New Space

New Space

From Low Earth Orbit to the Moon and Beyond

Edited by

Stephen W. Cheung
Former Adjunct Professor, Department of Physics
The University of Hong Kong, Hong Kong, China

Chih-Yung Wen
Department of Aeronautical and Aviation Engineering
The Hong Kong Polytechnic University, Hong Kong, China

IEEE Press
WILEY

Published by John Wiley & Sons, Inc., Hoboken, New Jersey.

Library of Congress Cataloging-in-Publication Data Applied for

Hardback ISBN: 9781394292929

Cover Design: Wiley
Cover Images: © SN VFX/Shutterstock, © Jiojio/Getty Images

Set in 9.5/12.5pt STIXTwoText by Straive, Chennai, India

Contents

About the Editors

Stephen W. Cheung earned his master's and doctoral degrees in physics from the University of Virginia.

His interest in space sciences began at Stanford University, where he was a research scientist on Gravity Probe B, a NASA-funded satellite program. Since 2010, he has remained a passionate advocate for the New Space era. As an adjunct professor of physics, he taught a space entrepreneurship course in the master's program in Space Science at the University of Hong Kong from 2020 to 2023. He emphasizes the practical applications of space technologies and enjoys communicating these ideas in accessible terms.

He also has an accomplished career in the medical device industry, particularly in radiation therapy systems. He has held both technical and managerial positions at two Silicon Valley companies—Varian Medical Systems and Accuray Incorporated. At Accuray, he led development efforts in medical and industrial X-ray systems and served as Principal Investigator on high-energy X-ray projects for cargo security. His work earned the company the prestigious Tibbetts Award and was awarded several US patents.

He spent a decade in senior management roles at a consulting organization and a major electronics firm in Hong Kong. He is the author of *Microwaves Made Simple: Principles and Applications*, a frequent speaker at industry and nonprofit events, and a strong advocate for integrating space technology into academic curricula.

Chih-Yung Wen received his Bachelor of Science degree in mechanical engineering from National Taiwan University in 1986. He subsequently earned his Master of Science and Doctor of Philosophy degrees in aeronautics from the California Institute of Technology (Caltech), USA, in 1989 and 1994, respectively. He commenced his academic career at the Department of Mechanical Engineering, Da-Yeh University, where he was promoted to full professor in

2002. During his tenure at Da-Yeh University, he held several key administrative positions, including Chairman of the Department of Mechanical and Vehicle Engineering and Provost.

In 2006, he joined the Department of Aeronautics and Astronautics at National Cheng Kung University. In 2012, he was appointed professor in the Department of Mechanical Engineering at The Hong Kong Polytechnic University. From 2019 to 2025, he served as Interim Head and subsequently Head of the Department of Aeronautical and Aviation Engineering. He currently holds the position of Chair Professor of Aeronautical Engineering and serves as Director of the Research Institute for Large Aircraft, Director of the Research Centre of Unmanned Autonomous Systems, and Associate Director of the Research Institute for Sports Science and Technology.

He has authored or coauthored more than 300 scientific publications, including journal articles, conference proceedings, and book chapters, and is the holder of 14 patents. He is a Fellow of the American Society of Mechanical Engineers (ASME), the Royal Aeronautical Society (RAeS), and the Hong Kong Institution of Engineers (HKIE), as well as an Associate Fellow of the American Institute of Aeronautics and Astronautics (AIAA). He is actively involved in professional and academic activities in the fields of mechanical and aerospace engineering at both national and international levels and serves on several prominent professional boards and committees related to aerospace engineering.

List of Contributors

Xu Bing
Department of Aeronautical and Aviation Engineering
The Hong Kong Polytechnic University
Hong Kong, China

Stephen W. Cheung
Former Adjunct Professor
Department of Physics
The University of Hong Kong
Hong Kong, China

Wang Tianqi
Department of Aeronautical and Aviation Engineering
The Hong Kong Polytechnic University
Hong Kong, China

Xue Xiaopeng
School of Aeronautics and Astronautics
Central South University
The People's Republic of China

Guan Yu
Department of Aeronautical and Aviation Engineering
The Hong Kong Polytechnic University
Hong Kong, China

Preface

Space inspires us to dream.

The words of Astronaut Neil Armstrong in 1969 still resonate today:

That's one small step for man, one giant leap for mankind.

For our generation, the space missions of the 1960s were a source of awe—the thunderous roar of rockets ascending, and the boundless possibilities they symbolized. Today, our fascination continues as we witness boosters landing with pinpoint precision on drone ships and launch pads mere minutes after liftoff. Space exploration is no longer driven solely by scientific curiosity; economic ambitions now share the spotlight. And thanks to social media, the public has a front-row seat to this extraordinary journey.

The past decade has brought disruptive advancements and groundbreaking records after records in space technology, making access to space more affordable and launches more frequent. The 21st century promises even greater milestones: a sustained human presence on the Moon. Decades after Armstrong's historic first step, history will be made again as the first woman and a person of color set foot on the lunar surface. By the 2030s, lunar outposts may be buzzing with activities, paving the way for humanity's next giant leap—becoming a spacefaring civilization. It all begins with an integrated vision for the Cislunar economy. After the Moon, on to Mars we go.

The New Space era has democratized access to the cosmos. Nations worldwide now have space agencies and low-Earth orbit programs. Middle school students have the same opportunity as university researchers to propose, design, and launch nanosatellites, and test innovative ideas.

We hope to share this excitement with a broader audience, especially young minds interested in science, technology, engineering, and mathematics (STEM).

This book grew out of a course taught by the author Stephen W. Cheung at the University of Hong Kong (2021–2023), part of the Master's program in Space

Science and another undergraduate course by the author Chih-Yung Wen at the Department of Aeronautical and Aviation Engineering, The Hong Kong Polytechnic University (2022–2025). The commercial focus of the course formed the foundation for this work.

Our deepest gratitude goes to Dr. Nim Cheung of the Institute of Electrical and Electronics Engineers (IEEE), whose support and encouragement turned this book project from concept to reality.

On the education front, the Orion Astropreneur Space Academy (OASA), founded by Professor Gregg Li and his associates has been instrumental in fostering space enthusiasm in Hong Kong. The authors are frequent participants. The seminars refine the contents of this book.

The author Stephen W. Cheung dedicates this book to his beloved parents, siblings Yuk-Lan and Dora, and nieces and nephews. His children, Carolisa and Quincy, are his endless source of motivation. He is especially indebted to Miss Nora Pong for her steadfast support during the months of writing. His former student, Miss Zijing Zhang, also helped out with the early artwork.

The author Chih-Yung Wen of the Department of Aeronautical and Aviation Engineering, The Hong Kong Polytechnic University, wishes to express his deepest gratitude to his family, especially his wife Shu-Hui Chang, for their hearty encouragement and constant support.

We also thank faculty members Drs. Guan Yu, Wang Tianqi, and Xu Bing of the Department of Aeronautical and Aviation Engineering, The Hong Kong Polytechnic University. Our appreciation also goes to the department's support staff Miss Samantha Sin, Miss Athena Cheung, and Mr. Patrick Lo in helping this one-year project smoothly cross the finish line. Dr. Xue Xiaopeng of the Central South University, China, contributed to the majority of the chapter on Mars.

As humanity prepares for deep-space exploration this century, we close with President John F. Kennedy's timeless words from 1962:

> We choose to go to the moon. We choose to go to the moon in this decade and do the other things, not because they are easy, but because they are hard.

Stephen W. Cheung
Mountain View, CA, USA

Chih-Yung Wen
Hong Kong, China
24 September 2025

1

Introduction

1.1 Introduction

1.1.1 Preamble

> 2022 was a record year for space launches 180 rockets lifted off successfully with SpaceX driving the pace
>
> Nature, 11 November 2023 [1]

> Elon Musk's Starship booster captured in world first
>
> BBC News, 13 October 2024 [2]

> How the Inspiration4 Mission Fits Into the Long History of Civilian Space Travel
>
> Time Magazine, 9 August 2021 [3]

> Jeff Bezos reaches space on Blue Origin's first crewed launch
>
> CNBC, 20 July 2021 [4]

Space launch is now routine and rarely makes headline news. We have come a long way since the 1950s.

In 1945, British Royal Air Force officer and later famous writer Sir Arthur Clarke suggested a "rocket station" orbiting the Earth at an altitude of about 36,000 km above Earth's surface, which would move at the same rotational speed as the Earth. The altitude suggested by Clarke, now known as geosynchronous Earth orbit (GEO), would become the de facto orbit for communications satellites for decades.

The space race began in the 1950s between the two superpowers, the United States and the (former) Soviet Union. Sending astronauts and spacecraft into space and going to the Moon were central themes of the race. In 1962, the American Telephone and Telegraph Company launched the world's first active communications satellite, Telstar-1, to test communications via space.

New Space: From Low Earth Orbit to the Moon and Beyond, First Edition.
Edited by Stephen W. Cheung and Chih-Yung Wen.

Since then, satellites of various applications have been conceived and launched in different orbits.

Today, we routinely cheer on professional sports on television broadcasts via satellite. Thousands of satellites now orbit our planet. Much of the current space industry is about satellites. Space stations such as the International Space Station (ISS) and China's Tiangong, as well as their predecessors, are also orbiting Earth in low orbit.

The thundering sound of a rocket lifting off remain just as intoxicating to space enthusiasts. Now these enthusiasts have the added excitement of the rocket boosters nailing perfect vertical landings on drone ships several hundred kilometers downrange. In October 2024, the world witnessed the huge Super Heavy booster caught by a pair of "chopsticks" on the launch tower in the Starship Integrated Flight Test #5. Another new chapter on reusable rockets began.

Countries have active programs to land humans on the Moon, with the intention of extended stays and establishing a long-term presence. We can envision the Moon as just the beginning of our civilization stepping outside of our planet.

1.1.2 Why Go to Space?

Space benefits us on Earth in many ways. From early explorations to the numerous satellites now serving different purposes, feeding our economy, we simply cannot live without space.

Going to space serves several purposes:

1) Spacecraft orbiting Earth provide vital information to the ground
2) Find sources of energy and materials to benefit Earth and future space settlers
3) Utilize the low-gravity environment
4) Satisfy our curiosity

Purposes 1) and 2) are tied to society's economic well-being and our quality of life. Purpose 3) is a platform that prepares humans to leave Earth. Purpose 4) has been in our DNA since humanity first gazed at the sky and wondered what was up there.

Expanding on Purpose 2), the natural resources on Earth are rapidly depleting while demand continues to rise. By the end of the 21st century, the supply of fossil fuels will likely run out, at least becoming too expensive to extract. Energy and raw materials are always needed to produce goods and services for citizens to keep up with our standard of living. The search for new sources of energy and materials has never been more pressing. Space is a viable option. The solar system has an infinite supply of energy and materials for our use for many centuries. The Sun is a super nuclear fusion reactor. The Moon has raw materials similar to Earth, and some of which are absent in Earth. Asteroids are a potential threat to Earth's safety but are also a rich source of raw materials.

The perspectives above imply that space is more than for exploration. Some applications are ready to scale up. Commercial interests need to participate to bring these efforts to an industrial scale. Commercialization is an integral part of the New Space era and the emerging space economy. Twenty-first-century space is a combination of exploration and industrialization. To commercialize and industrialize space, we need space infrastructure.

Governments recognize the need to change their roles in space to facilitate participation by the private sector. Governments will continue to support exploratory missions. They have changed from being in the driver seat controlling space agendas to being paying customers. They allow the private sector to build up the infrastructure where commercialization is ready.

1.1.3 About This Book

This book covers a range of topics that are happening in the New Space era. We focus on applications such as reusable rockets and asteroid mining. The topics show commercial promises.

The book targets audiences with some background in STEM. New Space topics are explained in simple terms using very few equations and without engineering details. Our purpose is to convey insights and further readers' interest in the subject.

1.2 New Space

The New Space era is characterized by the changing role of governments, commercialization and expanding applications. Going to space is democratized in countries large and small; commercial firms, and educational institutions can all participate.

Commercialization nurtures the growth of a space economy to serve the public at large and not just government agendas. It begins with orbits close to Earth, known as the low Earth orbit (LEO), as it is the "front door" to space. LEO is the volume of space between 200 and 2,000 km above sea level. This volume of space is buzzing with space activities because all space launches must go through it, and most satellites operate there.

The New Space economy begins with space ventures in LEO that benefit users on Earth. Examples of innovative use of technologies include miniaturized satellites and reusable rockets. These innovations drive down the costs of access to space and space missions, which in turn drive up the number of missions and participation by nations that would otherwise not have joined the space community. Affordable access increases space activities such as the demand for launch vehicles and services. More space missions translate to increased commercial opportunities and therefore strengthen the space ecosystem on Earth.

New Space is also about accessing large amounts of space resources. These resources will help address serious shortages facing humanity on Earth. In the long run, it is envisioned that humans will establish settlements in orbital spacecraft and on neighboring celestial bodies such as the Moon and Mars. These settlements will develop new technologies to extract materials for their survival and "export" them back to Earth and other space communities.

To a space enthusiast, space is the only option for civilization's long-term survival. There is no reason to limit our existence to planet Earth. We could be a spacefaring civilization traveling to or settling on other celestial bodies in the solar system and beyond.

In summary, the New Space era paves the way to transition from space exploration of the 20th century to one of proliferation of space.

1.2.1 Infrastructure

Chapters in this book often examine three infrastructures to be built in space: transportation, energy, and information. Rockets and space stations are examples of space transportation infrastructure. Transportation delivers humans and cargo from Earth to different orbits and other celestial bodies. Everything begins with the cost of access. Low transportation cost translates to affordable access to space and more users.

Satellites are part of the space information infrastructure. Present space information infrastructure primarily serves Earth users. This space infrastructure needs to expand beyond Earths orbits to cover deep space exploration and future space settlements.

Space energy infrastructure has not yet been developed. Energy powers an economy. Capitalizing on energy and energy-producing materials from space will be a major part of the space economy.

1.2.2 Destinations

Space infrastructures connect space communities of the future. That is what space commerce is about. Currently, we have a handful of space stations, the Moon, and Mars as potential communities. We need more man-made destinations in space. More destinations raise the demand for transportation, energy, and information infrastructure.

Initially, the new destinations will be in LEO. Commercial space stations, warehouses, hotels, factories, and universities alike can be in LEO and higher orbits in the Cislunar space (see below).

1.2.3 Payload Capacity

Payload refers to crew and cargo. High payload capacity is most desirable. For example, a jetliner has large passenger and cargo capacity. For commercialization to proliferate, more payloads should be delivered in one launch to keep the average cost per unit weight low. High payload capacity requires more lifting power from rockets.

High payload capacity allows more astronauts and cargo to be transported in one launch. We envision tens to hundreds of astronauts, together with machinery and mobile vehicles, being transported to the Moon and Mars.

1.2.4 Refueling

Vehicles on Earth do not go long distances on one tank of gasoline without refueling. Refueling is a vital part of terrestrial transportation infrastructure. The same is true with space transportation. Overcoming Earth's gravity to reach just the edge of space consumes tremendous amounts of fuel. Payload capacity is sacrificed in order for rockets to carry more fuel. Refueling is the only solution for high-capacity vehicles carrying passengers and cargo to the Moon and beyond. LEO is the first stop for refueling before going further.

1.3 Cislunar Space

The word Cislunar originally refers to the region of space encircling Earth and the Moon. The word's meaning has been expanded by space enthusiasts to cover the larger volume of space from Earth to Mars and everything in between. Cislunar space is the "sandbox" space enthusiasts have now set their sights on.

The Smithsonian, in a 2017 article "Cislunar Space: The Next 30 Years," conveys the vision articulated by the company United Launch Alliance (ULA) about the Cislunar economy. Cislunar economy refers to economic activities involving the Earth, the Moon, Mars, near-Earth asteroids, and future space settlements in different orbits. Each of these communities will have its own economy. ULA believes trade values worth trillions of US dollars will be realized when sizable populations and industrial activities take root beyond Earth in the future [5–7].

ULA offers a visual of what the Cislunar transportation infrastructure would look like in Figure 1.1. Here, space railroads represent transportation infrastructure connecting Earth, the Moon, Mars, and near-Earth objects (NEOs). Hubs, depots, and various services are located in different orbits along the

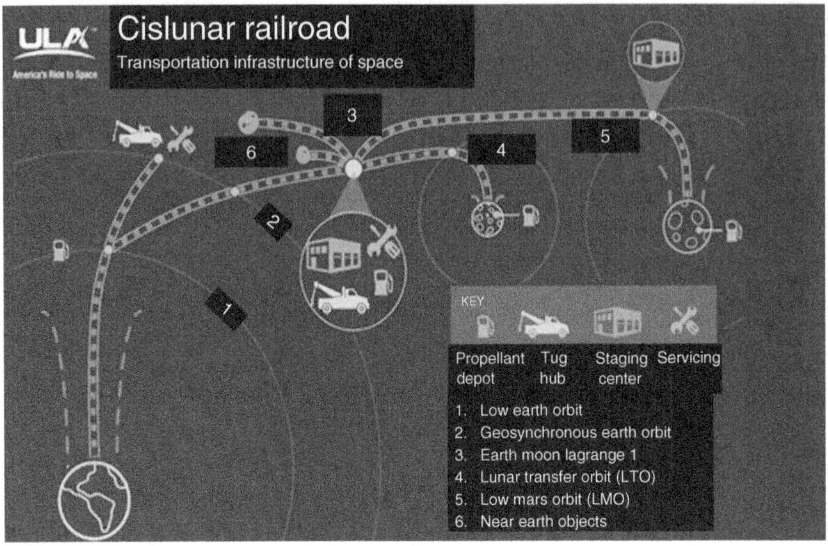

Figure 1.1 Cislunar railroad concept representing transportation infrastructure in space. *Source:* Ref. [8] / Multiverse Media Inc.

Cislunar railroad. These depots symbolize Cislunar commerce made possible by the space railroads. Refueling stations are shown in various orbits in Figure 1.1 and illustrate the significance of refueling.

The middle of Figure 1.1 shows a point labeled Earth-Moon Lagrange 1 (EML-1). Lagrange points are covered in Chapter 3. Next to EML-1 in Figure 1.1 are the NEOs. These objects traverse Cislunar space and are sources of raw materials. Asteroid mining will be covered in Chapter 10. Landing on asteroids and scooping up samples have already been demonstrated several times and found to be technologically doable using flyby spacecraft to intercept the target asteroid millions of kilometers away. Asteroid mining has attracted entrepreneurs and speculators for the promise of incalculable fortunes. These ventures are almost entirely commercial in nature. To be commercially viable, special-purpose vehicles will be needed to reach asteroids, mine them, and transport the tons of materials to designated locations for industrial processing. That needs space transportation infrastructure.

1.3.1 The Moon

None of the planets and celestial bodies in the solar system are identical to Earth. Without exception, the environment is harsh on these celestial bodies. The Moon will be our first test destination.

Chapter 11 describes how to reach the Moon and where to land. Lunar programs include NASA's Artemis program and China's *Chang'e* program. Space vehicles and flightpaths of the initial Artemis missions will be described. A space station known as Gateway will be a key piece of the lunar infrastructure. Chapter 12 looks at settling on the Moon and covers key issues such as the source of water, solar power, and means of mobile transport around the lunar settlement.

1.3.2 Mars

Chapter 13 is about Mars. The Moon is the training ground prior to going to Mars. Earth and Mars are millions of kilometers apart and go around the Sun at different speeds, resulting in a 26-month launch window for Martian missions. A round-trip human mission will take over 3 years. A variety of propulsion options are under investigation to ensure initial human missions have sufficient resources (fuel and food) for a round trip without relying on any local resources.

Generally speaking, some of the skills and technologies developed for the Moon will be transferable. Other technologies will need modification and localization.

1.4 Economics and Business Considerations

For decades, venturing into space was entirely driven by government objectives. Such missions did not necessarily have or need economic justification. Exploratory missions soon led to satellites that have proved to be of tremendous economic value to Earth.

New Space came about when conditions were ripe for a paradigm shift. Traditional government-centric practices gave way to government-private sector collaborations. The private sector can accelerate the commercialization of space with public capital. Young companies are springing up to compete with mature players.

1.4.1 Morgan Stanley's Projections

In February 2021, investment bank Morgan Stanley published an article "*A New Space Economy on the Edge of Liftoff*" [9]. The company projected the space economy will pass the coveted trillion US dollar bar by 2040. Several financial institutions also came out with bullish reports around that time. Their difference lies in the size of their projections, between US$1 trillion and $2 trillion by 2040.

> .. space is existential, from the future of the planet to the future of commerce
>
> Adam Jonas [10], Head of Morgan Stanley, Research Space Team

The Morgan Stanley article identified 10 drivers of the New Space ecosystem. The number one driver was Satellite Internet with a short description: "These companies focus on improving connectivity through low-earth-orbit satellites."

The Morgan Stanley article further projected satellites will constitute the largest share of the space industry by 2040. Within the "satellite pie" itself, satellite internet will take up a lion's share of revenues. This projection is based on the anticipation of thousands of satellites being deployed in low orbit. Satellite internet will be discussed in Chapter 6. Several constellations of thousands of communications satellites are at different stages of development or deployment. Constellations of tens of remote sensing satellites will soon join the crowd as well.

Another of the 10 drivers was Satellite Launch. Morgan Stanley stated, "one of the largest subsectors, satellite launch companies focus on technology and infrastructure to send satellites to near space and low earth orbit." The economics of rocket launches determines the cost of access to space. The unprecedented numbers of satellites planned by the private sector are driving up demand for launch vehicles and launch services. Discarding rocket stages after a single use is simply an unacceptably costly option in deploying thousands of satellites. Affordable launch cost and high vehicle payload capacity become inevitable requirements. Reusable rockets are the answer to this problem.

1.5 Conclusion

Space is the new frontier of the 21st century. Commercial interests will coexist with government objectives. The exploratory phase of Earth's orbits is now transitioning to the infrastructure-building phase. Nations are going to the Moon to explore, not just out of curiosity as before, but to settle with industrialization in mind. Lunar infrastructure will then follow. Mars and asteroids are next.

New technologies have been developed, which include reusable rockets, miniaturized satellites, and on-orbit refueling toward the foundation of transportation and information infrastructure. New commercial activities are imminent. Among them, the more visible activities include space tourism, asteroid mining, and space manufacturing. Reusable spacecraft will make access to space affordable and more "user-friendly" to the general public.

Commercialization and entrepreneurial companies will drive the New Space era forward. Soon, governments will still push their national agendas. They will be customers of space missions, mostly by subcontracting commercial firms to do the work.

Commercialization starts with the low orbits around Earth, forming the base of space infrastructure. Earth populations are the beneficiaries of the initial infrastructure in Earths orbits. Extending from Earth as the base, infrastructure going further into the Cislunar region will come along.

In the more distant future, as longer-term habitation of space becomes a reality, Cislunars infrastructure will take shape to accommodate such expansion.

Reading List

Anderson C. *The Space Economy: Capitalize on the Greatest Business Opportunity of Our Lifetime.* Wiley; 2023.

National Academies of Sciences, Engineering, and Medicine. *NASA Space Technology Roadmaps and Priorities Revisited.* The National Academies Press; 2016.

National Academies of Sciences, Engineering, and Medicine. *Thriving on Our Changing Planet: A Decadal Strategy for Earth Observation from Space.* The National Academies Press; 2018.

National Academies of Sciences, Engineering, and Medicine. *Assessment of Commercial Space Platforms for Earth Science Instruments: Report Series – Committee on Earth Science and Applications from Space.* The National Academies Press; 2023.

National Academies of Sciences Engineering, Medicine. *Thriving in Space: Ensuring the Future of Biological and Physical Sciences Research: A Decadal Survey for 2023–2032.* The National Academies Press; 2023.

Seedhouse E. *Mars via the Moon: The Next Giant Leap* (1st Edition). Springer; 2016.

References

1. Witze A. 2022 was a record year for space launches. Nature. 2023;*613*(7944):426–426.
2. Stallard E. Elon Musk's Starship Booster Captured in World First. BBC [updated 2024 Oct 13; cited 2025 Feb 27] [online]. Available from: https://www.bbc.com/news/articles/c8xe7exjy1go.
3. Waxman OB. How the Inspiration4 Mission Fits Into the Long History of Civilian Space Travel. TIME [updated 2021 Aug 9; cited 2025 Feb 27] [online]. Available from: https://time.com/6083973/history-civilian-space-flight/.
4. Sheetz M. Jeff Bezos Reaches Space on Blue Origin's First Crewed Launch. CNBC [updated 2021 Jul 20; cited 2025 Feb 27] [online]. Available from: https://www.cnbc.com/2021/07/20/jeff-bezos-reaches-space-on-blue-origins-first-crewed-launch.html.
5. Spudis PD. Cislunar Space: The Next 30 Years. Smithsonian Magazine [updated 2017 Feb 22; cited 2025 Feb 27] [online]. Available from: https://www.smithsonianmag.com/air-space-magazine/cislunar-space-next-30-years-180962249/.

6 New Space Economy. What is Cislunar Space and its Implications for the Space Economy?https://newspaceeconomy.ca/2023/04/01/what-is-cislunar-space-and-why-is-it-important/.

7 Scott DW, Curreri PA, Ferguson CK, Nall ME, Tinker ML, Wright GM. Germinating the 2050 cis-lunar econosphere. In: *2015 IEEE Aerospace Conference*. IEEE; 2015. p. 1–16.

8 CisLunar Railroad. United Launch Alliance. Cited by Boucher M. Panel: Cis-Lunar Self-sustained Space Economy. Space Ref [updated 2017 Apr 4; cited 2025 Feb 27] [online]. Available from: https://spaceref.com/space-commerce/panel-cislunar-self-sustained-space-economy/.

9 Creating Space. Morgan Stanley [cited 2025 Mar 20] [online]. Available from: https://www.morganstanley.com/Themes/global-space-economy.

10 Dasgupta A. Earth Observation: The Growth Story. Geospatial World [cited 2025 Feb 21] [online]. Available from: https://geospatialworld.net/prime/special-features/earth-observation-the-growth-story/.

2

Rockets

Guan Yu[1] *and Stephen W. Cheung*[2]

[1]*Department of Aeronautical and Aviation Engineering, The Hong Kong Polytechnic University, Hong Kong, China*
[2]*Former Adjunct Professor, Department of Physics, The University of Hong Kong, Hong Kong, China*

2.1 Introduction

Rockets have been the most reliable vehicles to reach space since the beginning of the space age. There are different types of propulsion systems. Our emphasis is on chemical propulsion in this chapter. Other propulsion systems will be briefly covered. It comes down to the power source that energizes the propulsion process.

The gigantic rockets we see on television are composed of two or more stages based on chemical combustion. The first stage (called Stage 1 or booster) constitutes the majority of the whole rocket by weight and volume due to its size and the rocket propellants in the tanks. Passenger and cargo payloads are carried to space by the upper stage(s).

We will use Δ (delta) to denote an increment of a physical quantity, such as mass and time. For example, Δt represents an increment of time t.

In this chapter, we shall skip the history of rocket development. Readers are referred to many books and articles to learn about the evolution of rockets.

2.2 Tsiolkovsky Rocket Equation

The principle of propulsion is to create a change in velocity by ejecting mass from the host. The Tsiolkovsky rocket equation is the foundation of propulsion. We start with a rocket with an initial mass M_0 that is initially at rest in free space. When a small amount of mass Δm is ejected at the rear of the rocket with an exhaust speed V_{exh} relative to the rocket, the ejected mass carries away momentum $(\Delta m) * V_{\text{exh}}$. Momentum conservation suggests that the same amount of

New Space: From Low Earth Orbit to the Moon and Beyond, First Edition.
Edited by Stephen W. Cheung and Chih-Yung Wen.

momentum will act on the remaining mass ($M_o - \Delta m$) in the forward direction. The vehicle is thrust forward because of the momentum change.

In rocketry, the ejected mass is the propellant. By continuously discharging mass, the vehicle continues to propel forward and gain speed.

The Tsiolkovsky equation is one of momentum conservation in accordance with Newton's laws of motion. If the vehicle has an initial speed, then ejecting mass as described above results in a change of forward speed V. In mathematical form, the Tsiolkovsky equation can be stated as follows:

$$\Delta V = V_{exh} * \ln(M_o/M_f) \tag{2.1}$$

ΔV is the change in velocity of the rocket body, M_o is the initial mass of the rocket before mass discharge, M_f is the mass momentarily after ejecting Δm, and ln is the natural logarithmic function.

We can express Equation (2.1) in a format used in kinematics.

$$V_f = V_o + \Delta V = V_o + V_{exh} * \ln(M_o/M_f) \tag{2.2}$$

where V_o and V_f are the initial and final velocities of the rocket in the discharge process, respectively.

We make an important observation from Equation (2.1): Either discharging propellants or high exhaust speed can result in a large change in the forward speed ΔV. A large ΔV is therefore the result of ejecting large amounts of high-speed propellants.

2.2.1 Thrust

The rocket equation gives us ΔV in response to mass being ejected. However, it contains no time element about how fast this ΔV will happen. Intuitively, if mass is discharged fast, ΔV will increase at a rapid rate.

We have used the term "thrust" loosely so far. Thrust is a force and follows Newton's laws of motion. Newton's Second Law states that force is the rate of change of momentum. The Tsiolkovsky equation can express thrust (force) if the rate of momentum change is known.

According to Equation (2.1), either V_{exh} or Δm or both can have a rate of change over time. Propulsion systems usually work by discharging propellants at a constant exhaust speed. Therefore, thrust primarily comes from the rate at which propellant Δm is ejected.

$$\text{Force (Momentum Thrust)} = V_{exh} * (\Delta m/\Delta t) \tag{2.3}$$

Equation (2.3) shows that the thrust propelling the rocket forward is equal to the product of the rate of propellants discharged by propellants ($\Delta m/\Delta t$) and the speed of the ejected mass V_{exh}.

Momentum thrust is produced by the chemical energy released from combusting propellants and ejecting the resultant gaseous byproducts. Before exiting the chamber, the hot gases exert pressure, pushing on the walls of the chamber and are the source of the thrust on the rocket. This gas pressure multiplied by the cross-sectional area of the chamber is the momentum thrust.

According to Equation (2.3), there are three options to produce a large thrust in free space:

a) Large V_{exh}
b) Large $\Delta m/\Delta t$
c) Both a) and b)

Momentum thrust is due to momentum exchange made possible by discharging propellants at high speeds. In a later section, we will introduce the "pressure-area thrust." The momentum thrust and the pressure-area thrust together constitute the total thrust acting on the rocket. The momentum thrust is the dominant thrust.

2.2.2 Including Local Gravity

The rocket equation assumes the vehicle is in free space, devoid of any external forces. Reality is we must leave Earth to reach space. Upward thrust is needed to overcome gravity and atmospheric resistance (drag). Local gravity also exists in other celestial bodies such as the Moon and Mars.

We shall focus on local gravity on Earth and ignore atmospheric drag due to the complexity of the latter. Gravity is a decelerating force against an ascending rocket. Therefore, the net upward force is the difference between the rocket's upward thrust and local gravity.

Net Force upward = Upward Thrust – Gravity

$$M * a = V_{exh} * (\Delta m/\Delta t) - M * g \tag{2.4}$$

$$a = V_{exh} * (\Delta m/\Delta t)/M - g \tag{2.5}$$

where "a" is the net acceleration, M is the mass of the rocket at a given moment, and g is the local gravity nominally equal to 9.8 m/s^2. Both M and "a" change with time due to mass being discharged from the rocket.

Equation (2.4) becomes more complex when atmospheric drag is introduced. Such solutions are usually done by calculus and computer programs and will not be covered in this book.

2.2.3 Delta-V (ΔV)

Velocity change ΔV in Equation (2.1) is a convenient and very important term in rocketry and orbital mechanics. ΔV is the product of two algebraic components: V_{exh} and ln (M_o/M_f). Let us look at what it takes to generate a large ΔV.

(1) V_{exh}: High exhaust velocity of the propellant produces large ΔV. V_{exh} from rocket engines is often on the order of km/s.
(2) ln (M_o/M_f): We can look at three mathematical possibilities:
   (a) ln $(M_o/M_f) < 1$: Natural logarithm is negative when the quantity M_o/M_f is less than 1. This means the initial mass is less than the final mass, which is impossible.
   (b) ln $(M_o/M_f) = 0$: This happens when M_o/M_f equals 1. When the final mass is the same as the initial mass, no propellants are expelled, so there is no velocity change.
   (c) ln $(M_o/M_f) > 1$: This happens when $M_o/M_f > 1$, i.e. the final mass is less than the initial mass, so there is a momentum exchange that results in a velocity change. The larger the M_o/M_f, the larger the ΔV.

The logarithmic function is not linear. Table 2.1 shows the ln (M_o/M_f) values over a range of M_f/M_o. It is more intuitive to use M_f/M_o because it is the final mass as a fraction of the initial mass and should be between 1 and 0 as shown in Figure 2.1.

It can be seen in Figure 2.1 that the ln (M_o/M_f) values start out linearly until M_f drops below 30% of M_o. Ln (M_o/M_f) equals to 1 when M_f is about 37% of M_o. Therefore, when M_f drops to below 37% of M_o, ΔV increases at a sharper slope.

Conclusion: Discharging a significant portion of the initial mass results in high ΔV and large thrust.

Looking ahead, a rocket launched from the Earth's surface to enter low Earth orbit requires a ΔV of nominally 8 km/s. Earth's gravity needs to be overcome, and then the orbital speed must be built up. As this book focuses on low Earth orbit (LEO), it is important to remember that an object needs to travel at 8 km/s to orbit Earth in LEO.

Table 2.1 ln (M_o/M_f) values versus M_f/M_o. The latter ratio means M_f as a fraction of the original mass M_o.

M_f/M_o	ln (M_o/M_f)	M_f/M_o	ln (M_o/M_f)
1	0	0.4	0.92
0.9	0.11	0.3	1.20
0.8	0.22	0.2	1.61
0.7	0.36	0.1	2.30
0.6	0.51	0.05	3.0
0.5	0.69	0	Infinite

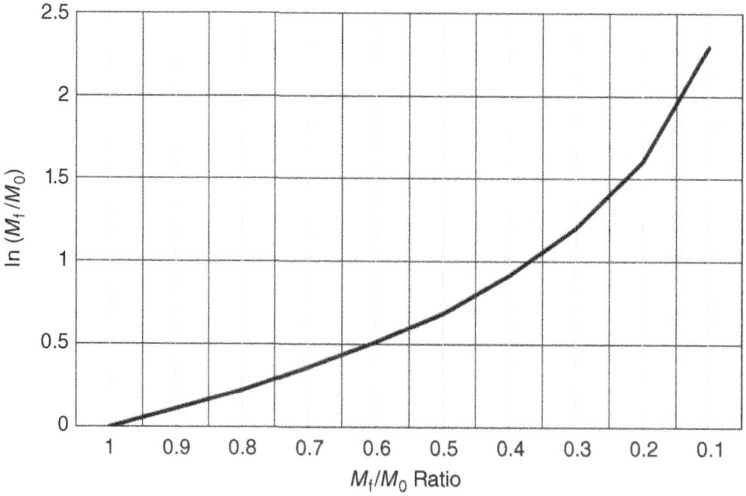

Figure 2.1 Logarithmic behavior of the ratio of the rocket's final mass to initial mass.

2.2.4 Thrust-to-Weight Ratio (TWR)

A common term to characterize a rocket and its engines is the thrust-to-weight ratio (TWR). An engine's TWR compares two forces: thrust generated by the engine versus the weight of the engine on Earth. Expressed differently, TWR is how many times the thrust is that of the engine's Earth weight.

We will use the Merlin engine used in the SpaceX Falcon 9 rocket to illustrate the concept of TWR. Using information on the Merlin engine [1], the "dry weight" of the engine is 470 kg (without fuel), and its thrust at sea level is 845 kN (kilo Newton). We will address the term "sea level" next. Expressed in Newtons, the weight of the engine is $470 * g = 470\,\text{kg} * 9.8\,\text{m/s}^2 = 4{,}606\,\text{N}$.

The TWR of the Merlin engine is therefore $845{,}000\,\text{N}/4{,}606\,\text{N} = 183.5$, which means a Merlin engine can deliver a thrust of 183.5 times its weight.

The calculation can be extended to the whole Falcon 9 rocket. The rocket has nine identical sea-level Merlin engines. When these engines are on full throttle at launch,

Total thrust by nine engines $= 845\,\text{kN} * 9 = 7{,}605\,\text{kN}$

The maximum total weight of a fully loaded Falcon 9 rocket with fuel and payload is 549,000 kg based on information obtained online.

Therefore, the TWR of the entire rocket at launch $= 7{,}605{,}000\,\text{kN}/(549{,}000 * 9.8\,\text{N}) = 1.41$

There are a few points to note about TWR when Earth's gravity is present.

(a) Lowest TWR is 0 when the rocket engines produce no thrust at all.
(b) $0 < \text{TWR} < 1$: This means the rocket engines' thrust is less than the rocket's weight, so the rocket does not get off the ground. And if the rocket is going up with thrust, gravity will decelerate it.
(c) $\text{TWR} = 1$: Upward thrust is the same as gravity. There is no net thrust. If the rocket is in the air, it will either continue its vertical velocity or remain still (known as "hovering") like a helicopter hovering above.
(d) $\text{TWR} > 1$: There is a net upward thrust and therefore upward acceleration. The net upward acceleration = $(\text{TWR-1}) * \text{g}$.

2.2.5 Concept of Specific Impulse (Isp)

The primary performance of a rocket engine is its efficiency in producing thrust. An engine's efficiency is a measure of maximum momentum change possible per unit weight of propellant used in the engine.

Rocket efficiency is analogous to the gasoline mileage of an automobile. Automobile efficiency is expressed as either miles per gallon or kilometers per liter. Car A is said to be more efficient than car B if car A has higher gas mileage using the same amount of gasoline.

While automobile gas mileage is about how far the vehicle can go on a unit amount of gasoline, rocket engine efficiency is about how long in time a unit weight of propellant can last delivering thrust. Engine efficiency is therefore expressed as the time a unit weight is being just lifted by the same unit weight of propellant.

In physics, impulse I is the change in momentum due to the application of a force F over a period Δt. Impulse is the product of the force and the duration of the force application, i.e. $\text{I} = \text{F} * \Delta t$. Dimensionally, impulse is momentum, not force. It represents the momentum change on the object when the force F is applied during Δt. The expression can be rearranged to obtain

$$\Delta t = \text{Impulse}/\text{Force} \tag{2.6}$$

If the momentum change due to the impulse is on an object of mass 1 kg and the thrust is from 1 kg of propellant, then Δt is the duration or engine efficiency. This Δt is known as specific impulse (Isp). Isp is commonly expressed in seconds.

Specific impulse (Isp) is a common quantity to characterize engine efficiency. A propulsion system with a high Isp uses a unit of propellant to produce more momentum change than one with a low Isp.

Isp is linked to ΔV. Without going into details, we state the result as:

$$\Delta V = \text{g} * \text{Isp} \tag{2.7}$$

Hence, knowing the Isp of an engine or a rocket, the velocity change can be easily estimated.

For example, a liquid hydrogen and liquid oxygen (LH_2/LOX) mixture as propellants in a particular rocket engine has an Isp of 400 seconds. It is equivalent to a ΔV of $9.8 * 400 = 3{,}920$ m/s or 3.92 km/s. That means 1 kg of LH_2/LOX in this engine can thrust 1 kg of the engine for 400 seconds and deliver a ΔV of 3.92 km/s.

ΔV and Isp are therefore not independent quantities but linked to each other. Isp is an important concept in rocketry. In this book, we limit ourselves to using ΔV.

2.3 Chemical Propulsion and Main Components

Propulsion makes a rocket lift off from Earth and accelerate into outer space. We first look at the most important item of a propulsion system: the rocket engine. Associated components of a rocket engine will be discussed.

2.3.1 Combustion Process

Combustion involves chemical reactions that are exothermic (producing heat). Combustion systems release the energy trapped in chemical compounds during this process. The amount of heat released is large and turns the reaction byproducts into hot gases. The combustion process is carried out inside a rocket engine's combustion chamber. The gaseous byproducts greatly increase the pressure inside the combustion chamber. The pressure inside the chamber ranges from 10 to 300 atm, where atm is a more intuitive unit representing 1 standard atmosphere of gas pressure and is numerically equal to 14.7 psi or about 100,000 Pascal (the metric unit of Newton per square meter).

In chemistry terms, combustion is a reduction–oxidation (or redox) process. An oxidizer is a type of chemical that fuel needs to burn. Oxygen is an oxidizer, as the name suggests. Other chemicals can also function as oxidizers. A reducing agent, on the other hand, reacts with an oxidizer to complete a chemical reaction. In combustion, fuel will not combust without the oxidizer. The terms fuel and propellant are often used interchangeably. A propellant is a mixture of fuel and oxidizer, i.e. the final product.

Near Earth's surface, combustion in automobiles and aviation utilizes oxygen in the atmosphere. In a rocket engine, both the fuel and the oxidizer must be available internally.

When ignited, propellants react and combust inside a combustion chamber, forming hot gaseous byproducts. The fore end of the chamber is closed while the aft steadily narrows to form a throat for the hot gases to escape. The pressure of the hot gases acts on the fore end, resulting in a force pushing the rocket in the forward direction. There is pressure on the sidewalls as well, but they have no bearing on the forward force.

The hot gases escape from the chamber through the throat. The aft of the chamber is connected to a nozzle where the opening widens steadily in certain geometric shapes.

2.3.2 Propellants

There are two categories of propellants: liquid and solid. They are shown in Figure 2.2.

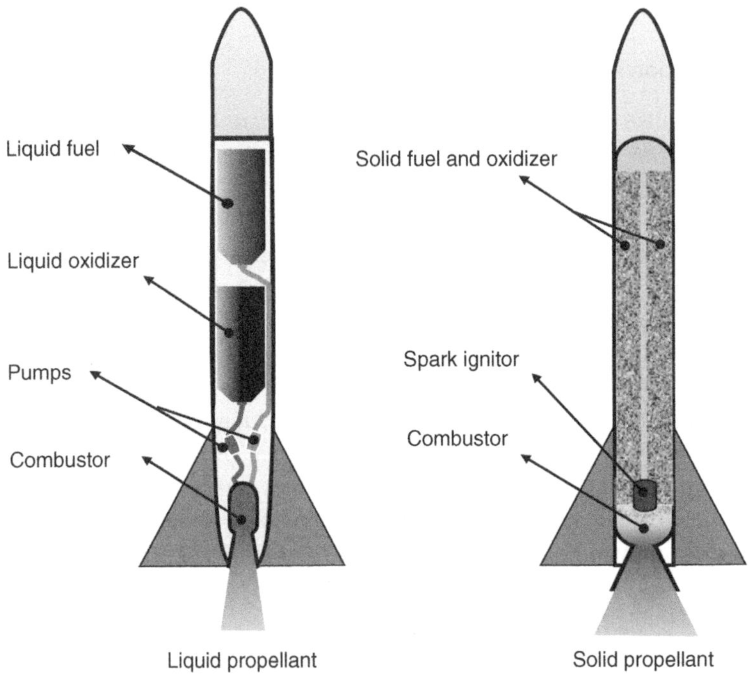

Figure 2.2 Propellants can be liquid or solid. Liquid propellants are stored separately in tanks and injected into the combustion chamber separately. Solid propellants are premixed and placed inside the chamber. *Source:* Adapted from Ref. [2].

2.3.2.1 Solid Versus Liquid Propellants

In a solid fuel rocket, the fuel and oxidizer have been premixed and pressed into a stack of annulus disks inside a cylindrical combustion chamber as shown on the right of Figure 2.2. The ignitor is located on the fore end of the chamber. Once ignited, solid propellants combust from the fore toward the aft, and the hot gaseous byproducts exit from the nozzle. Upon ignition, solid propellants continue to burn until no propellant remains. Rockets using solid propellants are relatively simple without complex pumping systems. They are easy to store. Missiles use solid propellants. The side boosters of the Space Shuttle also used solid propellants.

The disadvantage of solid propellants is that they cannot be throttled and lack the flexibility of multiple stop-and-restart.

In a liquid fuel rocket, the fuel and oxidizer are liquids stored in separate pressurized tanks. Pumps are used to inject the liquids into an engine combustion chamber, where the mixture is then ignited. This is shown on the left of Figure 2.2. The ignitor is usually like a spark plug.

The advantage of liquid propellant systems is throttling control and multiple stop-and-restart, which are crucial in upper-stage engines and reusable boosters. The disadvantages are the complex plumbing and the need for special storage tanks as the liquids are often in a cryogenic state. For a rocket with multiple engines, a complex plumbing system has the task of distributing the liquids to selected engines.

Liquid propellants generally have higher V_{exh} than their solid counterparts. Table 2.2 lists several propellants for reference.

Special Note According to Tsiolkovsky's equation, high V_{exh} serves to increase ΔV. Liquid hydrogen and liquid oxygen combust to yield ΔV near 4 km/s and are used in many missions. However, hydrogen is the lightest element, so large quantities of liquid hydrogen are needed, thereby requiring a much larger storage tank.

Table 2.2 Exhaust velocities of three solid and liquid propellants.

Propellants	Approximate effective exhaust velocity (km/s)
Solid	2.4
Kerosene/liquid O_2	3.0
Liquid H_2/liquid O_2	4.1

Extreme caution is needed to ensure no leaks due to material fatigue, as hydrogen can escape through the smallest cracks.

The mass of the discharged products is also important because thrust comes from the propellant's mass as well as the rate of mass discharge $\Delta m/\Delta t$. For this reason, petroleum-based fuels are often used in the first stage because they are heavier despite their lower V_{exh}, and they produce higher thrust to lift the rocket when it is the heaviest at launch.

2.3.3 Rocket Stages (Staging)

From Figure 2.1 and Table 2.1, a large amount of mass needs to be ejected to have a large $\ln(M_f/M_0)$. In other words, the liftoff process consumes a lot of propellants to overcome Earth's gravity and accelerate upward. That leaves little room for the rocket structure, propellant tanks, equipment, and payload.

The goal for a rocket is always to deliver the largest payload mass possible and have the least non-payload mass. Non-payload mass is defined as the structural weight of the rocket without the propellants and payloads. Non-payload mass is often called dry mass.

If a large portion of the rocket structure can be removed after large amounts of propellants have been consumed, there is less dry weight for the propulsion system to propel. This is known as staging. There are two types of staging:

1) Serial staging: Each stage is ignited serially after separating from its previous stage, which has exhausted its propellants and is now "dead weight."
2) Parallel staging: Usually applies to Stage One. All boosters are ignited during liftoff. They are jettisoned after they burn out.

There are three primary considerations for the staging of the launch system:

1) The first stage should focus on generating liftoff thrust to overcome gravity and accelerate upward to build up some speed to be ready for entering orbit.
2) The second stage is to build up ΔV.
3) Each successive stage should be smaller in mass than the previous stage.

The first stage drops away after having exhausted its propellants. The remaining stage(s) with payload is now much lighter without the burden of the now-empty first stage. The second stage can now focus on building up ΔV.

In parallel staging, a rocket's first stage is composed of a core booster in the center with side boosters as shown in Figure 2.3. The side boosters are symmetrically located relative to the core booster. These boosters are often fired together to provide maximum lift at launch. This design is often used for launching heavy payloads.

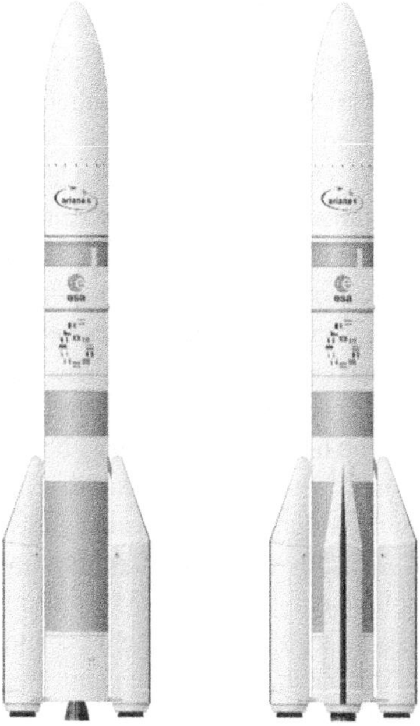

Figure 2.3 Both variants of Ariane 6. The image on the left has two solid side boosters. The image on the right has four solid rocket side boosters. *Source:* Ref. [3] / Wikimedia Commons / CC BY 3.0.

Side boosters usually use solid fuels. They drop out after their solid fuels are exhausted while the core booster continues to burn. The core booster usually uses liquid propellants, so throttling can be executed during ascent.

2.3.4 Combustion chamber

As the name suggests, the combustion chamber is where the fuel and the oxidizer combust to generate thrust to push the rocket forward. A rocket's first stage usually has several engines, each of which has its own combustion chamber.

A combustion chamber is cylindrical with a length between half a meter and 2 m. The physical environment during combustion includes high pressure and high temperature. The pressure inside the chamber can reach several hundred atmospheres and temperatures over 3,000 °C.

The chamber is made from materials that can withstand high pressure and temperature. Copper alloys are typically used. A portion of the heat generated by the combustion process heats up the chamber walls. Cooling tubes are often wrapped around the exterior of the chamber to transfer heat away. Super cold liquid propellants are routed through the cooling tubes to be "warmed up" before being pumped into the chamber for combustion.

2.3.5 Nozzle

The nozzle is an integral part of the combustion chamber design. It is the extension of the aft section of the chamber.

The nozzle can be understood by first looking at an example of a balloon. If you fill a balloon with air and let go, the air pressure inside the balloon is higher than outside and will escape through the lip of the balloon. It will fly around in uncontrolled, random directions; the escaped air scatters in all directions as soon as it exits the lip. If a straw is attached to the balloon lip, the air-filled balloon now moves in an almost fixed direction. The straw guides the exiting air along the straw.

In a combustion chamber, the high-pressure, hot gases move through the throat and reach hypersonic speeds in the order of kilometers per second. At the throat's exit, the hot gases collectively become the "exhaust plume." As the plume moves through the nozzle, it steadily expands along the nozzle wall, whose geometry slowly widens, until it reaches the nozzle's exit end. The pressure of the plume also steadily drops during this transition along the nozzle. The nozzle serves to contain and guide the exhaust plume like the straw in the balloon example above. Otherwise, the plume might expand and scatter in many directions coming out of the chamber's throat. We want the exhaust plume to exit exactly in the opposite direction of the rocket's motion.

The nozzle shape is designed to "squeeze" more forward thrust from the exhaust plume. This thrust is known as "pressure-area thrust" and is based on the plume pressure multiplied by the cross-sectional area of that part of the nozzle. This extra thrust comes from the residual pressure of the plume pushing the walls of the nozzle. The pressure-area thrust accounts for less than 10% of the total thrust on the rocket engine, while over 90% of the total thrust is from the momentum transfer of massive amounts of propellants being discharged.

Two nozzles of the Merlin rocket engine are shown in Figure 2.4. We will explain the difference between "sea-level" and "vacuum" nozzles. Notice the geometric shapes of these nozzles. They come in different shapes, ranging from conical to bell-shaped and more complex designs.

2.3.5.1 Sea-Level Versus Vacuum-Level Nozzle

The exhaust plume exits a nozzle and enters the ambient environment. The pressure of the plume drops from super high (in the combustion chamber) toward that of the ambient as it expands along the nozzle in accordance with thermodynamics. There are three scenarios of how the ambient pressure shapes the exhaust plume as illustrated in Figure 2.5.

1) Ambient pressure is at sea level: this ambient pressure is the highest among the three scenarios. The ambient pressure limits the expanding plume. When the ambient pressure is at sea level, it "contains" the plume toward the nozzle's

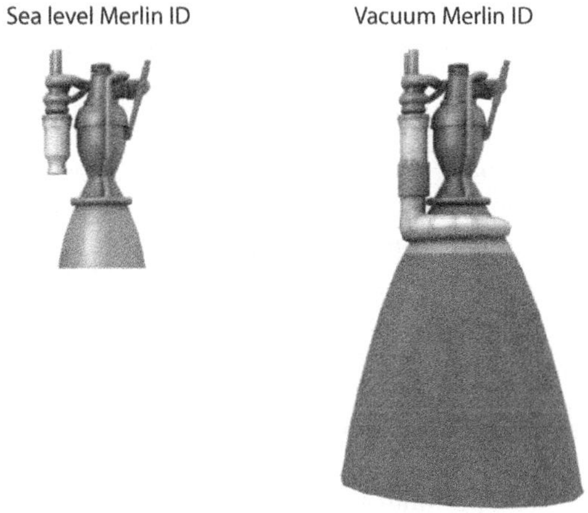

Figure 2.4 The Merlin engine with a large nozzle for vacuum and a small nozzle for sea level. *Source:* Ref. [4] / Everyday Astronaut.

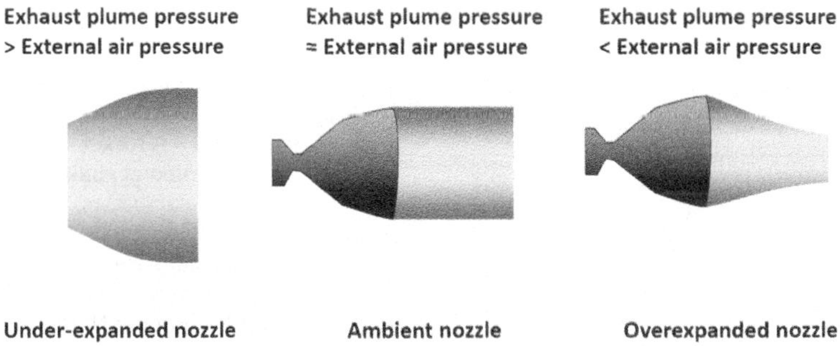

Figure 2.5 Vacuum-optimized rocket nozzles. *Source:* Adapted from Ref. [5].

central axis. This containment can be at the edge of the nozzle or inside. The nozzle is known as "over-expanded." In this scenario, there is no benefit in making the nozzle too big.

2) Ambient pressure less than sea level: the vehicle is at a higher altitude where the ambient pressure is sufficiently less than in scenario 1. The plume is perfectly contained within the nozzle rim at this moment. The nozzle is an ambient nozzle at this moment.
3) Ambient pressure is nonexistent: the rocket is now at high altitude or in space. There is no external pressure to contain the plume, so it continues to expand beyond the nozzle rim. The nozzle is known as "under-expanded."

The nozzle size of a rocket's first stage is smaller in scenario 1 above since the plume is being squeezed back by the ambient pressure at launch and at low altitudes. The nozzle size for high ambient pressure is optimized to be small. This works well in that more engines can fit in the first stage to build up thrust.

One can visually see the plume change shape from online videos of rocket launches. The plume begins to broaden as the rocket ascends beyond 20 km, when the ambient pressure steadily decreases.

A much bigger nozzle is used in an upper stage because it operates at a significantly higher altitude or in space. The much bigger nozzle allows for the most thrust on the upper stage possible. Usually, one or more engines are in the upper stage, and their nozzles take up a lot of room.

2.3.6 Steering

The rocket, as a vehicle, needs steering or changing its pointing direction. Gimballing the engine and the nozzle provides steering. Gimballing is achieved by actuators pushing on the engine as illustrated in Figure 2.6. The actuators are usually driven by electromechanical, electrohydraulic, or pneumatic means.

Typically, gimballing is done two dimensionally within a range of 10° in either direction.

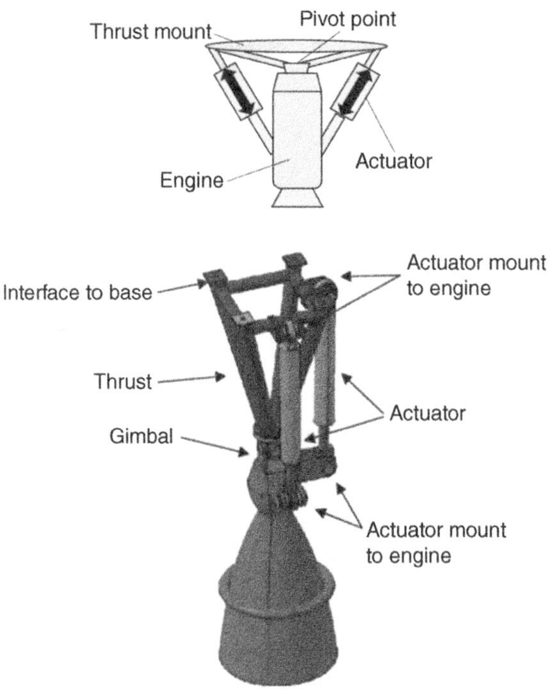

Figure 2.6 A rocket engine can be gimballed by one or two actuators, depending on the number of dimensions we want the engine to move. *Source:* Ref. [6] / Springer Nature / CC BY 4.0.

2.3.6.1 Fins

A flying vehicle can move in any of three ways: yaw, pitch, and roll. Gimballing works in all rocket stages but only when the engine is firing. Fine directional adjustment is often necessary when the engine is not firing.

Fins offer fine directional adjustments as well as stabilization. They function only in the presence of some atmosphere. Rotational motion of a fin results in a difference in aerodynamic pressure between the two sides of the fin, and the vehicle steers accordingly. Fins take on different shapes. Figure 2.7 shows the fins of a missile. They are usually moved by electromechanical means inside the vehicle.

2.3.7 Payload Fairing (Nose Cone)

The fairing is made of two halves and is located at the top of the rocket. It serves as a shroud to protect the payload in the upper stage from exposure to large aerodynamic pressure during ascent.

The fairing is an integral part of the whole rocket design. It should have sufficient interior volume to house a range of payloads. The fairing of the SpaceX Falcon 9, for example, is 13 m long, 5.2 m in diameter, and each half weighs approximately 1,900 kg. The fairing is made of robust materials such as composites. It is light so as not to take away weight from the payload. Its shape and exterior surface are aerodynamically streamlined to minimize drag while taking on the aerodynamic pressure. Figure 2.8 illustrates the payload inside the fairing.

The fairing usually opens by a release mechanism soon after the upper stage clears much of the atmosphere during ascent. The payload is then exposed to an almost space environment. The fairing drops away from about 100 km altitude and falls to the ground or the sea.

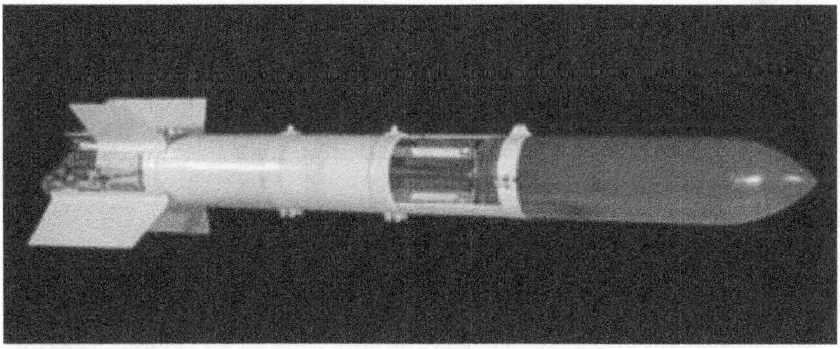

Figure 2.7 Fins on a missile. They stabilize and steer the missile. *Source:* Ref. [7] / Smithsonian Institution.

Figure 2.8 The payload, which is the Sentinel-2C satellite, is encapsulated in the two fairing halves of the Vega rocket. *Source:* Ref. [8] / European Space Agency (ESA).

2.3.8 Aerodynamic Pressure *Q*

The atmosphere presents friction to a moving object. That drag is mathematically linked to the object's speed. The atmosphere exerts pressure on both stationary and moving objects. Static pressure is related to the density of the air column weighing on the object. Relative to the sea level, the static pressure decreases when altitude increases as the air gets thinner.

Aerodynamic pressure occurs when an object moves through a fluid such as air. It is an added pressure to the static pressure. A typical example of aerodynamic pressure in our daily life is evident inside a fast-moving automobile or train. One can hear the sounds and vibrations on the windshield and side windows. Such extra pressure is aerodynamic and is absent when the vehicle is stationary.

Aerodynamic pressure Q is described by Rolle's theorem:

$$Q = \frac{1}{2}\rho v^2$$

where ρ is the density of the fluid and v is the relative speed between the moving object and the fluid.

The object's speed and fluid density counteract each other when applying Rolle's theorem to a rocket launched at sea level. As the rocket rises vertically and gains speed, the initial aerodynamic pressure on the rocket body goes up due to increased speed faster than the air density drop due to altitude.

The aerodynamic pressure Q reaches a maximum at some speed and altitude beyond which it begins to drop as air gets thinner further. This turning point is known as Max Q. For typical rocket launches, Max Q happens between 10 and 15 km above sea level. Max Q is usually called out during a rocket launch.

Aerodynamic pressure "crushes" the rocket body and is considered a survival test of the rocket's structural integrity. Such integrity ensures the payload, and the rocket structure are protected during launch.

To avoid exceeding Max Q, a rocket usually throttles down its thrust to about 80% of its full power as it approaches Max Q to decrease the dynamic pressure on its structure. It then throttles back up to full power soon afterward.

Max Q also happens when a fast-moving descending vehicle moves from high altitude to low altitude as the atmosphere gets thicker.

2.3.9 Kármán Line

The Kármán line is the official "front door" to outer space. It is 100 km above sea level that borders Earth's upper atmosphere and is considered the beginning of outer space. However, the line where space begins is never clear. Earth's atmosphere does not disappear abruptly. There is simply no definitive boundary to Earth's atmosphere.

Wings are crucial for airplanes to fly. Lift is based on the difference in air pressure above and below the wings. The lift decreases as the air becomes too thin at high altitudes. Therefore, a plane would need more wingspan or more speed to maintain lift at high altitudes. There comes an altitude where the residual air can no longer provide lift. In the absence of air in space, wings have no meaning. Therefore, winged aircraft cannot take us to space.

Thrust based on the principle of momentum exchange does not depend on the environment the vehicle is in. It works in the presence of an atmosphere as well as in space.

2.4 Other Thrusters

Rocket engines based on chemical combustion generate large thrust due to the large amounts of high-speed byproducts of exhaust. There is always a need for thrusters to deliver small thrust. Such thrusters serve to push the rocket body sideways or to thrust on the much lighter payload. This section covers several of these thrusters.

Payloads do not have the luxury of having large thrusters and tanks of propellants. Once the payload reaches LEO, it can generate small thrust to slowly increase its orbital velocity to rise to its destination. Satellites, for example, also use small thrusters to make small adjustments to their orbital altitudes and attitudes (their orientations). These thrusters should last as long as the life of the satellites with the capability of multiple start-restart.

2.4.1 Ion Thruster (Ion Drive)

Referencing the Tsiolkovsky equation, one can get ΔV from either exhaust velocity or ejecting a lot of mass or both. One option is therefore to find mechanisms that produce high exhaust velocities but do not eject much mass. Ion thrusters serve that purpose.

The principle of ion thrusters is to use electric means to turn an amount of gas into ions by stripping their electrons. The ions are then accelerated to high velocity and ejected. The gas needs to be relatively easy to ionize and sufficiently heavy to yield thrust. Inert gases fit the bill.

Ionization is often done by subjecting the gas to high-voltage electrodes. Acceleration can be achieved by electrostatic or electromagnetic forces depending on the design. A number of designs exist to make these thrusters more compact. Special hardware is usually installed in place such that the ions pick up electrons they have lost right before exiting the spacecraft, so that the spacecraft stays electrically neutral overall.

Many ion drives in satellites use xenon gas, which has an atomic number of 54 and is the second heaviest inert gas. Other inert gases, such as krypton (atomic number 36), are also used.

Ion thrusters deliver higher exhaust velocities than chemical combustion, but the total mass of gas discharged is much smaller. Therefore, their thrust is usually measured from milli-Newtons (mN) to a few hundred Newtons, depending on the thruster's size and amount of electric power consumed. As a measure of propellant efficiency, a thruster's Isp is often quoted.

Table 2.3 shows three ion thrusters from a list of thrusters. We can see the Isps are high, i.e. almost 10 times those of chemical propellants. However, their thrust is small.

Table 2.3 Ion thruster characteristics.

Thrusters	Weight (kg)	Propellant	Isp (s)	Max thrust (mN)
NEXT	13.5	Xenon	4,200	236
Starlink Gen2	2.1	Argon	2,500	170
BHT-8000	25	Xenon	2,210	449

Ion drives are used when the spacecraft has already reached orbits where local gravity has decreased. Their small thrust will still accelerate the spacecraft to a higher speed, but it takes time to get to the destination.

2.4.2 Cold and Hot Gas Thrusters

Cold gas thrusters produce thrust by discharging a gas through a small nozzle. The gas may be cold or at room temperature. Cold gas stored in a pressurized tank is routed to the thruster. The extent of pressurization in the tank determines the thrust a given amount of gas can deliver. Nitrogen is commonly used due to its low cost and straightforward storage. Other gases such as argon and helium are also used.

Cold gas thrusters are popular in small space missions due to their simplicity. Nitrogen thrusters are also used as the reaction control systems (RCS) to reorient a rocket booster, as will be shown in Chapter 5.

Hot gas thrusters produce thrust by the combustion of chemicals. They can be thought of as mini rocket engines. Such combustion can be a "mono prop," namely, one type of material ignited by passing it through a catalyst. The other option is "dual prop" by igniting a mixture of propellants such as methane and oxygen. Hot gas thrusters produce more thrust due to higher pressure and exhaust speeds of the gaseous byproducts.

2.4.3 Nuclear Thrusters

Nuclear energy is an energy source to discharge propellants at high exhaust velocity to generate propulsion. Nuclear energy has the advantage of being a continuous energy source for an extended period—often measured in years. It is therefore a more reliable source than solar power and batteries.

Nuclear energy in propulsion is used in two ways: Nuclear Thermal Propulsion (NTP) and Nuclear Electric Propulsion (NEP). The concept of NEP is relatively straightforward and will be explained first.

NEP involves using either a nuclear fission reactor or the natural decay of a radioisotope to generate heat. The heat is then turned into electricity to operate an ion thruster. When using radioisotopes to generate heat, the heat cannot be turned on or off. Nuclear fission generates plenty of heat energy and can be turned on or off by controlling the chain reaction. An energy of about 200 MeV (million electron volts) is released in each reaction, which is plenty in atomic energy terms. Such energy manifests as heat at very high temperatures.

NTP utilizes the heat from a nuclear reactor to heats up liquid propellant directly. The propellant is often pumped through the reaction core, where nuclear fission generates plenty of heat. The propellant gains kinetic energy and is then routed through a nozzle to be discharged. There is no chemical combustion in this process. The propellants can be liquid forms of hydrogen (LH_2), water (LH_2O), methane (LCH_4), as well as denser liquids.

Nuclear fission has been terrestrially used for decades. NTP was proposed in the early stage of the US space program. It is now being actively investigated again for lunar missions (see Chapter 12).

Note that the ln (M_0/M_f) factor of the Tsiolkovsky equation comes into play for ion drives because the mass being discharged is far less than chemical propellants. Hence, NTP and NEP do not have the thrust to lift payloads off the Earth's surface. Rather, the current thinking is to use conventional rockets to deliver the payload with built-in NTP or NEP to LEO and activate from there.

We can look at Equations 2.4 and 2.5 to explain the points above. The "g" factor is important, i.e. the presence of gravity from any celestial body. The thrust-producing component is $(V_{exh} * \Delta m/\Delta t)$.

If V_{exh} can be increased 10 times by a new drive, then $\Delta m/\Delta t$ can be decreased by a factor of 10 to keep the thrust-producing component the same. The determining factor is Mg in Equation (2.5). If it is Earth's gravity well for a fairly large spacecraft weighing tens of tons, the Mg factor is very large, relying on contributions from both V_{exh} and $\Delta m/\Delta t$. On the other hand, if the "g" factor is much reduced such as on an asteroid's surface or away from the gravitational influence of large celestial bodies, the spacecraft has no gravity well to overcome, and a "supercharged" drive with high V_{exh} will be of great value.

Therefore, the Tsiolkovsky equation dictates the $\Delta m/\Delta t$ factor in climbing up and down the gravity well of large celestial bodies. One commonly mentioned possibility is to set up transfer stations in LEO, like terrestrial airports to where traditional chemical rockets deliver passengers and payload. They are then transferred to spacecraft more suitable for high orbits and deep space missions.

2.5 Conclusion

This chapter covers the basics of rocketry and related components. Propulsion is the result of momentum exchange as dictated by the Tsiolkovsky rocket equation.

Chemical combustion provides the energy source to generate thrust for the liftoff of a rocket and its payload. Large thrust is generated by ejecting large amounts of mass as well as a sufficiently high exhaust speed. We are currently bound by this means of propulsion to lift off from Earth's surface to reach space.

Electric propulsion usually comes in after the spacecraft is in orbit after having gained sufficient speed. Their thrust is smaller and ejects much less mass. They can work for prolonged periods.

When launching from Earth, the consequence of the rocket equation leads to that at least 70% of a rocket's size and weight is in the first stage, thereby leaving not much room for the payload. Rocket staging is now the common practice to shed weight at high altitudes.

Reading List

Mishra D. *Fundamentals of Rocket Propulsion*. CRC Press; 2017.

Rogers L. *It's Only Rocket Science*. Springer; 2008.

Sutton GP, Biblarz O. *Rocket Propulsion Elements* (9th Edition). John Wiley & Sons; 2016.

References

1 Space X. Falcon 9. Space X [cited 2025 Feb 17] [online]. Available from: https://www.spacex.com/vehicles/falcon-9/.

2 Rodić V, Petrić M. The Effect of Additives on Solid Rocket Propellant Characteristics. FCAD [cited 2025 Mar 20] [online]. Available from: https://www.fcad.com/solid-rocket-propellant/.

3 Skywalker PL. Ariane 62 and 64. Wikimedia [cited 2025 Feb 27] [online]. Available from: https://commons.wikimedia.org/wiki/File:Ariane_62_and_64.svg.

4 Dodd T. Are Aerospike Engines Better Than Traditional Rocket Engines? https://everydayastronaut.com/aerospikes/.

5 Wessels W. Why Nozzles On Vacuum Optimized Rocket Engines Are Bigger Than Those On Sea Level Engines. Head for Space [cited 2025 Feb 27] [online]. Available from: https://headedforspace.com/vacuum-optimized-rocket-nozzles/.

6 Krammer A, Blecha L, Lichtenberger M. Fin actuation, thrust vector control and landing leg mechanisms design for the RETALT VTVL launcher. CEAS Space Journal. 2022;*14*(3):577–591.

7 Tiny Tim Missile. National Air and Space Museum [cited 2025 Feb 21] [online]. Available from: https://airandspace.si.edu/collection-objects/tiny-tim-missile/nasm_A19660030000.

8 European Space Agency. Sentinel-2C Being Encapsulated in the Vega Fairing. European Space Agency [updated 2024 Aug 23; cited 2025 Feb 27] [online]. Available from: https://www.esa.int/ESA_Multimedia/Images/2024/08/Sentinel-2C_being_encapsulated_in_the_Vega_fairing.

3

Orbits

Wang Tianqi[1] *and Stephen W. Cheung*[2]

[1] *Department of Aeronautical and Aviation Engineering, The Hong Kong Polytechnic University, Hong Kong, China*
[2] *Former Adjunct Professor, Department of Physics, The University of Hong Kong, Hong Kong, China*

3.1 Introduction

Going to space usually means reaching an orbit of the desired altitude. This chapter covers orbits around Earth with a focus on the low Earth orbit (LEO). Most LEO spacecraft are below 1,500 km altitude. For example, the International Space Station (ISS) and the China Tiangong Space Station currently in orbit are at a nominal altitude near 400 km.

As of May 2024, there are over 8,000 satellites in LEO. These satellites range from communications (largest share), Earth observations, intelligence gathering, to education and research. LEO is the first stop a spacecraft goes to before moving on to higher orbits or other celestial bodies and beyond. Rocket stages usually deliver payloads to enter LEO, where the payloads use their own propulsion systems to reach their destined orbits.

3.2 Orbit Dynamics

All spacecraft motions in space can be described as orbital motions. The gravitational force from a massive body such as Earth is the source of pull on the spacecraft. In this section, orbits are explained using Earth as the massive body.

Significance of Altitude on Earth's Gravity

The radius of the Earth R_e is 6,370 km, which we approximate as 6,400 km. Gravitational acceleration on the Earth's surface is expressed as "g."

$$g = GM_e/R_e^2 \tag{3.1}$$

New Space: From Low Earth Orbit to the Moon and Beyond, First Edition.
Edited by Stephen W. Cheung and Chih-Yung Wen.

Table 3.1 Gravitational acceleration as a function of altitude h above sea level. The orbital speeds at these altitudes in km/s are shown.

Altitude (km)	$g(h)/g$ in %	Orbital speed v in km/s
50	99	7.89
100	97	7.86
300	91	7.74
600	84	7.57
1,000	75	7.37
2,000	58	6.91
36,000	23	3.08

where G is the universal gravitation constant, and M_e is the mass of the Earth. The value of g is 9.8 m/s^2 in metric units.

There is a misconception that there is no gravity in space and that it is weak in low orbit. The answer comes from the inverse square law of distance. Say, an object is at altitude h above Earth's surface, the distance between the object and the center of Earth is $(R_e + h)$. The gravitational acceleration $g(h)$ as a function of h is now

$$g(h) = GM_e/R_h^2 = GMe/(R_e + h)^2 \tag{3.2}$$

Equation (3.2) can be expressed as

$$g(h)/g = (1 + h/R_e)^{-2} \tag{3.3}$$

From the above equation, we see that the larger the h is, the smaller the $g(h)$ will be. Table 3.1 shows the value of $g(h)$ and orbital speed as a fraction of g for different altitudes of h.

The numbers in Table 3.1 suggest that Earth's gravitational influence drops to under 60% beyond 2,000 km altitude. Gravity continues to taper off with increasing altitude but is never zero. Therefore, one can surmise that Earth's "gravity well" is the steepest from initial ascent to LEO. The orbital speeds in Table 3.1 will be used in the next subsection.

3.2.1 Orbital Velocity

Orbits take on different geometric shapes from circles to ellipses to hyperbolas. We will use circular orbits to develop the basics. For our purpose, speed and velocity are sometimes used interchangeably.

An object in a circular orbit means its orbital speed is uniform. Orbital speed is the instantaneous tangential speed of the circular motion and is altitude dependent. In simple terms, an observer on the ground would call the tangential speed "horizontal." A spacecraft reaching altitude vertically alone is insufficient to enter orbit. It must attain the needed tangential speed to enter and maintain orbit. Without sufficient orbital speed, the spacecraft would simply fall back to Earth.

The altitude, or more accurately, the object's distance to the center of Earth, determines the orbital speed needed to stay in that orbit. Gravity provides the centripetal force the object needs in circular motion. In the absence of an atmosphere, an object can theoretically orbit just above the Earth's surface. The orbital speed works out to be about 7.9 km/s as shown in Table 3.1. This speed gets progressively smaller at higher altitudes. In Chapter 2, we use $\Delta V = 8$ km/s as the nominal value to go from Earth's ground to enter LEO so the numbers are consistent.

The orbital speed at 1,000 km altitude, for example, is less than that at 100 km in Table 3.1. This can be explained by conservation of total energy. An object at 1,000 km altitude has higher potential energy relative to 100 km but less kinetic energy. The total energy (potential + kinetic) remains the same, i.e. conserved.

Therefore, launching an object at rest from sea level to LEO needs to gain about 8 km/s in speed. The underlying assumption is there is no atmospheric drag acting on the object. It is customary to tag an extra 1.0–1.5 km/s to the ΔV to account for such drag. Together with some margin, the working value for ΔV to LEO orbit is between 9 and 10 km/s.

For an elliptical orbit around Earth, the object is sometimes close to and sometimes far from Earth. Its orbital velocity is thus distance dependent. But the total energy (potential + kinetic) also remains the same.

3.2.2 Orbital Geometry

There are two broad categories of orbits: open versus closed. An object in an open orbit means it approaches Earth once, travels with too much speed to be captured, and then flies away and never returns. In trigonometrical terms, an open orbit is either hyperbolic or parabolic. An object on a closed orbit comes around the Earth repeatedly. A closed orbit is either a circle or an ellipse. Planets in our solar system move in closed orbits around the Sun, as does the Moon orbit around Earth.

In terms of speed and energy, if the said object travels at a speed so fast that Earth does not have enough gravitational pull to "capture" it, the object can escape from Earth's gravitational influence. If the object's speed is not sufficiently fast, it gets captured and bound to some form of closed orbit.

Six parameters, known as classical orbital elements (COEs), completely describe the instantaneous position of an object in a closed orbit. One of the COEs is a length measurement, and the others are angles between 0° and 360°. We will only describe parameters applicable to the contents of this book.

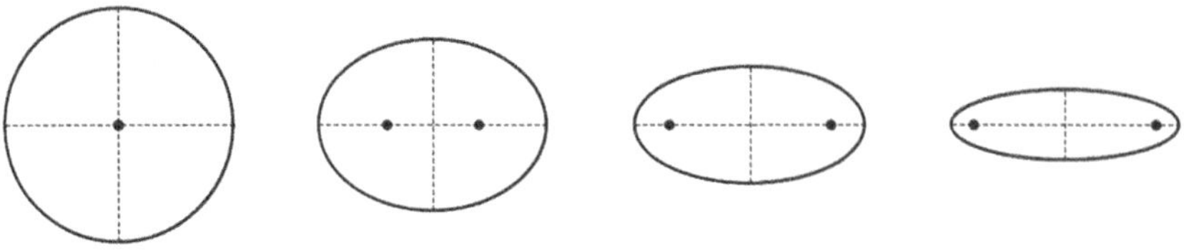

Figure 3.1 Geometric shapes with different eccentricities. The heavy dots are the foci. Eccentricity = 0 is a perfect circle. High eccentricity means a flat ellipse. The center of the perfect circle "splits' into two foci along the major axis.

3.2.2.1 Major Axis

An ellipse is a trigonometrical figure with two foci along a line that is dimensionally the longest. The length is known as the "major axis." A circle can be thought of as a special case of an ellipse where the two foci converge into one. The dimension of a circle is characterized by its diameter. The size of an ellipse is similarly measured by the length of its major axis. The source of gravity, viz. Earth, for an object in an elliptical orbit, is located at one of the two foci along the major axis.

3.2.2.2 Eccentricity

Eccentricity describes an ellipse's shape and represents the degree of deviation from a perfect circle. Eccentricity is a dimensionless value. The eccentricity of a closed orbit varies between 0 and 1. A circle's eccentricity is zero. A small eccentricity value such as 0.1 means the ellipse looks close to a circle, while a high value such as 0.8 means the ellipse looks like an American football. Eccentricity equal to 1 is the boundary between closeness and openness, and the orbit is parabolic. A hyperbolic orbit has eccentricity >1. Figure 3.1 shows four shapes from eccentricity 0 up to 0.8.

3.2.2.3 Perigee Distance

Perigee is an important parameter although it is not a COE. A satellite describing an elliptical orbit goes between the closest and farthest point from Earth and back. Perigee (also known as periapsis) is the closest distance an object is to the center of Earth, while apogee (also known as apoapsis) is the farthest. Perigee is commonly chosen to represent this aspect of the ellipse. The perigee can be calculated using the values of the major axis and eccentricity ε.

The Earth's orbit around the Sun is an ellipse with eccentricity varying from 0.0034 to 0.058. Its current value is 0.017. The difference between the aphelion and perihelion at this moment is 220 km. Planets in the solar system have eccentricities less than 0.1 except for Mercury, whose eccentricity is about 0.2. Similarly, the distances between the centers of Earth and the Moon are 363,396 km (perigee) and 405,504 km (apogee), giving an eccentricity value of 0.055.

It is useful to note that a satellite's speed is slow at a far distance due to weak gravity and fast at a close distance. The satellite moves the fastest at the perigee and the slowest at the apogee. The satellite appears to hang around in the sky longer at the apogee. This feature is useful for a satellite to spend more time viewing a particular region.

3.2.2.4 Inclination

An orbit is on a two-dimensional plane immersed in the three-dimensional world around Earth. Inclination is the angle between an orbital plane and the equatorial plane using the center of Earth as the origin. This definition applies to both circular and elliptical orbits. The equator's inclination is set at angles 0° and 180°. The north pole is +90°, and the south pole is −90° or 270°. A satellite in a polar orbit moves from pole to pole along the north–south direction. A 45° orbit, for example, means a satellite only reaches as far north as 45° latitude and as far south as −45° latitude.

A special case is the Sun-synchronous orbit (SSO). It is a near-polar orbit, usually with an inclination between 98° and 99°. This inclination compensates for the Earth's slight equatorial bulge that disturbs satellite orbits. The SSO ends up with a precession that manifests itself as a satellite always crossing the Equator at the same time of day. Earth observation satellites are usually sent to SSOs.

Prograde and Retrograde The Earth rotates eastward. A satellite moving along the same rotational direction is said to be on a prograde orbit regardless of the hemisphere it is in. Its inclination is between 0° and 90°. Conversely, a satellite on a retrograde orbit means it is moving westward. A retrograde orbit has an inclination between 90° and 180°.

The general preference is for prograde orbits because the Earth's rotation gives a spacecraft an inherent eastward velocity component at launch. At the Equator, this inherent velocity is about 0.46 km/s. This inherent velocity is beneficial to a rocket launch to save fuel. Therefore, if a rocket is launched from the Equator pointing exactly eastward, it gets a "free" ΔV of 0.46 km/s. This "free" velocity component decreases with higher latitudes. Conversely, extra energy is required to launch to a retrograde orbit to erase the inherent velocity component.

3.3 Launching to Orbit and Orbital Transfer

From Chapter 2, a rocket is launched vertically to pass the point of maximum aerodynamic pressure. After Max Q, the next task is to start gaining horizontal speed to be ready for orbit entry. The objective is for the rocket thrust to provide as much horizontal acceleration while also having enough vertical acceleration to overcome gravity and keep ascending.

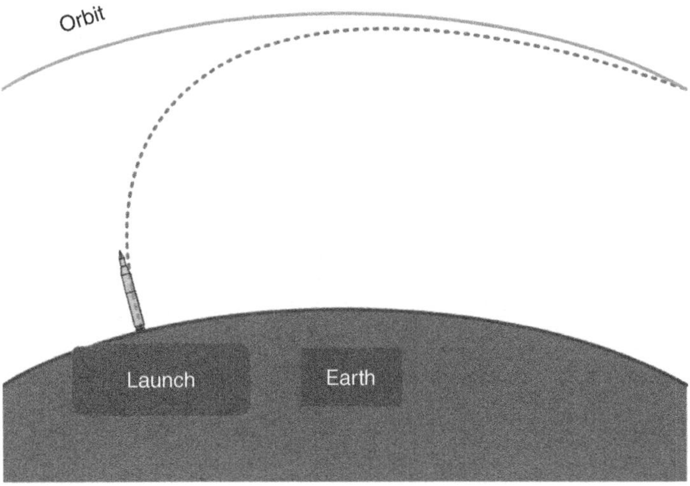

Figure 3.2 Flight path from initial vertical launch to entering orbit, demonstrating gravity turn.

3.3.1 Gravity Turn

The nominal path is for the rocket to begin a "pitch-over maneuver" soon after Max Q by gimballing one or more of the rocket engines. Gravity pulls on the leaning rocket's center of gravity to continue the pitching action. As the horizontal velocity component increases, the rocket begins to describe a segment of a circular motion utilizing the gravitational pull as the centripetal force. The vertical component of the rocket thrust effectively has less gravity to overcome.

The path just described is known as gravity turn. Figure 3.2 illustrates this motion. Gravity turn is like an "aid" to going to orbit.

3.3.2 Orbital Transfer

Standard launch practice is to lift off from Earth and enter LEO. Then, the spacecraft will execute one or more orbital transfers from LEO to other orbits.

Orbital transfer is achieved by adding or subtracting ΔV. Such ΔV is assumed to be achieved instantaneously by a thrust applied for a negligible amount of time. Though unrealistic, this instantaneity is reasonable because a thrust is usually applied for a few minutes, which is negligible compared to the long flight time of the orbits.

We start with orbital transfers being coplanar and co-apsidal. Coplanar means the initial and final orbits are on the same plane, i.e. they have the same inclination. Co-apsidal means the major axes are aligned before and after a transfer.

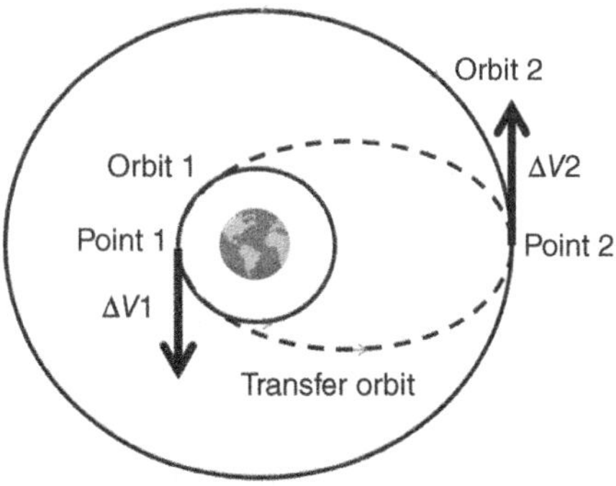

Figure 3.3 A regular Hohmann Transfer, which is from one circular orbit to another circular orbit via half an elliptical transfer orbit (dotted line). Two burns are needed.

Therefore, coplanar and co-apsidal mean thrusts are applied at the two ends of the same axis, so the initial orbit changes size but does not move away from the major axis.

As we will see in later chapters, there are two basic options to execute an orbital transfer: fuel (energy) efficient versus time efficient. Fuel efficient means that minimal fuel is used in executing the transfer. Time efficient means the time from the beginning to the end of the transfer is the shortest possible. A mission completed faster is not necessarily fuel efficient and vice versa.

3.3.3 Hohmann Transfer

Hohmann transfer is the most common mode of orbital transfer from one circular orbit to another circular orbit. It is energy efficient and is a two-burn process as illustrated in Figure 3.3.

Burn 1: A thrust is executed, giving the spacecraft a velocity change of $\Delta V1$ at Point 1. This increased velocity at Point 1 puts the spacecraft on the perigee of a new elliptical orbit with a longer major axis than the initial orbit. This elliptical orbit is called a transfer orbit.

Burn 2: Upon completing the first half of the transfer orbit, the spacecraft reaches the apogee (Point 2) and moves at a slow speed. Another boost is executed, adding $\Delta V2$ to the velocity at Point 2. The right amount of $\Delta V2$ turns the transfer orbit into a circular orbit that is larger than the original circular orbit. The total burn (total ΔV) of the two maneuvers is $(\Delta V1 + \Delta V2)$.

Hohmann transfer is energy efficient. A possible limitation is that some transfers aim to intercept another spacecraft or a distant celestial body. Such a requirement constrains timing.

3.3.4 Bi-elliptic Hohmann Transfer

This transfer takes three burns: start with a circular orbit and finish in a circular orbit via two half-elliptical transfer orbits of different sizes, TO-1 and TO-2.

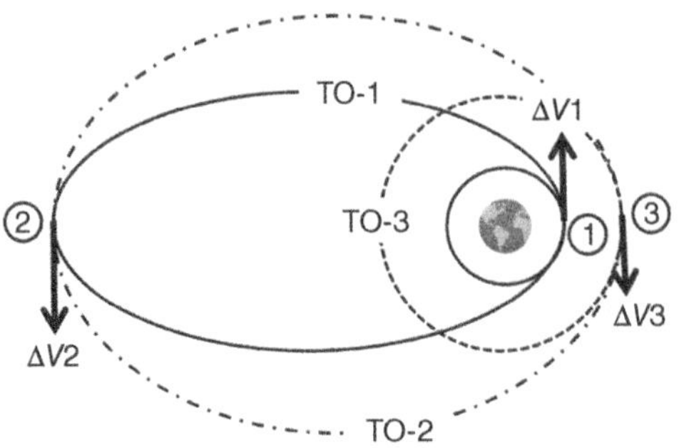

Figure 3.4 The bi-elliptic Hohmann transfer. Two burns $\Delta V1$ and $\Delta V2$ are applied in sequence at the original circular orbit to result in two progressively larger transfer elliptical orbits TO-1 and TO-2. Then a decelerating burn $\Delta V3$ is applied to end up with the final circular orbit TO-3.

Figure 3.4 illustrates this transfer. Two thrusts are first executed, resulting in a half-elliptic transfer orbit TO-1 followed by a larger, second half-ellipse TO-2. A third thrust is applied to result in the final target orbit.

Maneuver 1: A thrust is executed, resulting in ΔV at Point 1 and an elliptical transfer orbit TO-1 toward Point 2 as a Hohmann transfer.

Maneuver 2: The spacecraft is now at apogee (Point 2). A second boost is executed by adding $\Delta V2$ at Point 2. This second boost is insufficient to form a larger circular orbit like the regular Hohmann orbit above. Instead, another elliptical transfer orbit TO-2 is formed. TO-2 is larger than TO-1. This brings the spacecraft to Point 3, which is the corresponding perigee of TO-2.

Maneuver 3: Point 3 is farther from Earth than Point 1 but is closer to Earth than Point 2. A third thrust $\Delta V3$ in the opposite direction at Point 3 to reduce the spacecrafts speed will turn the second transfer orbit TO-2 into the final circular orbit.

The total burn of the three maneuvers leads to a net ΔV of $(\Delta V1 + \Delta V2 - \Delta V3)$. In terms of energy efficiency, we can compare the total burn of the bi-elliptic Hohmann versus the regular Hohmann. The regular Hohmann would have different values of $\Delta V1$ and $\Delta V2$.

In practice, it is the size of four orbits (initial, TO-1, TO-2, and final) that matters. It can be shown that the ratio of the radius of the final target r_3 to the initial orbit r_1 that determines which orbital transfer approach is more energy efficient.

$r_3/r_1 > 15.58$ (Regular Hohmann is more energy efficient)

$r_3/r_1 < 11.94$ (Bi-elliptic Hohmann is more energy efficient)

3.3.5 Ellipse to Ellipse Hohmann Transfer

Ellipse to Ellipse (E2E) transfer works similarly to the two previous cases. We will work on the case of moving from an initial ellipse to a larger ellipse. Thrust is applied to the "perigee" and the "apogee" like the regular Hohmann.

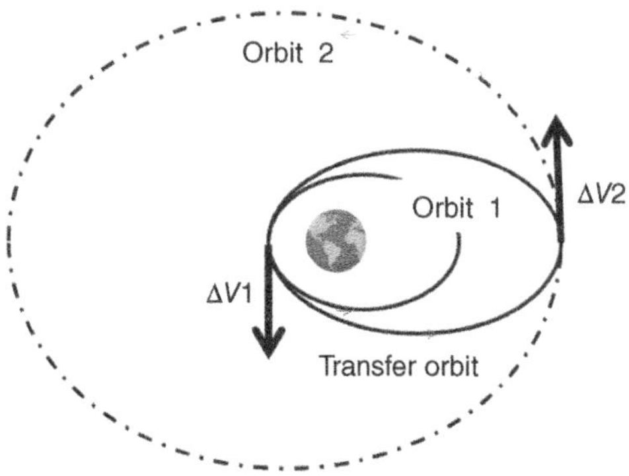

Figure 3.5 Ellipse to Ellipse Hohmann transfer from initial elliptical Orbit 1 to a half-elliptical transfer orbit and then reach the final Orbit 2. Two burns are needed.

Maneuver 1: At the perigee of the initial ellipse, a thrust is executed to result in a $\Delta V1$ in Figure 3.5. This leads to the first elliptical transfer orbit.

Maneuver 2: When the spacecraft reaches the apogee of the transfer orbit, its speed is the slowest. A second thrust is then executed to result in an added $\Delta V2$. The higher speed leads to a larger elliptical orbit. The apogee remains the same, but the perigee has now moved further from Earth.

To summarize, Hohmann transfer works to raise or lower orbits. The examples above are for raising orbits further away from Earth. In lowering orbit, an example is a spacecraft from the space station returning to Earth. It must go to lower altitudes and eventually deorbit. It is not practical for the spacecraft to come to a halt because that would take almost 8 km/s to lose all that speed. Instead, the spacecraft's orbital speed is brought down enough that it begins to describe an elliptical orbit around Earth that would put the perigee near the edge of space to utilize the higher atmospheric drag to slow down further. This deorbit process is essentially done by progressively shrinking elliptical orbits.

In coplanar orbital transfers, thrust is applied along the same line as the direction of travel, i.e. the spacecraft's velocity. Changing orbital velocities is just a matter of adding and subtracting the velocities. There are cases where orbital changes require the thrust to be applied at an angle to the direction of travel. Below is an example.

3.3.6 Inclination Change

Inclination change can be pictured as follows. Imagine we start with a satellite in an inclined orbit. To move to another orbit of different inclination, both orbits still need to use the Earth as the source of gravitational pull. The two inclined orbits intersect at the Equator in either direction (north-going-south or south-going-north). Hence, inclination change must be executed when the satellite crosses the Equator.

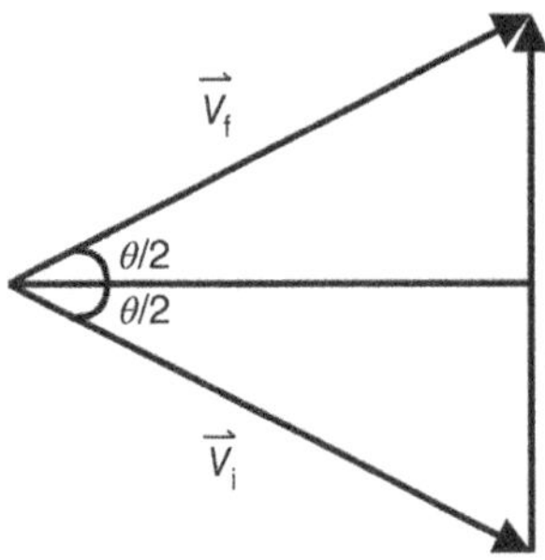

Figure 3.6 Some orbital changes require the velocity vectors to change in magnitude or direction. An external vector ΔV_s is needed to change the initial vector V_i to the final vector V_f to result in a net inclination change of angle θ.

We will demonstrate an inclination change of a circular orbit. For simplicity, only the inclination is changed but not the size of the orbit, i.e. the magnitude of the orbital velocity stays the same. The previous cases have the before and after orbit on the same plane. Therefore, the velocity vectors are aligned in the same direction as addition and subtraction.

With inclination change, there is a net change in orbital inclination by an angle θ. Hence, the initial and final velocity vectors are not aligned, and we need to use trigonometry. Figure 3.6 shows the trigonometry of the initial and final velocities. The change in direction from V_i to V_f is the result of a ΔV_s applied at an angle $(90° - \theta/2)$.

Here V_i and V_f are assumed to be the same in magnitude as an illustration. From trigonometry, we have the formula

$$\Delta V_s = 2V_i \sin\left(\frac{\theta}{2}\right) \tag{3.4}$$

One can look at a few special cases to test Equation (3.4) with different angle θ.

1) $\theta = 0°$, $\sin(\theta/2) = 0$, $\Delta V_s = 0$. The case is reduced to coplanar, and there is no orbit change. There is no ΔV_s because the initial and final velocities are the same.
2) $\theta = 180°$, $\sin(\theta/2) = 1$, $\Delta V_s = 2V_i$. This is also a coplanar case. The angle of ΔV_s must be applied at an angle $(90° - 180°/2 = 0°)$. The spacecraft must come to a full stop and then reverse its direction by a velocity of the same magnitude. Therefore, this maneuver requires $2V_i$.
3) $\theta = 30°$, $\sin(\theta/2) = 0.26$, $\Delta V_s = 0.52V_i$. The amount of thrust needed to create a ΔV_s equal to $0.52V_i$ at an angle $= 90° - 15° = 75°$ to effect this change. If this plane change is to happen in LEO, where the orbital velocity is about 8 km/s, a net change of inclination by 30° would need over 4 km/s, which is substantial.

Inclination change like the examples above is usually done at the moment the spacecraft crosses the equator to the north or the south.

3.4 Lagrange Points as Orbits

So far, satellite orbits are fairly close to a massive body such as Earth. The presence of other massive bodies such as the Sun or the Moon nearby is usually ignored due to their distances. The primary gravitational pull comes from Earth. But there are cases where the gravitational pulls from two massive bodies must be considered together.

Take Earth and the Moon as an example. Both are massive enough to have gravitational pulls. A requirement is that the Moon orbits Earth. There are locations where Earth's gravitational pull is exactly counteracted by that of the Moon. These locations are known as Lagrange points. There are five Lagrange points for two massive bodies, M1 and M2, in which M2 revolves around the more massive M1. At these Lagrange points, a small object will follow M2 to orbit M1 in a synchronized manner. These five points L1, L2, L3, L4, and L5 are shown in Figure 3.7.

Lagrange points are about the orbital behavior of a small object (spacecraft) in the presence of the gravitational tug of war between a pair of massive bodies. It is theoretically known as a three-body problem. Lagrange points apply to all the planets orbiting the Sun. Between Earth and Mars, for example, no Lagrange points are formed because neither planet orbits around the other.

Not all the five Lagrange points are equal. L1, L2, and L3 are inherently in unstable equilibrium between the two forces of gravity. Any small disturbances will cause the spacecraft to drift further away toward either M1 or M2. Hence, natural objects such as debris and asteroids do not gather there. A spacecraft placed there would require regular station keeping via onboard thrusters.

Only L4 and L5 are stable, which means that a spacecraft on these two orbits is less sensitive to disturbances and would return to its position naturally. The stability of L4 and L5 rests on the condition that the more massive object M1 is at least 24.59 times more massive than M2. Less than this ratio, L4 and L5 also turn unstable.

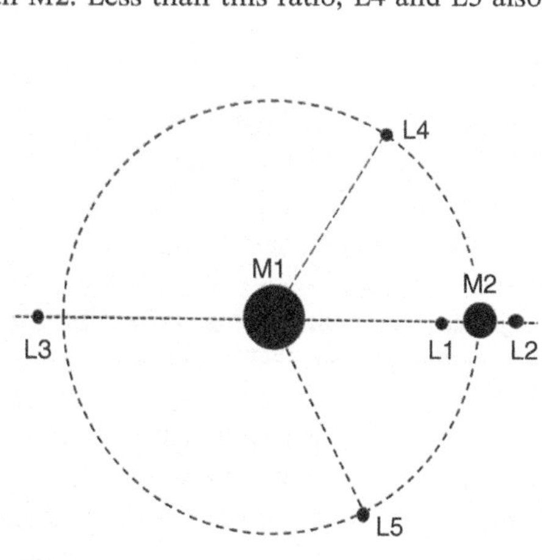

Figure 3.7 There are five Lagrange points between two massive bodies, M1 and M2 with M2 orbiting around M1. M2 has less mass than M1.

The ratio M1/M2 larger than 24.59 holds true with the Earth-Moon pair and all the Sun-planet pairs. One point to notice with Lagrange points is that L1 and L2 are usually close to the lighter mass of M2. The larger the M1/M2 ratio, the closer L1 and L2 are to M2.

In the case of the Earth-Moon pair, spacecraft at L1 and L2 are "within reach" from the lunar surface. China places the Queqiao communications satellite in L2 to communicate with missions on the far side of the Moon. L4 and L5 are stable and collect interplanetary dust known as the Kordylewski cloud. A large number of Trojan asteroids gather around L4 and L5 of the Sun-Jupiter Lagrange points.

In the case of the Sun-Earth pair, L1 and L2 are located about 1.5 million km from Earth. An observatory placed at Sun-Earth L1 would have an unobstructed view of the Sun and is good for observing solar activities. India's Aditya-L1 mission is a solar observatory located at L1. IMAP, which stands for Interstellar Mapping and Acceleration Probe, is a mission planned for 2025 by NASA, sending a satellite to L1 to study the interaction of solar wind on radiation and materials from intergalactic sources.

On the Sun-Earth L2 side, the Earth blocks the solar influences. Therefore, the James Webb Space Telescope placed there would be good for observing the universe. L3 is located behind the Sun, well out of the line of sight with Earth. We have little knowledge about if and what is located there. Dust and asteroids gather around the Sun-Earth L4 and L5.

The dimensions of the Earth-Moon and Sun-Earth Lagrange points are shown in Tables 3.2 and 3.3, respectively.

Table 3.2 Lagrange points in the Earth-Moon system.

Lagrange point	Distance from Earth	Distance from Moon	Application/example
L1	≈323,000 km	≈61,000 km	SMART-1 mission
L2	≈448,000 km	≈64,000 km	Queqiao satellite
L3	≈381,000 km	≈768,000 km	Not commonly used
L4	≈384,000 km	≈384,000 km	Kordylewski cloud
L5	≈384,000 km	≈384,000 km	

Table 3.3 Lagrange points in the Earth-Sun system.

Lagrange point	Distance from Earth	Distance from Sun	Application/example
L1	$\approx 1.5 \times 10^6$ km	$\approx 148.5 \times 10^6$ km	Aditya-L1, IMAP
L2	$\approx 1.5 \times 10^6$ km	$\approx 151.5 \times 10^6$ km	James Webb Space Telescope
L3	$\approx 299 \times 10^6$ km	$\approx 149.6 \times 10^6$ km	Not commonly used
L4	$\approx 149.6 \times 10^6$ km	$\approx 149.6 \times 10^6$ km	Trojan asteroids
L5	$\approx 149.6 \times 10^6$ km	$\approx 149.6 \times 10^6$ km	

3.5 ΔV Map

We have shown in this chapter that ΔV is intimately connected to orbits. Orbital changes require burns (energy), a simple measure is ΔV. Missions requiring multiple orbital changes can be characterized by adding or subtracting ΔV of each maneuver.

Numerous ΔV charts are available online. Some maps are quite extensive. The ΔV values are approximate and often depend on the many options a spacecraft can go from Point A to Point B. For example, extra ΔV would be needed if a spacecraft goes through one or more inclination changes. Hence, the actual ΔV is path dependent.

Figure 3.8 is a good illustration of how ΔV values shown in the figure are approximate. Let us embark on a journey to leave Earth in Figure 3.8.

(i) From Earth to LEO, $\Delta V = 9.5$ km/s. About ~1.5 km/s is allocated to overcome atmospheric loss during ascent.
(ii) From LEO to GEO (geosynchronous orbit), a boost of $\Delta V = 4.3$ km/s is needed to overcome Earth's gravity to reach the high orbit
(iii) From LEO directly to Lagrange point EML1 (skipping GEO), $\Delta V = 3.8$ km/s
(iv) From EML1 to lunar transfer orbit (LTO), $\Delta V = 0.7$ km/s; then to Moon's surface, $\Delta V = -2.5$ km/s (deceleration boost)
(v) From EML1 to low Mars orbit (LMO) (skipping the Moon), $\Delta V = 2.4$ km/s; from LMO to Mars's surface, $\Delta V = -4.1$ km/s (deceleration boost)

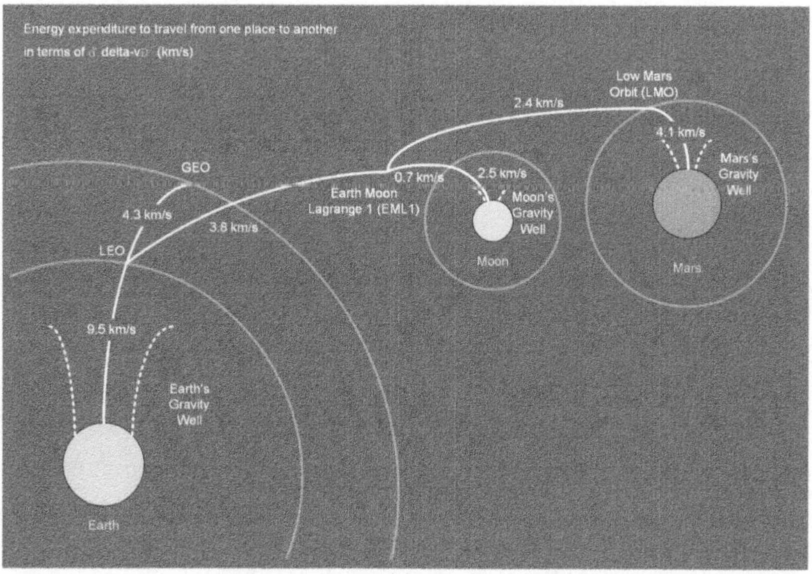

Figure 3.8 This is a ΔV map of the Cislunar railroad previously shown in Chapter 1. *Source:* Adapted from Ref. [1].

3.6 Conclusion

Space flights are about the mechanics of orbital changes. A spacecraft never travels in a straight line toward its target due to the constant presence of gravitational force from one or more massive bodies. Going from one orbit to a higher orbit away from the source of gravity requires energy consumption, or ΔV, by the spacecraft. Energy would be expended in the opposite direction of travel in order to reach lower orbits. ΔV is therefore a convenient indication of energy consumption to undergo orbital change.

The simplest and most energy-efficient way to make orbital change is the Hohmann method, which is to go from one orbit to another under certain conditions, such as coplanarity. Significant ΔV is needed whenever the target orbit is not on the same plane.

Lagrange points are special orbits that can be leveraged. We will see in Chapter 11 that a special orbit is chosen for the proposed lunar space station Gateway. The orbit leverages the Earth-Moon Lagrange points.

Reading List

Bate, R.R., Mueller, D.D., and White, J.E. (2020). *Fundamentals of Astrodynamics*, (Second Edition). Dover Publications.

Cicci, D.A. (2022). *Orbital Mechanics: The Essentials*. DS Publishing.

Logsdon, T. (1997). *Orbital Mechanics: Theory and Applications*, (1st Edition). Wiley-Interscience.

Wiesel, W.E. (2010). *Spaceflight Dynamics*, (Third Edition). CreateSpace Independent Publishing Platform.

Reference

1 CisLunar Railroad. United Launch Alliance. Cited by Boucher M. Panel: CisLunar Self-sustained Space Economy. Space Ref [updated 2017 Apr 4; cited 2025 Feb 27] [online]. Available from: https://spaceref.com/space-commerce/panel-cislunar-self-sustained-space-economy/.

4

Satellites and Space Stations

Xu Bing

Department of Aeronautical and Aviation Engineering, The Hong Kong Polytechnic University, Hong Kong, China

4.1 Introduction

Launching a spacecraft to low Earth orbit (LEO) is the first step of a space mission. It requires less propellant and thrusts to reach compared to medium and high Earth orbits. In 2024, the total number of operational satellites in space has reached 10,000 satellites, 84% of which are in LEO. About 7,000 of these satellites belong to the Starlink constellation (Chapter 6) [1].

LEO is the only zone of space that humans have set foot in apart from the lunar surface. Space stations are technically also satellites. They are megastructures assembled in orbit for humans' long-term presence in microgravity. They are national and international projects so far. Commercial stations will enter the picture in the next two decades.

Given its significance, this chapter will cover satellites in LEO, with an emphasis on crewed space stations and their commercial development. This chapter provides an overview of spacecraft and explores the future of long-term human presence in LEO to be driven by the commercial sector.

4.2 Types of LEO Satellites

4.2.1 Earth Observation Satellites

Earth observation (EO) satellites are designed to monitor activities on Earth and collect scientific data on the atmosphere, land, and oceans in understanding geographic sciences and assessing the state of climate, natural resources, and human activities. This category is considered one of the most important and versatile applications.

New Space: From Low Earth Orbit to the Moon and Beyond, First Edition.
Edited by Stephen W. Cheung and Chih-Yung Wen.

EO satellites are equipped with various instruments, such as optical sensors, radars, and LiDAR (light detection and ranging), to capture images and measurements across different wavelengths using tools such as multispectral and hyperspectral imaging (see Chapter 7).

Meteorological satellites are designed to observe and analyze atmospheric and weather conditions. LEO is one of the most common orbits, with Sun-synchronous orbit (SSO) being particularly advantageous for missions requiring global coverage and frequent revisits to targets due to its high inclination and short orbital period. Because SSO passes directly over Earth's poles, it is an ideal choice for missions focused on polar observations. Beyond the LEO, inclined geosynchronous orbits (IGSO) and geostationary orbits (GEO) are also popular for EO satellites, as they enable continuous monitoring of specific regions on Earth's surface.

Sun-synchronous orbit is "near polar" with a typical inclination of 98–99° at an altitude of 600–800 km with an orbital period of about 100 minutes. SSOs can be circular or elliptical. It is unique in that the satellite always comes back to visit a spot on Earth at the same local "solar time," which means the view of the same spot can be compared today with yesterday at the same time of the day. This facilitates tracking under the same lighting conditions. As will be seen in Chapter 7, many EO satellites are in SSO.

Meteorological satellites focus on targets such as weather, storms, cloud formations, wildfires, pollution levels, auroras, and lightning activity. They are equipped with various sensors, including imagers that capture visible and infrared light, sounders and GPS radio occultation that provide temperature and humidity profiles of the atmosphere, radiometers that measure Earth's radiation levels, LiDARs that detect atmospheric particles, spectrometers for chemical composition analysis, and altimeters for surface topography. Notable examples of LEO meteorological satellites include the US's Defense Meteorological Satellite Program (DMSP) and China's Fengyun series.

Environmental monitoring and mapping satellites are primarily used to collect data on a wide range of terrestrial and environmental factors, including geological structures, topographical features, agricultural productivity, oceanological conditions, ecological health, and climatic changes. This data is essential for understanding and managing natural resources, disaster response, and climate change mitigation. These satellites often operate in LEO (including SSO) for global coverage and in GEO for regional observation. Prominent examples of LEO satellites in this category include the Landsat program jointly by NASA and the US Geological Survey (USGS) and ESA's Sentinel missions.

4.2.2 Communication Satellites

Communication satellites have the most impactful commercial benefits. These satellites relay signals from ground stations and transmit directly to users, making them an integral part of the global communication infrastructure. Instruments on board must be capable of facilitating the transmission, reception, and relay of signals across vast distances, ranging from hundreds to thousands of kilometers. Power comes from solar arrays and batteries.

Traditionally, a small fleet of large satellites formed global communication constellations due to high launch costs. To maximize each satellite's coverage, these constellations are placed in higher orbits, such as medium and geosynchronous orbits.

A significant reduction in launch costs in recent years has enabled the deployment of low-cost satellites in LEO to create mega constellations such as the Starlink global communication constellation (see Chapter 6). Communication satellites can be categorized into two main subcategories: telecommunication satellites and navigation satellites.

Telecommunication satellites are designed to relay television, radio, telephone, and Internet signals from and back to Earth's surface. Various applications are supported by these satellites, including mobile broadband, television broadcasting, maritime and aviation communications, military and governmental hotlines, and emergency services. Satellites for television broadcasting and radio communication are typically placed in MEO or GSO, where they benefit from extensive coverage and prolonged access times. These characteristics are ideal for delivering direct-to-home television, long-distance communication, and regional media distribution services. For broadband Internet services, LEO is preferred and has seen the most rapid development in recent years. The LEO altitude provides strong signal strength and low latency, both of which are critical for high-speed Internet access.

Navigation satellites are designed to provide positioning, navigation, and timing (PNT) services to users on Earth or in space. By making use of the high predictability of satellite position and velocity, the user receiver's position, velocity, and time can be computed. Traditionally, navigation satellites are deployed in MEO as part of global navigation satellite systems (GNSS), such as GPS, Galileo, and Beidou. LEO-based satellite navigation is also becoming an emerging concept. Mega LEO constellations will significantly increase satellite visibility from the ground, enabling positioning with improved accuracy and convergence time.

4.2.3 Scientific and Experimental Satellites

Scientific and experimental satellites are designed to collect scientific data, conduct in-space experiments, and demonstrate or test new technologies. The key distinction is that their main objectives are focused on academic research and engineering development rather than commercial services.

Technology demonstration satellites validate new engineering concepts, test technologies in the microgravity environment of space, or examine the launch vehicle's lifting capability. In recent years, it has become increasingly common for research institutes to launch small satellites as part of research projects aimed at promoting and facilitating education and research in aerospace engineering. LEO is the first choice for these smaller-scale missions due to budget constraints. Notable programs for technology demonstration satellites in LEO include JAXA's Innovative Satellite Technology Demonstration Program and NASA's Technology Demonstration Missions (TDMs).

Research and experimental satellites utilize the space environment to collect scientific data that is otherwise difficult or impossible to obtain on Earth in the fields of space physics, astronomy, atmospheric science, Earth science, life science, and materials science. For example, the microgravity environment is highly desirable for the development of high-purity crystals, fibers, and nanomaterials.

Astronomical satellites, commonly known as space telescopes, operate beyond the interference and obstruction of Earth's atmosphere, which massively improves astronomical observations across all frequency bands. The orbit selection for this class of satellites is highly diverse, ranging from LEO to pseudo-orbits at Lagrange points and trajectories to outer space. LEO remains the most widely used orbit due to its cost-effectiveness. Notable examples of research satellites in LEO include the famous Hubble Space Telescope.

4.2.4 Crewed Spacecraft

Crewed spacecraft are special satellites that orbit Earth but for a limited time. Then, they will return to Earth. These spacecraft transport astronauts to space, maneuver in orbit, and safely return them to Earth. Usually, these spacecraft transport humans and cargos to space stations.

Such spacecraft is equipped with life support systems, propulsions and thrusters, technologies for docking, undocking, and extravehicular activity (EVA), to name a few. As of 2024, only the United States, China, and Russia (formerly the Soviet Union) possess manned spaceflight capability.

To date, all crewed space station missions have been conducted in LEO. The trend of crewed and cargo missions is for governments to subcontract such services to commercial vendors, who would develop their own launch and transport systems. NASA's Commercial Crew Program (CCP) is one such example that transports humans and cargo to and from the International Space Station.

4.3 Miniaturized Satellites in LEO

Commercialization of LEO and low cost of access led to the emergence of miniaturized satellites. These satellites weigh much less than the big satellites in high orbits. This development has made satellites affordable and accessible to a broader market, including smaller organizations, research institutes, private companies, and even individual customers.

Today, nanosatellite missions to LEO use commercial off-the-shelf (COTS) components with mission durations from a few months to several years. A nanosatellite weighs 1–10 kg with a volume of a few thousand cubic centimeters and costs under US$50,000 to build. It is equipped with subsystems, together with scientific payloads and solar arrays. The short development timeline of miniaturized satellites results in more frequent launch opportunities and faster iterations of technology, making smaller satellites suitable for constellations.

Reduced satellite fabrication costs, availability of small-lifting rockets, and breakthroughs in reusable rockets have significantly lowered project costs. CubeSats started out being launched as secondary payloads attached to other missions on small rockets. Examples of CubeSats in LEO include the AAU CubeSat by Aalborg University in Denmark and the QuakeSat by Stanford University.

Nanosatellites can also be released directly from orbiting space stations, allowing for even more flexible deployment. SpaceX's Transporter program now regularly launches multiple small satellites in one Falcon 9 launch.

For a small satellite project, the launch cost remains to be the primary financial expenditure. Therefore, LEO is often the only affordable orbit to reach. Most rockets designed for quick-launch-readiness are small-lifting vehicles, such as Rocket Lab's Electron and China Aerospace Science and Technology Corporation's Long March 6. These vehicles are incapable of delivering payloads to higher orbits, making LEO the only viable option.

CubeSat is a class of nanosatellites and a standardization concept that has become the most popular and successful small satellite model. The basic unit of a CubeSat is 1U (which stands for "one unit") and serves as a modular building block that can be assembled into a larger satellite of several standard units. 1U is an aluminum cube, dimensions $10 \times 10 \times 10\,\text{cm}^3$, weighing less than 1.33 kg and providing an interior volume of 1,000 cc.

Developed by California Polytechnic State University (Cal Poly) and Stanford University, CubeSats are intended to enable students to learn the design, fabricate, test, and also operate spacecraft. CubeSats have evolved into a global standard for nanosatellites. As of May 2024, about 2,400 CubeSats have been launched.

The education sector has benefited the most from standardized CubeSats. Students from middle schools to universities can experience the design of flyable satellites for various scientific objectives. Using COTS components has greatly reduced the cost and scope of development from design to launch.

A significant milestone was achieved in 2018 when 2 MarCO CubeSats (MarCO-A and MarCO-B) were deployed in NASA's InSight mission to Mars, marking the first time CubeSats ventured into deep space.

Concepts similar to CubeSats have also been offered. For instance, PocketQubes are similar in design but measure $5 \times 5 \times 5\,cm^3$ per unit. PicoSats are "do-it-yourself" kits that can operate as a satellite swarm and rely on mother satellites for communication. They cost under US$ 12,000. Another alternative is the TubeSats, which are cylindrical, measuring 8 cm in diameter and 13 cm in height, and weigh less than 1 kg. Lastly, ChipSats in the femtosat category are credit card size and are made from printed circuit boards (PCBs), including solar cells.

4.4 Space Stations

A space station is an orbiting, crew-occupied habitat. A station lacks major propulsion or landing systems and has docking ports to allow other spacecraft to transfer crew and supplies. Space stations are multifunctional including microgravity research, Earth and astronomical observation, and a variety of research areas. Space stations play a crucial role in studying the long-term effects of the weightlessness environment on human health.

A range of systems is onboard, including life support systems, thrusters, technologies for extravehicular activity, docking, and research facilities.

In recent years, commercial space stations have become a point of interest in the space industry. Multiple concepts have been proposed and actively developed. Currently, all operational space stations are in LEO as will most future commercial stations. Governments will focus their priority and resources on deep space ventures. The Lunar Gateway will be an orbiting station around the Moon, see Chapter 11.

This section will discuss the current state and prospects of space stations, from government-led initiatives to cutting-edge commercial developments.

4.4.1 International Space Station and Tiangong Space Station

Two space stations are currently in LEO: the International Space Station (ISS) and China's Tiangong Space Station. The ISS is an international collaborative project. It was assembled and is now maintained by five space agencies, including NASA, Roscosmos, European Space Agency, Japan Aerospace Exploration Agency (JAXA), and Canadian Space Agency. The Tiangong space station is a crewed structure constructed and operated solely by China.

Table 4.1 compares several key information between the ISS and Tiangong. Both stations have been fully utilized to fulfill their functions.

Missions of resupply and crew changes take place regularly. Stations are also the platform for next-generation transportation vehicles, microgravity experiments, and technology demonstrations for future missions.

Table 4.1 A quick statistics comparison between Tiangong and ISS.

	Tiangong space station	International space station
Size	Length: ~55 m Diameter: ~39 m	Length: ~109 m Width: ~73 m
Weight	~100,000 kg	450,000 kg
Number of modules	3	16
Max. crew	6	13
Average orbital altitude	289.1 km	400 km
Starting launch date	April 2021	November 1998
Mission status	Operational	Operational

As the ISS is to be phased out by 2030, the view is that commercial space stations in LEO will steadily emerge.

4.4.1.1 International Space Station (ISS)

Operational since 1999, the ISS shown in Figure 4.1 has been the largest man-made satellite in orbit to date. The space station provides a habitable volume of 1,005 m^3. The ISS is a platform for long-term human habitation and scientific research in space. As of 2018, the construction cost was around US$155 billion. At a given

Figure 4.1 Image of the International Space Station taken in 2021. *Source:* Ref. [2] / NASA public domain.

time, the ISS supports a crew of seven on average, although it has a maximum capacity of 13 during shift change.

The structure of the ISS comprises 43 major elements (modules), including pressurized modules, trusses, solar panels, docking ports, airlocks, and a wide array of sensors, electronics, and robotic systems. Among these 43 elements, 16 are pressurized modules where the crew members reside and conduct experiments. The modules belong to the five space agencies who are directly involved in the ISS: US (8), Russian (6), ESA (1), and Japanese (1).

To date, several vehicles support ISS activities. Launch vehicles from different countries have been used to transport cargo and crew. SpaceX's Falcon 9, for example, is a frequent flyer to the ISS on its Dragon capsule.

The ISS is a microgravity laboratory. Since becoming operational, astronauts on the ISS have conducted around 3,000 experiments, averaging three per week. The ISS also enables engineers and scientists to study space construction techniques, conduct astronaut training, and deploy small satellites from the station. Due to several factors such as uneven distribution of Earth's gravity, residual atmospheric drag, and floor vibration due to machinery, the ISS cannot maintain an absolute zero-g environment. Residual atmospheric drag requires regular boosting to raise its orbit to the nominal altitude.

The ISS is scheduled to retire by the end of 2030 after three decades of service. SpaceX will assist in deorbiting the megastructure. However, before moving to complete retirement, the ISS will be expanding its role in commercial activities such as commercial research, manufacturing in space, tourism, and even film production. Additionally, the ISS continues to support the release of nanosatellites and new technology testing, such as the BEAM inflatable module, which could shape the future of space habitation.

4.4.1.2 Tiangong Space Station

The Tiangong Space Station shown in Figure 4.2 is the first operational and permanent crewed space station constructed by China. The name "Tiangong" translates to "Heavenly Palace." The space station was under the China Manned Space (CMS) program.

The construction of Tiangong Space Station took around 1.5 years between 2021 and 2022. The CMS program is also preparing to launch the Xuntian telescope – a co-orbiting telescope currently under development – by 2026.

Tiangong has similar scientific objectives to the ISS, focusing on research across multiple disciplines, such as space life sciences and biotechnology, that involve exploring the effects of microgravity on living organisms and human health. Research on microgravity physics and material science is being prioritized as well, exploring how materials behave in the unique space environment. Earth observation is also carried out to enable high-resolution environmental

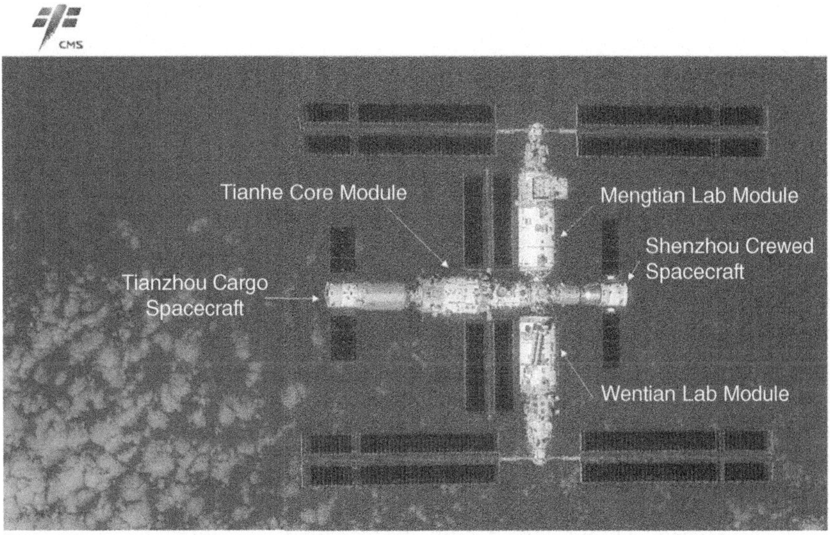

Figure 4.2 Tiangong Space Station with all its modules labeled. The Tianzhou Cargo Spacecraft serves to transport cargo from Earth to the space station and return to Earth. *Source:* Ref. [3] / China Manned Space Engineering Network/Public domain.

monitoring and climate studies and testing of space robotics and new space systems. Educational outreach is also integral with live interactions and experiments shared with students around the world.

The microgravity environment is known to affect human health during a lengthy stay. Examples of such effects include loss of bone density, dehydration due to excess fluid excretion, space sickness due to a weakened immune system, and balance disorders on returning to Earth.

4.4.2 Commercial Space Stations

Commercial space stations are launched and operated by the private sector with little to no government funding. Presently, space agencies worldwide plan to design and fabricate miniaturized space station components and modules. Small stations are considered more cost-effective for dedicated purposes. Private companies such as Axiom Space, Sierra Space, and Nanoracks are looking to provide these structures. Concepts of a new generation of commercial space stations in LEO have been proposed. A wide range of markets include commercial R&D, tourism, manufacturing, biotechnology, media and film production, astronaut training, transfer and logistics, satellite services, education, and many others.

Compared to government-led projects, public-private and pure private projects enjoy multiple advantages such as motivation for cost reduction and sustainability, economies of scale, fast development and iteration, lower risk for governments, and utilization of the capital market.

New ideas such as miniaturization, inflatability, expandability, and modularity are driving the advancement of commercial space stations. Below are a few examples of commercial space stations under development.

As a sign of the paradigm shift, NASA has the Commercial LEO Destinations (CLD) program in which contracts are awarded to encourage companies looking to build and send space stations to LEO.

4.4.2.1 Bigelow Expandable Activity Module

The Bigelow Expandable Activity Module (BEAM) is an expandable module developed by Bigelow Aerospace. The module was attached to the ISS as a demonstration unit as shown in Figure 4.3. Installed in 2016, the BEAM experimental program aimed to test and validate inflatable habitat technology, including launch and deployment of a commercial inflatable module, radiation protection, thermal and structural performance, mechanical durability, long-term leak resistance, and safe operation. BEAM provides an internal volume of 16 m^3.

4.4.2.2 Orbital Reef

Orbital Reef, also referred to as "mixed-use business park" in space, was proposed in late 2021 by Blue Origin. Multiple business and academic partners have joined the development of the station. Blue Origin was awarded a $130 million contract for the Orbital Reef project, one of three funded by NASA as the first phase of the agency's Commercial LEO Destinations (CLD) program.

Figure 4.3 The Bigelow Expandable Activity Module (BEAM) is attached to the Tranquility module of the ISS. *Source:* Ref. [4] / NASA/Public domain.

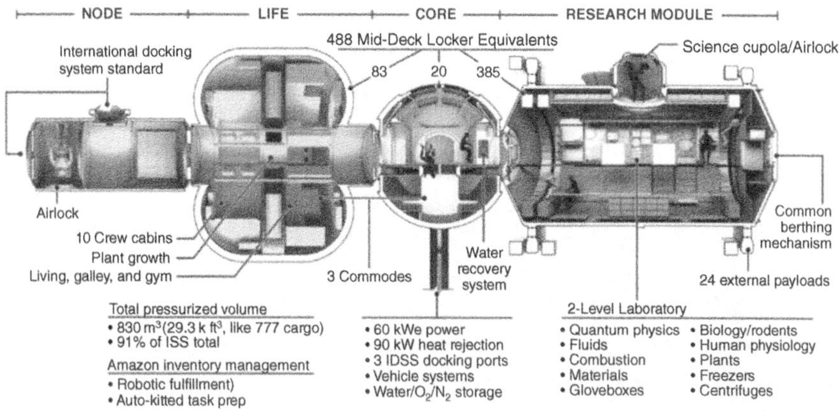

Figure 4.4 The Orbital Reef design has several modules connected together linearly. The station is therefore expandable by attaching more modules to the Airlock side. *Source:* Ref. [5] / Springer Nature / CC BY 4.0.

Orbital Reef features a modular design, as shown in Figure 4.4. The modular design can be scaled up with linear addition of new modules in support of growing demand and new functionality. The components of the station include a core, a node, a large integrated flexible environment (LIFE), a power/thermal mast, and a research module. The Core acts as the central hub for other modules and visiting vehicles. It is also the command-and-control center, data processing, and communications with parties on the ground. It offers 250 m^3 of habitable volume vs ISS's 388 m^3 and is equipped with an environmental control and life support system (ECLSS). Up to ten crew members can be supported.

The node, around 40 m^3 in volume, provides docks for visiting vehicles. It also supports extravehicular activity with an airlock.

The research module offers research opportunities such as payload experiments in a low gravity environment for universities or other research institutes. The LIFE habitat provides over 300 m^3 volume to host payloads and other facilities.

In March 2024, NASA reported that Blue Origin had passed four test milestones, including a trace contaminant control test, water contaminant oxidation test, urine water recovery test, and water tank test. Orbital Reef is scheduled to be operational by 2027.

4.4.2.3 Axiom Station

In 2020, NASA awarded a $140 million contract to Axiom Space to build at least one habitable modular spacecraft that can be attached to the ISS. Referred to as Axiom Orbital Segment, at least four Axiom modules are planned for attachment to the ISS. The project targets the late 2020s for completion. The modules and their assembly configuration are shown in Figure 4.5.

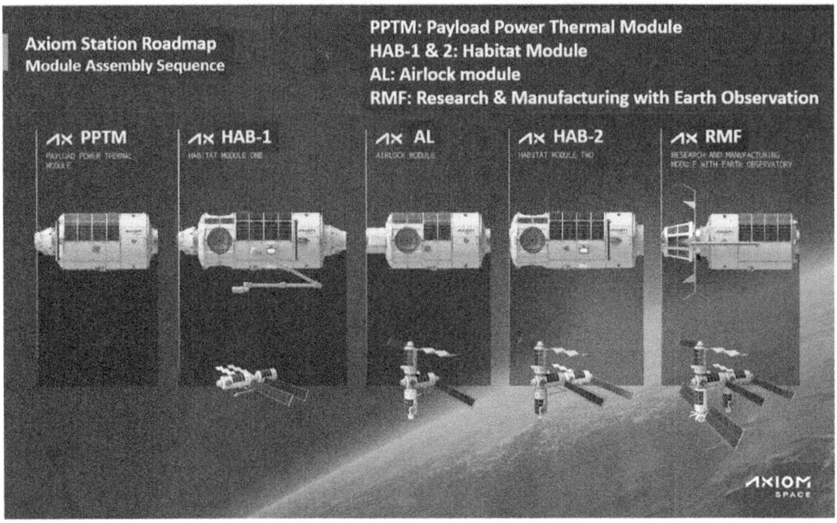

Figure 4.5 The Axiom Station is modular. Four modules plus an airlock module in the middle can be assembled. The HAB modules are habitats. The RMF is for research and manufacturing of high-end materials. *Source:* Ref. [6] / Axiom Space, Inc.

The first module, named Ax-H1 (Hab One), is capable of providing quarters for four crew members and room to accommodate research and manufacturing applications. Hab One is scheduled to be launched in late 2026. No information is available for the crew as of this writing. It will initially be attached to the ISS module Harmony, serving as the springboard for the remaining modules. Ax-H1 has a docking port for visiting vehicles and four "radial ports" for the assembly of future modules.

The second module Ax-H2 (Hab Two) is to be launched in 2028 and will provide quarters for an additional four crew members with complete ECLSS support. The third module Ax-RMF is scheduled for launch in the late 2020s and will provide a microgravity environment as a platform to conduct research and manufacturing.

The fourth module Ax-PPTM, which stands for Payload Power and Thermal Module, will be launched by 2030 and will provide expanded environmental, life support, storage, and payload capabilities, enabling the Axiom Station to detach itself from the ISS and become an independent commercial space station before the retirement of the ISS.

The Axiom Station will host crews, research, and manufacturing, which will lead to the proliferation of industrial R&D that can only be done in microgravity. The station will be a platform for private companies and national governments to conduct research.

4.4.2.4 Starlab

Starlab is being designed by Starlab Space, a joint venture subsidiary owned by Voyager Space and Airbus. The Starlab project is one of the three NASA's CLD contracts, with a grant of $160 million.

Starlab is a planned commercial space station for space agencies, researchers, and companies that aim for a continual human presence in LEO and a seamless transition from the ISS to the commercial era. Starlab is planned to launch in 2028.

The Starlab shown in Figure 4.6 will have two modules. The service module will provide propulsion and 60 kW of electric power collected by solar panels. The other module will function as a habitat and a laboratory. It also will have docking ports. The laboratory module is 8 m in diameter with a pressurized volume of 450 m^3.

Due to its size and mass, Starlab is planned for a single launch on the SpaceX Starship.

Figure 4.6 Starlab, a commercial LEO orbiting space station for use as a laboratory in 2027. *Source:* Ref. [7] / Lockheed Martin Corporation.

Table 4.2 Technical overview of the Gateway.

Guest and crew capacity:	1,250 guests, 150 crew
Perigee:	400 km
Apogee:	400 km
Orbital inclination:	28.5°
Width:	488 m
Depth:	76 m
Pressurized volume	11,906, 250 m^3
Atmospheric pressure:	1 atm, 101,300 Pa
Solar power generation:	56.1 MW

Source: Ref. [8] / Gateway Spaceport.

4.4.2.5 Gateway Spaceport

The Gateway Spaceport has several names: a destination hotel, a LEO cruise ship, or a city in space. It is a staging point for future spacecraft from Earth, other space stations, vehicles going to and returning from the Moon, and even other planets. The Gateway concept was proposed in 2012 by the Gateway Foundation to build the first rotating spaceport and space hotel as part of the construction industry in space. It is intended to be the first artificial gravity space station that can house over 1,400 people in orbit, including 150 crew members and 1,250 guests. Table 4.2 lists a technical overview of the Gateway design.

As the first step, the Von Braun Station (VBS) will be built. Also known as Voyager Station, the VBS is smaller. It is estimated to take more than 40 launches on the Starship to send all the pieces for assembly in space. The construction work is planned to start in 2025 and be completed in 2027. The VBS is a rotating station, composed of 24 cylindrical modules, each offering a total of 500 m^2 of habitable surface spread over three floors. Four hundred guests and over 100 crew can be hosted.

The diameter of the Gateway is 190 m, and it can be spun up to produce simulated gravity. This is done by using small engines placed at the two ends to spin up the structure. Figure 4.7 compares the Gateway and the VBS.

The Gateway station will be composed of the Hub, the Lunar Gravity Area (LGA), the Mars Gravitational Area (MGA), and the Bay. The Hub is the control room and centralized location of various facilities such as environmental processing. It will accommodate 50 crew members with a pressurized volume of 1,100,224 m^3.

The LGA and MGA represent locations where simulated gravity is like the Moon and Mars. The outer ring is the MGA and will provide accommodation for almost a thousand guests. The Bay is for passenger transfer for different spacecraft passing through.

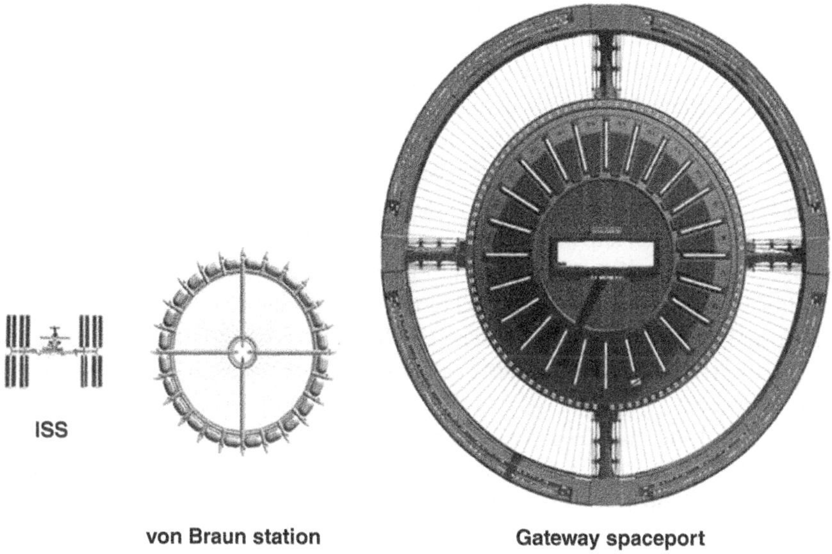

Figure 4.7 Comparison of the Gateway and the von Braun station by the Gateway Foundation. *Source:* Refs. [8, 9].

4.4.3 Other Space Stations

The abovementioned space stations are, or will be, in LEO. In the future, space stations will be deployed beyond Earth orbits around celestial bodies such as the Moon and Mars.

4.4.3.1 Lunar Gateway

The Lunar Gateway will be described in detail in Chapter 11. The Lunar Gateway will be the first space station beyond LEO and is part of the Artemis program to be integrated in orbit around the Moon. It will have multiple roles: providing a communication relay between the Moon and Earth, acting as a scientific laboratory, and offering habitat support for astronauts. It is a multinational collaborative project led by NASA, ESA, JAXA, CSA, and other space agencies and commercial partners. A wide range of experiments will be conducted across disciplines such as planetary science, astrophysics, Earth observation, heliophysics, space biology, human health, and spacecraft performance. Gateway's orbit is designed so it can spend a long time observing the lunar south pole.

4.4.3.2 Mars Base Camp

The Mars Base Camp (MBC) concept came from NASA. A conceptual design of the MBC is proposed by Lockheed shown in Figure 4.8.

The camp is a space station orbiting Mars. It is designed to support human missions for up to 1,000 days. It will be a part of a larger architecture that includes the Lunar Gateway. Scientific experiments will be conducted on board as well as serving as the interface for vehicles going to the two Martian moons and serving as a proving ground in preparing humans for future missions to the Martian surface.

The two camp elements are Habitat and the Mars Laboratory. Before going to Mars, the base camp elements will be assembled and tested in Cislunar space. The design of these elements will evolve from the Lunar Gateway. The two elements are to be connected as shown in Figure 4.8. Each element is connected to a hydrogen-oxygen fuel tank. Docked to the end of each fuel tank will be a crew vehicle similar to Orion (see Chapter 11). The Orion will have its own cryogenic propulsion stage. This MBC design assumes liquid hydrogen and liquid oxygen are the propellants.

The initial MBC crew, composed of six astronauts, will conduct studies of the two Martian moons and tele-robotically operate assets on the surface of Mars, including the return of Mars samples to the MBC. Subsequent missions will introduce a reusable, single-stage, crewed descent and ascent vehicle, enabling sortie missions to Mars' surface. The concept is similar to the Human Landing System of the Artemis program (see Chapter 11). Future missions will also transition toward producing propellants using water, either delivered from Earth or extracted from Mars.

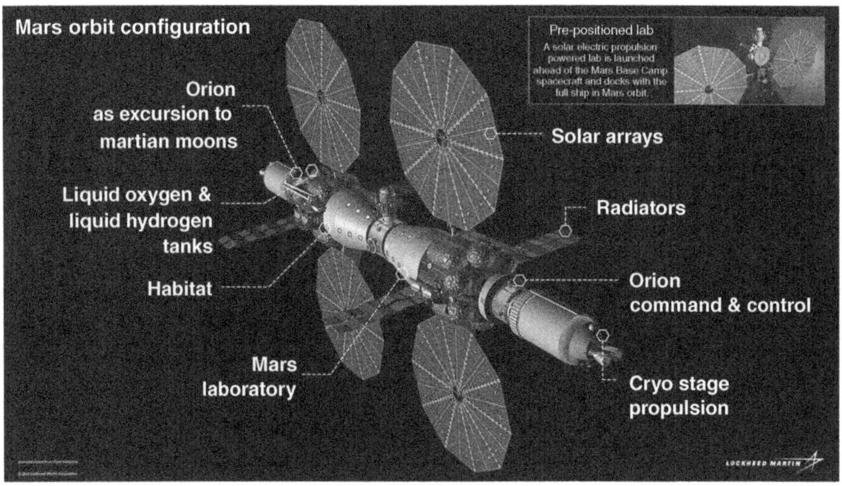

Figure 4.8 The Mars Base Camp concept. It is built up in modules [10]. *Source:* Lockheed Martin / https://en.wikipedia.org/wiki/File:Mars_Base_Camp_cover_view.jpg / last accessed Aug 09, 2025.

The MBC elements are built on a foundation of technologies developed for the Artemis program. They include the Space Launch System (rocket), Orion vehicle, Habitats, and solar electric propulsion (SEP). Chapter 11 will discuss in detail the series of Artemis missions to build the Gateway and land astronauts on the Moon. The MBC concept is quite similar to that of the Artemis program.

4.4.4 Development Trends of Future Space Stations

The New Space market is a driving factor for the commercialization of LEO. In turn, the number of satellites and commercial space stations is expected to increase. By sheer number, mega constellations of communication satellites in LEO will be the dominating force to populate LEO.

Soon, commercial space stations will begin to populate LEO as the ISS's retirement draws near. These commercial opportunities include hotels/motels, research facilities, education, factories, and warehouses. Key technologies such as reusable launch systems, high-capacity upper stages, crewed vehicles, and inflatable modules make LEO commercialization feasible. More experience will give all interested parties the confidence to go beyond LEO to high orbits and then further beyond.

It is likely that the new generation of space stations will be commercially built and smaller in size. Increased competition, together with reusable rockets, will drive down the cost of access to these stations. Governments will be paying customers instead of drivers.

In the intermediate term, space stations will orbit the Moon (the Lunar Gateway) and, in the future, Mars (the Mars Base Camp) to explore more scientific mysteries of deep space and enable further commercial scenarios. In the long term, space stations in the generic sense can be placed in selected Earth-Moon's and Sun-Mars' Lagrange points.

One of the foreseeable and key applications of space stations is their role in space transportation infrastructure. Currently, astronauts and cargos are directly sent to LEO space stations and the Lunar Gateway in the Artemis program. Therefore, LEO space stations can be part of the future transportation infrastructure. Some space stations can serve as transportation hubs for passengers to transfer between spaceflights, like airports, or as warehouses for space logistics servicing multiple commercial stations in LEO and higher orbits.

However, there are technical challenges. For instance, the delta-V requirement to move across different orbits (especially different inclinations) is high for large vehicles carrying passengers and cargos. Therefore, it will be a cost and energy effectiveness of single launches versus getting the materials delivered by small crafts from space warehouses akin to terrestrial delivery services.

4.5 Looking Ahead for the LEO Industry

LEO commercialization has opened new markets and revenue streams. LEO promises to revitalize many other sectors of the space industry, including satellite navigation, space tourism, space manufacturing, etc. LEO commercialization also pushes the advancement of innovation and technologies in satellite technology, launch systems, and on-orbit services, which, in turn will contribute to the commercialization of LEO and can enable further and broader applications beyond LEO. However, the development of the LEO industry faces two major challenges.

Launch limitations: Reusable launch vehicles have revolutionized the space industry by offering low launch costs. Mega constellations with thousands of satellites and their relatively short lifetime require frequent launches and multiple satellites in each launch. It is foreseeable that more private companies will be engaged in the LEO industry, offering satellite manufacturing, testing, launching, maintenance, and many related services. Limited choice of orbit, including altitude, inclination, and other orbital parameters, is a challenge to be addressed.

Space environment: LEO satellites are operating above the atmosphere, yet there is residual atmosphere that causes drag on satellites, hence leading to orbital decay over time. The rising problem with space debris is another challenge as the LEO industry develops. Other space environment challenges include radiation and orbital congestion.

Reading List

Kassas, Z.Z.M. (2021). Navigation from low Earth orbit, Part 2: Models, implementation, and performance. In: *Position, Navigation, and Timing Technologies in the 21st Century: Integrated Satellite Navigation, Sensor Systems, and Civil Applications*, 1ste (ed. Y.T.J. Morton, F. Van Diggelen, J. Spilker Jr., et al.). Wiley-IEEE Press.

Malik T. Success! SpaceX Launches Falcon Heavy Rocket on Historic Maiden Voyage [Internet]. 2018 [cited 2024 Dec 19]. Available from: https://www.space.com/39607-spacex-falcon-heavy-first-test-flight-launch.html.

Reid, T.G.R., Walter, T., Enge, P.K. et al. (2021). Navigation from low Earth orbit, Part 2: Concept, current capability, and future promise. In: *Position, Navigation, and Timing Technologies in the 21st Century: Integrated Satellite Navigation, Sensor Systems, and Civil Applications* (ed. Y.T.J. Morton, F. Van Diggelen, J. Spilker Jr., et al.). Wiley-IEEE Press.

References

1 Nano Avionics Kongsberg. How Many Satellites are in Space?. Nano Avionics Kongsberg [updated 2023 May 4; cited 2025 Feb 27] [online]. Available from: https://nanoavionics.com/blog/how-many-satellites-are-in-space/.

2 NASA Johnson. The Space Station is Pictured from the SpaceX Crew Dragon. flickr [updated 2021 Nov 8; cited 2025 Feb 27] [online]. Available from: https://www.flickr.com/photos/nasa2explore/51712323479/in/album-72157720187084178.

3 China Manned Space. China Manned Space [cited 2025 Feb 27] [online]. Available from: https://www.cmse.gov.cn/dmt/tj/tkyxz/index_4.html.

4 NASA. Bigelow Expandable Activity Module (BEAM). [cited 2025 Feb 27] [online]. Available from: https://www.nasa.gov/international-space-station/bigelow-expandable-activity-module-beam/.

5 Zea, L., Warren, L., Ruttley, T. et al. (2024). Orbital Reef and commercial low Earth orbit destinations—upcoming space research opportunities. npj Microgravity *10* (1): 43.

6 Axiom Station Roadmap. Axiom. Cited by Kuhr J. Axiom Space Adjusts Space Station Plans. Payload Space [updated 2024 Dec 18; cited 2025 Feb 27] [online]. Available from: https://payloadspace.com/axiom-space-adjusts-space-station-plans/.

7 Starlab. Cited by. Nanoracks, Voyager Space, and Lockheed Martin Teaming to Develop Commercial Space Station. Lockheed Martin [cited 2025 Feb 27] [online]. Available from: https://news.lockheedmartin.com/news-releases?item=129118.

8 Gateway Spaceport. The Gateway. Gateway Spaceport [cited 2025 Feb 27] [online]. Available from: https://gatewayspaceport.com/the-gateway/.

9 Jestin T. Meet the Von Braun Station: The Most Advanced Rotating Space Hotel Design To Date. Spacetech [updated 2019 Oct 22; cited 2025 Feb 27] [online]. Available from: https://www.spacetechasia.com/meet-the-von-braun-station-the-most-advanced-rotating-space-hotel-design-to-date/

10 humanMars net. Lockheed Martin's Mars Base Camp Concept. humanMars.net [updated 2017 Sep 28; cited 2025 Feb 27] [online]. Available from: https://www.humanmars.net/2017/09/lockheed-martins-mars-base-camp-concept.html.

5

Reusable Rockets

Stephen W. Cheung[1] *and Guan Yu*[2]

[1] *Former Adjunct Professor, Department of Physics, The University of Hong Kong, Hong Kong, China*
[2] *Department of Aeronautical and Aviation Engineering, The Hong Kong Polytechnic University, Hong Kong, China*

5.1 Introduction

The goal of reusability is that one or more stages and components of the launch system are recovered and reused with only routine maintenance. Up to a decade ago, rocket stages were not recovered and reused except for the Space Shuttle. The boosters usually fall into the ocean or desert areas and are often damaged. The upper stages, which take payloads to the LEO and higher orbits, are often left behind to slowly deorbit. The practice of single-use rockets began to change in the mid-2010s.

The common indicator of space flight economics is "cost per unit weight of payload" such as US$ per kg for cargo, which is a common reference for going to the low Earth orbit (LEO). The high cost of single-use rockets made space missions relatively infrequent events. Going to space and its commercialization can thrive only when the cost of getting there is affordable to attract more users. The launch cost and payload capacity are key. By recovering and reusing the rocket stages, the parts can be reused multiple times to spread the cost.

LEO is the first stop of rocket launches and is the base of all space activities. Its advantages are plentiful such as low signal latency and higher resolution by being close to the Earth's surface. Single-use and reusable rockets are all designed to reach LEO with different payload capacities. They are usually two-stage rockets. The booster, or Stage-1, does the heavy lifting and brings the whole rocket to the upper atmosphere at an altitude of around 70–80 km. The upper stage, or Stage-2, then takes the payload to a low LEO around 150–200 km altitude. The payload does the rest of the way in rising to its designated orbit using its own propulsion system.

New Space: From Low Earth Orbit to the Moon and Beyond, First Edition.
Edited by Stephen W. Cheung and Chih-Yung Wen.

This chapter covers the current state of reusable rockets by propulsive means, i.e. rocket stages that can return to Earth using rocket engines for landing. Reusability mainly applies to the booster stage as of this writing. Upon stage separation, the booster is to follow a flight path and land intact at a predetermined site with high precision. The perspective taken in this chapter is to describe the sequence of events for a rocket stage to return to a designated landing spot.

SpaceX is the farthest along in developing reusable rockets. There is a voluminous amount of media coverage and social media chat groups by space enthusiasts. The Falcon 9 system reached 390 successful and successive launches and booster returns by the end of 2024 – a success rate of 99.7%. In 2024 alone, 134 Falcon 9 launches took place, representing a 40% jump from the previous year and at an average of 2.5 launches per week. SpaceX is also advancing its next launch system, known as Starship. We will cover Falcon 9 and the Starship in this chapter.

There are three classifications of rockets by their payload capacity to LEO in accordance with NASA's classification: small lift (under 2,000 kg), medium lift (2,000–20,000 kg), and heavy lift (20,000–50,000 kg). Reusability applies to all these rockets. As payload capacity is vital for the commercialization of space, reusable medium-lift and heavy-lift rockets have a large impact on rocket economics.

5.1.1 Key Technology Components of the Reusable Rockets

There are several key components and subsystems that contribute to the operation of reusable rockets. This section focuses on reusable boosters, although these subsystems also apply to the upper stages.

1) RCS thrusters: RCS stands for reaction control system. RCS is covered in Chapter 2. In reusable stages, these small thrusters use gas to produce thrust to push on the booster. The push can change the booster's attitude and orientation by small or large amounts. Both hot and cold gases are used.
   RCS thrusters are also used to apply coarse adjustment of a booster's lateral position during the booster's return after stage separation, and will be described later in this chapter.
2) Rocket engine with range: During launch and return, the booster must be able to throttle, stop, and restart its engines multiple times in space and in the atmosphere at different phases of the space flight. The engines must work over a throttle range.
3) Different burns: The booster is to fly to its designated landing spot after stage separation. Therefore, booster engines also need to produce thrust on demand. There are several burns, each lasting a different duration.
   a) Boost-back burn: After the RCS thrusters reorient the booster's direction, boost-back burn is applied to slow down the booster's speed or simply reverse course and fly back to the launch site. The latter is known as the Return to launch site (RTLS) operation. Boost-back burn consumes

plenty of propellants. Alternatively, the booster can continue its course downrange, with or without any boost-back burn, and land on another platform such as a drone ship several hundred kilometers away.

Unlike single-use boosters, which exhaust their propellants before stage separation, a reusable booster needs a reserve of propellants to perform different burns. In general, RTLS consumes the most propellants to reverse course. Therefore, a RTLS operation works at the expense RTLS operation at the expense of the payload capacity. In the case of Falcon 9, as much as 30% of the booster's total propellants by weight is needed for the RTLS option and about 10% if landing on a drone ship [1].

Boost-back burn for drone ship landing is optional, depending on the fuel reserve set aside during mission planning. Without a boost-back burn, the booster will go further downrange. The drone ship placement needs to be adjusted accordingly.

b) Entry burn (sometimes called reentry burn): The entry burn reduces the booster's descending speed as it approaches the thicker part of the atmosphere, typically near the 50 km altitude. Only some booster engines need to be fired. Another objective of the entry burn is that as the booster tears through the thin atmosphere at hypersonic speeds; heat is generated by the induced plasma sheath. The entry burn pushes the plasma away from the engines and reduces the heat in the engine end. The pulse action is known as supersonic retropropulsion.

c) Landing burn: The landing burn is activated to reduce the downward speed in preparation for the final landing. The thrust is controlled just right for a soft vertical landing.

4) Fins for precision steering: The booster must land with precision within a few meters on a landing pad or on a drone ship. Fins provide fine steering to adjust the booster's orientation and lateral position. Sudden disturbances such as air currents in the atmosphere need to be compensated for in real-time during descent.

Emphasis should be made that fins work in the presence of some airflow. They do not work in a vacuum.

In essence, the RCS in 1) and the fins in 4) make a double-loop control system to adjust the booster's lateral position and orientation soon after stage separation and during descent. RCSs can work in both vacuum and atmosphere, but fins need some atmosphere to work.

5) Navigation and positioning: The booster follows a well-defined trajectory already programed in the booster's computer after separating from the upper stage. The booster has an onboard inertial navigation system (INS), gyroscopes, and sensors that constantly feed data to the computer about its position and orientation. Combined with instantaneous GPS data, the onboard computer controls various adjustments by the RCS, fins, and engine gimbal.

The landing spot of the RTLS is fixed. However, a drone ship at sea is a moving target until the final moment of touchdown. The onboard computer must account for the INS and GPS data of the booster and command the RCS and fins to steer the booster to coincide with the INS information of the landing spot on the drone ship.

6) Landing legs: A booster needs legs to land and stand after touchdown. Legs are not needed with single-use boosters. The landing legs of reusable boosters are stowed away during most of the flight and only unfold themselves when the booster is a short height away from landing. The legs also need mechanical means to absorb shocks during landing.

The above six categories of subsystems can be used to evaluate different reusable rocket systems. Their functions come in after the stage separation during ascent.

5.2 Falcon 9 Reusable Rocket

5.2.1 Rocket Description

SpaceX's Falcon 9 is a two-stage rocket. The payload is brought to the LEO by the upper stage and then released. Figure 5.1 provides a description of the rocket construction, showing the internal locations of the engines and fuel tanks. The booster has nine Merlin engines (sea-level), and the upper stage has one Merlin engine (vacuum-level). See Chapter 2 to review the sea-level and vacuum-level engines under the Nozzle section.

The Falcon 9 system stands 70 m long. The booster is almost 40 m without the interstage. The interstage is about 5 m long and connects the booster to the upper stage. The diameter of the booster and the upper stage is 3.7 m. The booster carries

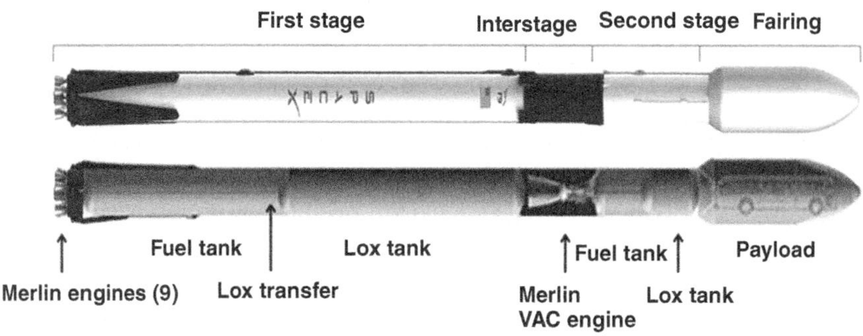

Figure 5.1 Key components of the Falcon 9 booster and upper stage. *Source:* Ref. [2] / SpaceX.

395 tons of propellants at launch. Its maximum payload capacity to the LEO is 22.8 tons if the booster is to be expended, i.e. not recovered. If the booster is to be recovered, the maximum payload mass to the LEO is around 18.5 tons, depending on the recovery via RTLS or drone ship. Falcon 9 is between the heavy-lift and medium-lift classes.

5.2.2 Merlin Engine

SpaceX developed the Merlin family of engines. Its dry weight is 470 kg. The full thrust is 854 kN (sea-level) and 980 kN (vacuum) with a throttle range of 57–100%. The engine runs on liquid oxygen LOX and RP-1 (RP stands for rocket propellant, a highly refined kerosene). The nine engines on the booster are in an 8-1 configuration, as shown in Figure 5.2.

5.2.3 Reaction Control System

The RCS is key in turning the booster to the needed orientation soon after stage separation and before any boost-back burn is applied. The Falcon 9 uses four sets of cold gas thrusters located around the top of the booster. Each thruster has three holes to eject pressurized nitrogen gas in three directions as shown in Figure 5.3.

Figure 5.2 Falcon 9 engines. *Source:* Ref. [3] / SpaceX.

Figure 5.3 One set of reaction control systems (in red circle) is between the two grid fins [4]. *Source:* Official SpaceX Photos / Flickr.

5.2.4 Fairing

The fairing measures 13 m in height and 5.2 m in diameter. Each half weighs 1,900 kg. The fairing is typically fabricated from lightweight carbon composites, making it robust and can withstand the aerodynamic pressure and other harmful effects from the atmosphere during launch and ascent.

Falcon 9's fairing, according to one of SpaceX's tweets in 2017, costs US$5 million and is about 10% of the cost of the Falcon 9 system. The fairing halves are designed for recovery and reuse. They parachute down to the ocean several hundred kilometers downrange from the launch site and are recovered by vessels, as shown in Figure 5.4.

Figure 5.4 shows a fairing half parachuting down to the ocean. Fairing reuse is near 100% for Falcon 9.

5.2.5 Gimballing

All nine Merlin engines are gimballed by pneumatic actuators. The center engine is two dimensionally gimballed, while the outer engines can do one dimension. The engines are gimballed to pitch the rocket sideways after Max Q to start the gravity turn maneuver.

5.2.6 Landing Legs

The Falcon 9 booster has four landing legs 90° apart near the engine end, as shown in Figure 5.5. Each leg unfolds by a long actuator rod. The legs are stowed away

Figure 5.4 Falcon 9 fairing is released soon after stage separation. The two halves parachute down to the ocean and are recovered by vessels. *Source:* Ref. [5] / SpaceX.

Figure 5.5 Falcon 9 booster has four landing legs. They are stowed away during the majority of the flight and are redeployed moments before landing. *Source:* Ref. [6] / SpaceX.

during launch and remain so until several seconds before the booster is about to touch down. Made of lightweight materials, the 8-m-long legs weigh around 1,000 kg each.

5.2.7 Grid Fins

The Falcon 9 booster has four grid fins located near the top of the booster before the interstage. The grid fins are nominally placed adjacent to the RCS, as shown in Figures 5.3 and 5.6. They are turned electrically by motors. The rotational

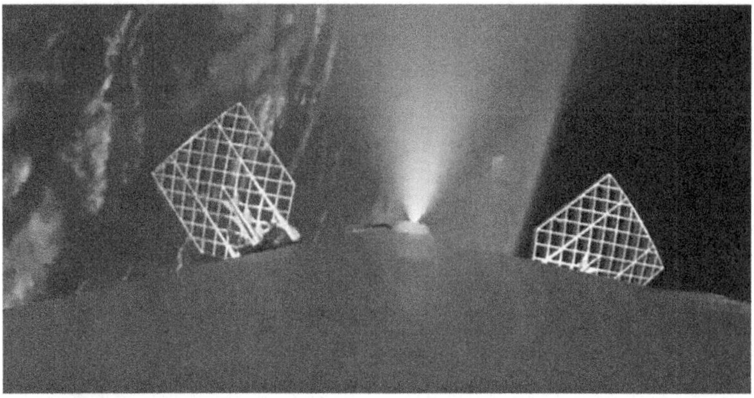

Figure 5.6 Cold gas RCS in action pushing and reorienting the booster. Two grid fins are also shown. *Source:* Ref. [7] / The Space Techie.

orientation of the grid fin alters the airflow, which results in different aerodynamic pressures pushing on the booster body. Steering works when the booster encounters the thicker part of the upper atmosphere. Figure 5.6 shows the grid fins as viewed from the top of the booster. The grid fins, which are folded-in until stage separation, are unfolded in conjunction with the RCS working.

5.2.8 Falcon 9 Flight Profile

The flight profile of the Transporter 6 mission in January 2023 illustrates the key events of the Falcon 9 launch. We focus on the booster and not the upper stage to understand the reusability perspective.

The Transporter 6 mission was launched from Cape Canaveral, Florida, and the booster returned to Landing Zone 1 approximately 14 km from the launch site. It is an RTLS operation. The general flight profile is shown in Figure 5.7 for a similar transporter mission.

The actual timeline and the speed and height of the booster at several key events are shown in Table 5.1 using telemetry data from online videos. The speed is the total speed, i.e. the vector sum of the horizontal and vertical speeds. The height and speed of the booster over time are plotted in Figure 5.8, in which the key events are numerically labeled. We further convert the speed from km per hour to km per second, keeping in mind that the orbital speed in LEO is about 8 km/s.

The outstanding feature of an RTLS operation is the significant drop in the booster speed when the boost-back burn is applied. This is clearly observed in Figure 5.8. The booster's speed reaches a minimum when it is at the apogee where the vertical speed is zero. The booster's speed picks up when freefalling back to Earth.

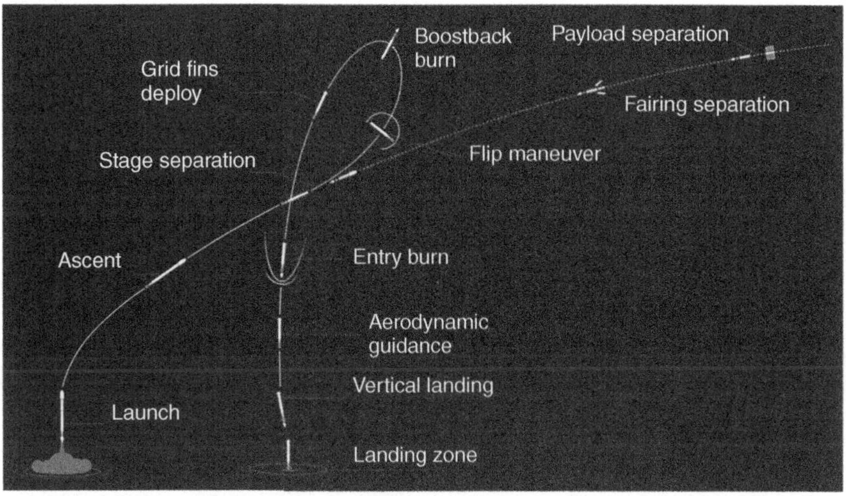

Figure 5.7 Diagram of the SpaceX Transporter-3 flight profile. This is a Return to launch site (RTLS) operation. *Source:* Ref. [8] / AmericaSpace.com.

Table 5.1 Timeline of key events for the Transporter 6 mission in 2023. Speed is the vector sum of the horizontal and vertical speeds.

Actual time	Altitude (km)	Booster speed (km/s)	Event
T + 00 : 00	0	0	Liftoff
T + 01 : 10	11.8	0.32	Max-Q
T + 02 : 23	69.2	1.65	MECO (main engine cutoff)
T + 02 : 26	71.9	1.64	Stage separation
T + 02 : 34	81.7	1.58	Second-stage engine ignition
T + 02 : 37	84.4	1.56	First-stage boost-back burn ignition (three engines)
T + 03 : 25	125	0.69	The first stage boost-back burn ends after 48 seconds. Grid fins deployed at 03 : 40
T + 03 : 53	139	0.46	Fairing jettison (upper stage)
T + 4 : 30	146	0.29	Apogee reached
T + 06 : 50	56.3	1.34	First-stage entry burns ignition (three engines)
T + 07 : 15	31.1	0.74	First-stage entry burns end lasting 25 seconds
T + 08 : 04	3.5	0.27	First-stage landing burn ignition (one engine)
T + 08 : 29	~0	0.03	Landing legs deployed
T + 08 : 35	0	0	Touchdown

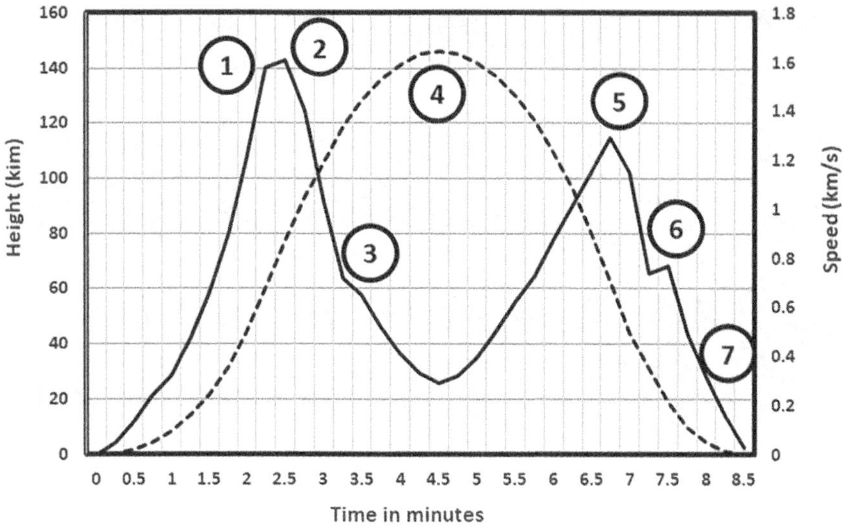

Figure 5.8 Height and speed of the Falcon 9 booster over time in Transporter-6 Return to launch site operation. The dotted line represents the altitude, and the solid line represents the speed of the booster. The booster's speed is greatly reduced when the boost-back burn is applied soon after the stage separation. The speed picks up again after the booster begins to fall from the apogee. 1: MECO, 2: boost-back burn starts, 3: boost-back burn ends, 4: apogee reached, 5: entry burn starts, 6: entry burn ends, and 7: landing burn starts. *Source:* Adapted from Ref. [9].

Figure 5.8 and Table 5.1 together are educational in following the key events that the booster goes through. Videos of the mission are available online [9].

5.2.8.1 Flight Profile of the Transporter 6 Mission

We will go through each event of the Transporter 6 mission play by play, as outlined in Table 5.1.

Event 0: Liftoff.

Event 1 Max Q: The booster throttles down to about 80% a few seconds before Max Q is called out. The Falcon 9 rocket passes the speed of the sound barrier (Mach 1) right before reaching Max Q. Max Q typically occurs between 10 and 12 km altitude about 1 minute after liftoff. After clearing Max Q, the booster resumes full throttle and begins a gravity turn in which the rocket slowly pitches sideways.

Event 2 MECO: MECO stands for main engine cut off. The rocket reaches an altitude of around 70 km and shuts down all the booster engines. The rocket continues to coast higher by inertia.

Event 3 Stage separation: A couple of seconds after MECO, the upper stage separates from the booster via a release mechanism. The booster and the upper stage continue to gain altitude by inertia while losing vertical speed due to gravity. The booster is now a projectile with both vertical and horizontal speeds.

Event 4 Second-stage engine ignition: Within 5 seconds after MECO, the upper stage fires its single engine and proceeds to speed up to a higher altitude and enters orbit. The upper stage is almost horizontal at this point. It gains speed from propulsion, while the booster loses vertical speed due to gravity.

Event 5 First-stage boost-back burn and grid fins deployed: The immediate action is to use onboard RCS thrusters to turn the booster around to point toward the launch site. The booster is still climbing due to inertia. The boost-back burn turns on with three engines firing for 48 seconds.

For RTLS operation, the booster reverses its direction of travel. Note that the booster speed rapidly drops as it loses both vertical and horizontal speeds, as shown in the middle of Figure 5.8.

Falcon 9's grid fins are folded in during the first part of the launch. After Event 3 and the boost-back burn (if applied), the grid fins unfold. There is no atmosphere for the grid fins to work yet.

Event 6 Apogee: The booster rises to its apogee over 140 km altitude with its horizontal direction already reversed and no vertical speed. The booster reaches its lowest speed around 0.3 km/s in the middle of Figure 5.8. The booster then starts to fall and gains downward speed while maintaining its horizontal speed toward the launch site.

Event 7 Fairing jettisoned: The fairing of the upper stage opens while climbing and exposes the payload to space. This usually occurs after the upper stage crosses the Kármán line. The fairing then falls back to Earth. Each half parachutes down in the lower atmosphere and hits the ocean downrange. They are recovered by vessels.

Event 8 First-stage entry burn (three engines): The booster freefalls while maintaining an inherent horizontal speed. It is also pushed by RCS thrust so that the booster is almost vertical with the engine-end pointing to the ground. The booster's speed is supersonic after falling for 60–70 km from the apogee to about 60 km altitude. The entry burn is turned on to reduce the booster speed before it enters the thicker part of the atmosphere.

Three engines are turned on for 25 seconds to provide enough reverse thrust to slow down the falling booster.

Event 9 First-stage entry burn ends: The booster's downward speed is much reduced by the entry burn. The atmospheric drag from the lower atmosphere now comes into play. After gaining some speed for a short moment, the booster loses speed to the atmospheric drag toward its terminal velocity.

Event 10 First-stage landing burn ignition (one engine): The booster is now 3–4 km from the ground, and its speed has been reduced to subsonic. The landing burn comes on and is controlled for the rest of the landing. Only the center engine is turned on to provide the decelerating thrust.

Event 11 Landing legs deployed: The landing legs are deployed when it is about 50–100 m above ground to ready the booster for landing.

Event 12 First-stage landing: The landing burn reduces the downward speed to exactly null when the booster touches down.

5.2.9 Landing on a Drone Ship

The flightpath of a booster landing on a drone ship downrange is shown in Figure 5.9.

Boost-back burn is optional after stage separation as the booster can travel further downrange to land on a drone ship. More payload margin can be gained. Therefore, choosing RTLS versus drone ship landing depends on the target orbit and the mass of the payload.

The Telkomsat Merah Putih 2 mission, launched from Cape Canaveral in 2024, calls for the booster to land on a drone ship. The payload is a satellite owned by Indonesia and is to be placed in a geosynchronous orbit. No boost-back burn after stage separation is executed in this mission.

We will replicate most of Table 5.1 for this mission as shown in Table 5.2.

Figure 5.10 shows the plots of the booster speed and altitude over the flight time with the key events numerically labeled. The speed drops after MECO. Its speed

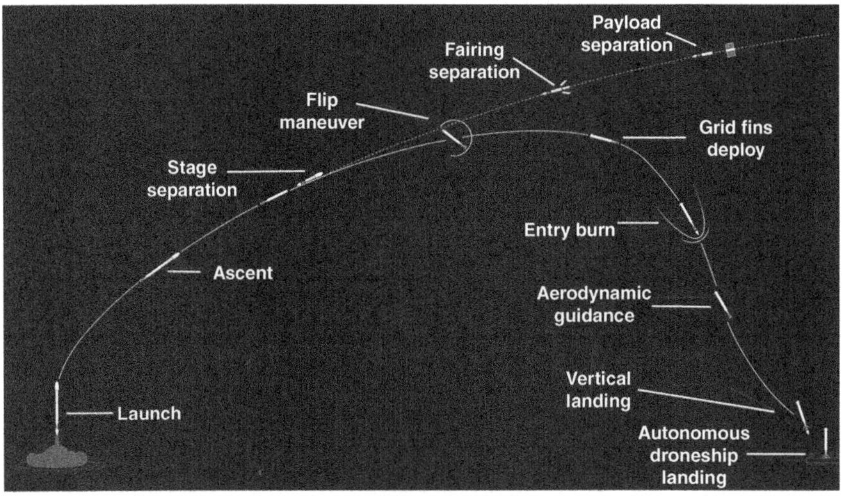

Figure 5.9 Flight path of a Falcon 9 mission where the booster lands on a drone ship. In this operation, the boost-back burn is eliminated and allows more payload mass to be delivered. *Source:* Adapted from Ref. [10].

Table 5.2 Telkomsat Merah Putih 2 mission, February 2024. The booster lands on a drone ship with no boost-back burn applied.

Actual time	Altitude (km)	Booster speed (km/s)	Event
	0	0	Liftoff
T + 01 : 10	11.4	0.43	Max Q
T + 02 : 35	66.1	2.29	MECO
T + 02 : 39	69.3	2.27	Stage separation
T + 02 : 46	81.7	1.58	Second-stage engine ignition
T + 2 : 56			Grid fins deployed No boost-back burn
T + 03 : 30	107	2.11	Fairing jettisoned
T + 06 : 18	67.1	2.28	First-stage entry burns ignition (three engines)
T + 06 : 46	42.7	1.38	First-stage entry burns end lasting 28 seconds
T + 08 : 11	2.4	0.25	First-stage landing burn ignition (one engine)
T + 08 : 33	~0	0.02	Landing legs deployed
T + 08 : 38	0	0	First-stage landing

Source: Adapted from SpaceX [11].

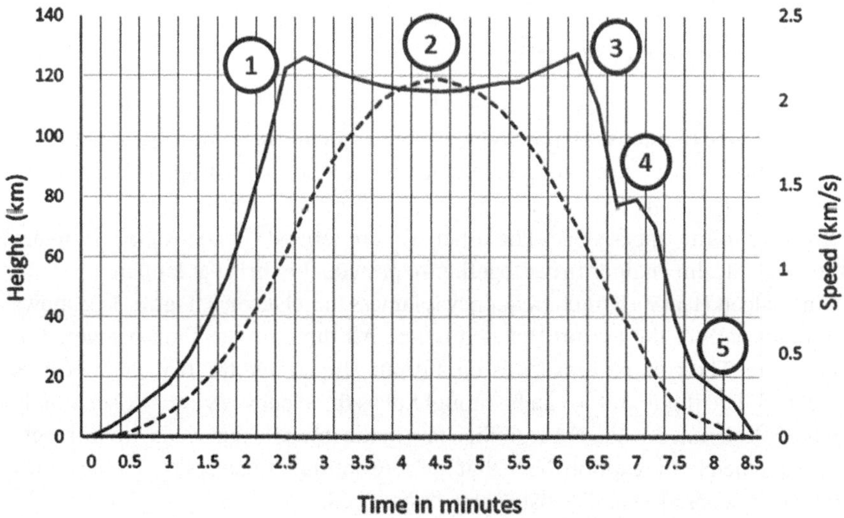

Figure 5.10 Height and speed of the Falcon 9 booster over time with the booster landing on a drone ship. The dotted line represents altitude, and the solid line represents speed. 1: MECO, 2: apogee, 3: entry burn begins, 4: entry burn ends, and 5: landing burn turns on.

reaches a low point at the apogee as expected and picks up again afterwards as it falls under gravity. The speed increases until the entry burns by three engines, which reduces the speed substantially as evident from the drop in speed from the entry burn start to end. Then, the atmospheric drag takes over to reduce the speed as the booster falls further.

Several observations can be made:

1) No boost-back burn is executed. The horizontal momentum carries the booster downrange, while the vertical momentum sends the booster to the apogee. The booster can therefore commit more fuel to impart more speed on the payload before stage separation. This is evident from the much higher booster speed (2.27 km/s) at the point of stage separation in Table 5.2 versus 1.64 km/s in Table 5.1. Such rocket speed is primarily horizontal.
2) The high rocket speed at stage separation also means that the horizontal velocity of the booster stays high for the rest of its projectile path.
3) Comparing the RTLS and drone-ship operation in Figures 5.8 and 5.10, we can see that the altitude curve looks the same. The speed curve for the RTLS, however, dips significantly in the middle of the figure. Otherwise, they have similar characteristics overall.

A daylight mission of a Starlink launch in 2024 shows a clear perspective taken from the interstage end. From stage separation at T + 02 : 50 onward, the engine end can be observed to swing across steadily by the action of the RCS thrusters, as seen from a video on the SpaceX website [12].

5.2.10 Falcon Heavy Launches

The Falcon Heavy is a heavy-lift system for heavy payloads going to high orbits or for deep-space missions. Figure 5.11 shows the Falcon Heavy during launch.

The Falcon Heavy uses three identical Falcon 9 boosters strapped together in a straight-line configuration. The center booster is the core with the interstage connected to the upper stage. The other two are termed side boosters. There are $3 \times 9 = 27$ Merlin engines firing together to provide the thrust at launch.

The Falcon Heavy's launch frequency is far less than Falcon 9. Figure 5.12 shows the flight path of the Falcon Heavy mission. All three boosters are recovered in this figure. The core booster lands on a drone ship, while the two side boosters are RTLS. The flight profile can be thought of as three boosters landing separately on their designated spots. For legibility, text in the image is replaced with numbers and explained in the caption. Note that the core booster executes a boost-back burn after stage separation to dissipate some of its speed.

The minor difference is that the side booster engine cut-off is termed BECO. BECO occurs first at a lower altitude, and the side boosters fall away from the core stage synchronously. The core stage continues to thrust forward with its engines

Figure 5.11 Image of the launch of the SpaceX Falcon Heavy. *Source:* Ref. [13] SpaceX / UnSplash.

firing. The two side boosters execute their own RTLS as illustrated in Figure 5.13. The core booster continues a flight path typical of the Falcon 9 and goes through MECO, stage separation, and the rest to eventually land on a drone ship.

Some customers of Falcon Heavy missions often need performance (payload mass or high orbits) over cost saving. The core booster and sometimes even the side boosters could be expended. We choose three Falcon Heavy launches and extract their available telemetry data to illustrate the key events. Time, altitude, and booster speed in each event are included.

Background of the three missions:

1) USSF-67 Space Force mission [16]
   - Launched in 2023
   - Reported payload weight 2,000–3,000 kg
2) ArabSat-6A [17, 18]
   - Launched in 2019
   - Geostationary communications satellite with a reported weight of 6,000 kg (with fuel)
3) ViaSat-3 Americas Mission [19]
   - Launched in 2023
   - Broadcast satellite with reported weight 6,400 kg

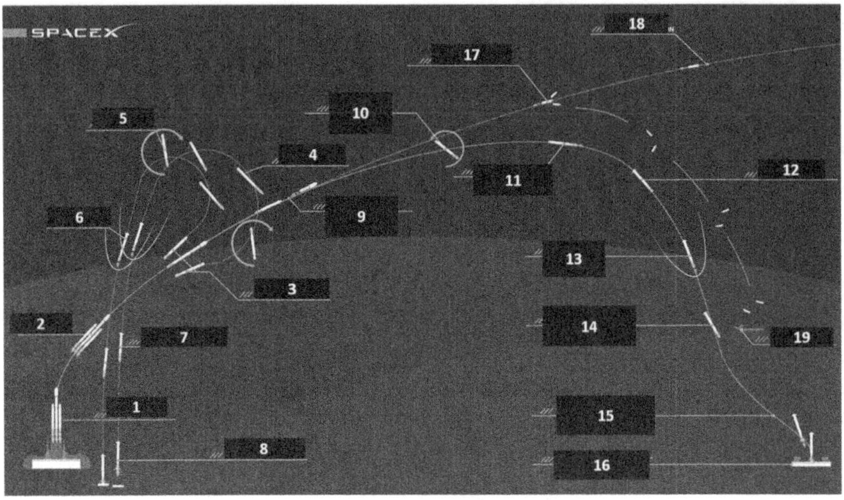

Figure 5.12 Launch profile of the Falcon Heavy. The side boosters drop out early in the "side-booster cutoff" and return to the launch site. The core stage lands on a drone ship. The meaning of the numbers are 1) launch, 2) ascent, 3) side booster separation, 4) side booster boost-back burn, 5) side booster flip maneuver, 6) side booster entry burn, 7) aerodynamic guidance, 8) side booster vertical landing, 9) core booster stage separation, 10) flip maneuver, 11) core booster boost-back burn, 12) grid fins deploy, 13) entry burn, 14) aerodynamic guidance, 15) vertical landing, 16) on drone ship, 17) upper stage fairing separation, 18) payload separation, and 19) fairing recovery. *Source:* Ref. [14] / Elon Musk.

Figure 5.13 Two side boosters of a Falcon Heavy flight successfully landed back in their designated landing zones. *Source:* Ref. [15] / SpaceX.

Table 5.3 Falcon Heavy data for three missions USSF-67, Arabsat-6A, and ViaSat-3.

Item	USSF-67	Arabsat-6A	ViaSat-3
Max Q	T + 0 : 1 : 00 8.9 km, 0.34 km/s	T + 01 : 11 12.6 km, 0.38 km/s	T + 01 : 05 11.0 km, 0.38 km/s
BECO	T + 02 : 32 59 km, 1.62 km/s	T + 02 : 36 59 km, 1.63 km/s	T + 03 : 12 80.6 km, 2.8 km/s
MECO (core)	T + 04 : 01 117 km, 4.03 km/s	T + 03 : 38 101 km, 2.98 km/s	T + 04 : 20 115 km, 4.74 km/s
Booster actions	Side boosters RTLS, core expended	All 3 recovered. Side boosters RTLS, core on the drone ship	All three boosters expended

Note the MECO data from the three Falcon Heavy launches in Table 5.3 compared with Falcon 9. All three core boosters rise to higher altitudes, and their speeds at MECO are higher. As all three boosters are expended in the ViaSat mission, the core booster altitude and speed at MECO are the highest.

5.2.10.1 Section Summary

The Falcon 9 launch system has proved to be a reliable workhorse for going to LEO with a strong track record and flight frequency, which passed 350 successful launch-and-recovery as of May 2024. It has disrupted the launch business by commoditizing space flights, such as having multiple launches a week.

Currently, the Falcon 9 does not have a reusable upper stage due to its orbital speed near 8 km/s. The upper stage is left in low LEO after releasing the payload. Reducing the upper stage's speed by a propulsive means to deorbit would take plenty of fuel and is impractical.

The economic factor is that as of 2021, the reported cost of the upper stage is US$10 million, which is about 20–25% of the total rocket cost. Launch costs will be reduced further if the upper stage is also reusable, particularly if a high payload capacity is included. Such a goal points to a large upper stage that can take many more passengers and cargo to orbit and return and land propulsively like the booster. A more powerful booster will be needed to get this upper stage off the ground.

We are paving the way for the Starship system to be discussed in Section 5.4.

5.3 Blue Origin Systems

Blue Origin's interests span from reusable launch vehicles to lunar landers and commercial space stations. The first working reusable launch system is the New Shepard.

5.3.1 New Shepard

The New Shepard is designed to go straight up to the edge of space, crossing the Kármán line, and return without entering orbit. The booster lands a short distance from the launch pad.

The New Shepard targets the space tourism market (Chapter 9) by offering paying customers the experience of rocket rides, the views of Earth from 100 km above, and about 3 minutes of weightlessness in space. It also offers a microgravity environment to do proof-of-concept tests for scientific payloads.

The rocket consists of a booster and a passenger capsule, as shown in Figure 5.14. The reusable booster consists of one BE-3PM engine that runs on LH_2/LOX.

As the only engine in the booster, its throttle range is substantially wider from 16% to 100%. The system's gross weight is 75,000 kg at launch and stands 18 m tall and 3.7 m in diameter. Passengers are seated in a pressurized cabin during the entire flight. The cabin can house six passengers and some cargo for the short flight, as shown in Figure 5.15.

For safety purposes, the capsule has an abort motor known as the Crew Capsule Escape Solid Rocket Motor (CCE-SRM). In case of an emergency, the solid-fueled motor fires and ejects the capsule away from the booster.

The launch process is shown in Figure 5.16. There is no gravity turn after Max Q because it is not an orbital flight. The MECO and stage separation events remain the same. MECO occurs at around 60 km altitude followed by stage separation at 70 km. Both the booster and the capsule continue their ascent by inertia, crossing the Kármán line to their individual apogees, which are about 110 km above Earth. The booster and the capsule then begin their own freefall.

The booster reenters the atmosphere, building up the speed. Reentry burn is not needed because the booster has no horizontal speed to dissipate. The booster reached its highest falling speed of 1.1 km/s at an altitude of about 27 km.

Passengers in the capsule begin to experience weightlessness at the moment of stage separation as inertia carries the capsule upward and then turns around at the apogee. The capsule's fall will be slowed by the built-in parachutes and will glide down to the ground about 3 km from the launch pad.

Three types of fins and brakes shown in Figure 5.17 are used on the New Shepard booster:

- The wedge fins provide stability
- Drag brakes are deployed in the lower atmosphere to slow down the booster's free fall. The speed drops significantly from this point on
- Aft fins are located near the bottom of the booster. They function like the grid fins on the Falcon 9 and steer and land the booster to the designated landing spot

Figure 5.14 Blue Origin New Shepard and a detachable passenger capsule at the top. *Source:* Ref. [20] / Blue Origin LLC.

Figure 5.18(a) and (b) shows the landing moments of the New Shepard booster and the capsule.

5.3.2 New Glenn

The New Glenn is a reusable, two-stage heavy-lift system under development by Blue Origin. It can deliver 45 tons of cargo to LEO. The system stands 98 m tall with

Figure 5.15 Blue Origin's first crewed capsule flight. The capsule is designed for civilians to experience weightlessness in space. *Source:* Ref. [21] / Blue Origin LLC.

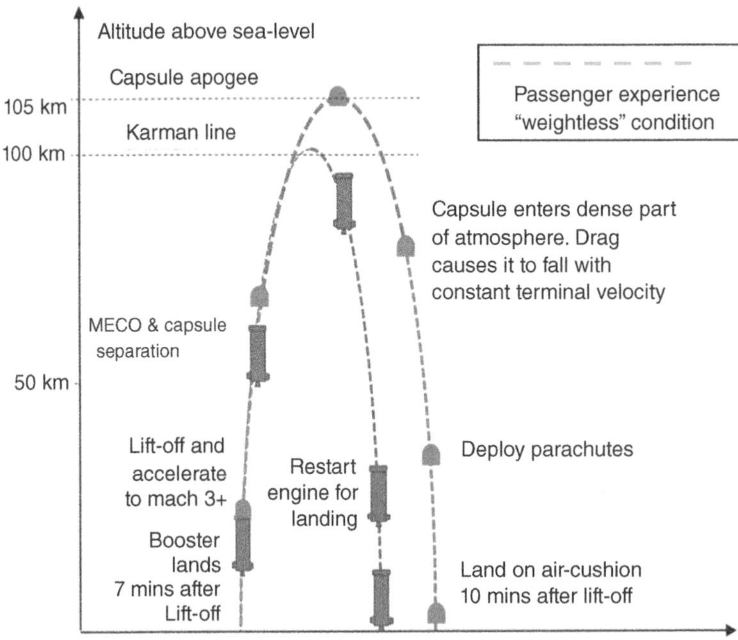

Figure 5.16 Flight profile of the New Shepard launch. The booster lands on the ground without going into orbit. The crew capsule reaches the Kármán line and parachutes down to the ground. *Source:* Ref. [22] / Design spark.

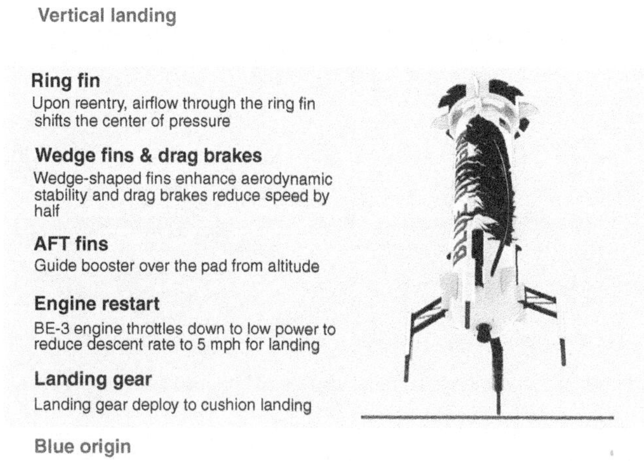

Figure 5.17 The New Shepard booster is equipped with fins and brakes for guidance and landing. *Source:* Ref. [23] / Blue Origin LLC.

a diameter of 7 m. The booster stage is 57.5 m tall, and the upper stage is 16.1 m excluding the nozzle. The booster has seven BE-4 engines running on a mixture of liquid methane and liquid oxygen, while the upper stage has two BE-3U engines running on LH_2/LOX. Two wing-like strakes are near the booster engine end to provide some lift during descent. At the top of the booster, there are four actuated fins that function like grid fins and provide attitude adjustment during descent and landing. The booster has six actuated legs for landing. These features are shown in Figure 5.19.

The flight profile is similar to that of the Falcon 9 drone-ship landing. This is shown in Figure 5.20. The booster lands on a drone ship nearly 1,000 km away from the launch site.

The first orbital test of New Glenn NG-1 took place successfully in January 2025 but did not demonstrate a booster landing.

5.4 SpaceX Starship Launch System

The Starship system attracts much public attention due to its enormous size, innovative features, and promise of lowering the launch cost by an order of magnitude from that of Falcon 9. It is designed to meet SpaceX founder Elon Musk's vision to take humans to Mars and return. As the Falcon 9 builds up a successful track record, the Starship aims to be a new workhorse servicing both LEO and deep-space journeys to the Moon and Mars.

Figure 5.18 (a) New Shepard booster touches down at the landing zone. *Source:* Ref. [24] / Blue Origin LLC. (b) The crew capsule of the New Shepard parachutes down to the desert not far from the launch site. *Source:* Ref. [25] / Blue Origin LLC.

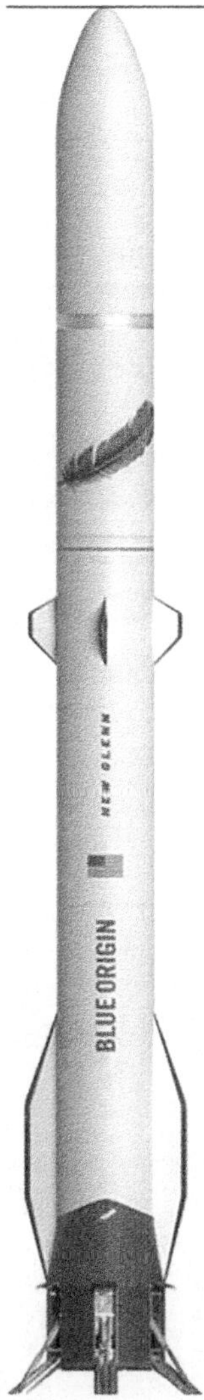

Figure 5.19 Fins and strakes on the New Glenn booster for stabilization and braking while returning and landing. *Source:* Ref. [26] / Blue Origin LLC.

Two features stand out for the Starship system: Payload capacity and full reusability (both booster and upper stage). Payload capacity and full reusability are key to low launch costs. A high-capacity vehicle can carry many crew members and cargos. A reusable upper stage, particularly with high capacity, makes the whole launch system fully reusable.

Starship is a two-stage system. The booster is called the Super Heavy and the upper stage is the Starship (or the Ship). The Starship system stands 120 m in height with a diameter of 9 m. The Super Heavy is 70 m tall and the Ship is 50 m tall, as shown in Figure 5.21.

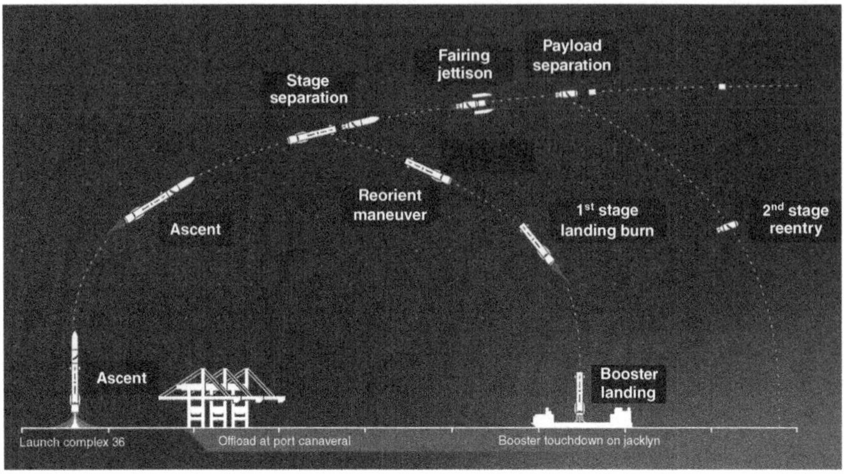

Figure 5.20 Flight profile of the New Glenn launch system being developed by Blue Origin. The booster is to land on a drone ship. *Source:* Ref. [26] / Blue Origin LLC.

Figure 5.21 The Starship upper stage is stacked atop the Super Heavy booster at SpaceX's Starbase test site [27]. *Source:* Ryan Chylinski / Adobe Stock Photos.

A quick comparison of the Starship system with the New Glenn in Section 5.3.2 shows that the Starship is taller and wider dimensionally. More importantly, the Starship upper stage aims to be fully reusable.

The Super Heavy is a scaled-up version of the Falcon 9 booster, while the Starship is a radical departure from conventional upper stages. A side-by-side comparison of the Starship system and Falcon 9 is listed in Table 5.4.

As of this writing, the Super Heavy has 33 sea-level raptor engines arranged in a 20-10-3 configuration, as shown in Figure 5.22. The Starship has six engines

Table 5.4 Comparison of the Falcon 9 launch system with the Starship launch system. Both the booster stage and the upper stage are compared. The Starship system is much bigger, and its upper stage (Starship) can return to the ground. Note (1): Only for variants that operate in the presence of some atmosphere.

Booster	**Super heavy**	**Falcon 9**
Height	71 m	40 m
Diameter	9 m	3.7 m
# of engines	33 Raptor-2 (sea-level)	9 Merlin-1D (sea-level)
Propellants	Liquid methane + liquid oxygen (methanox)	Liquid RP-1 + liquid oxygen
Fuel capacity	3,400 tons	396 tons
Grid fins	Yes (non-folding)	Yes (folding)
RCS	Hot methanox gas	Cold nitrogen
Landing legs	No	Yes, stowed away
Recoverable	Yes	Yes
Heat shields	No	No
Upper stage	Starship	Falcon 9
Height	40 m	13.8
Diameter	9 m	3.7 m
Fuel capacity	1,200 tons	92.7 tons
Payload capacity to the LEO	100+ tons	17.5 tons if the booster landing on a drone ship
# of engines	Raptor-2; 3 (vacuum) + 3 (sea level)	1 Merlin-D (vacuum)
Recoverable	Yes	No
Heat shield	Yes (1)	No
Fins/flaps	Yes (1)	No
Legs	Yes	No

Figure 5.22 The Super Heavy booster for SpaceX's Starship launch system is powered by 33 raptor sea-level engines in a 20-10-3 configuration. *Source:* Ref. [28] / SpaceX.

with three sea-level engines inside and three vacuum outside as shown in Figure 5.23.

Note that the number of engines on the booster and the Starship is subject to change after the launch system makes regular orbital flights. All the engines in Super Heavy are at sea level. For the Starship, sea-level engines are used in space as well as when operating in the presence of the atmosphere, such as launching from and landing on Earth and other celestial bodies.

The flight tests of the Starship system will be covered in Section 5.4.4.

5.4.1 Starship Launch Tower

SpaceX has built a special launch tower that allows both the launch and the return of the Super Heavy. This tower is nicknamed "Mechazilla." The Super Heavy and the Ship are assembled and integrated at the Mechazilla. The booster rests on a structure on the ground next to Mechazilla, known as the "launch mount." Under the launch mount is a deluge system that was installed after the first integrated flight test so as to dampen the large amount of energy released by the exhaust plume during the launch. Currently, about 1.3–1.5 million kg

Figure 5.23 The Starship upper stage has three sea-level and three vacuum engines. The vacuum engines are on the outside, and the sea-level engines are inside.
Source: Ref. [29] / SpaceX.

(350,000–400,000 gal) of water is used for each launch. The water all turns into vapor during launch.

Mechazilla has two pairs of robotic arms nicknamed "Chopsticks." The Chopstick pairs can lift the Super Heavy and hold it in place, pick up the Starship, and place it on top of the Super Heavy. The Chopsticks also serve as the "catcher" catching the Super Heavy when it returns after a launch.

The Super Heavy has no legs. Landing the Super Heavy at the Mechazilla suggests that RTLS is currently the only option. Alternatively, replicates of Mechazilla can be built in multiple locations. The Mechazilla-Chopsticks operation is considered an ambitious undertaking that takes the concept of precision landing to a new level.

The Ship has two landing options: caught by the Mechazilla or on the ground. Therefore, landing legs are required for the Starship so that the different variants can land on different destinations, including other celestial bodies.

Mechazilla services the two rocket stages before launch, such as stacking and fueling up. Both the Super Heavy and the Ship are expected to be "frequent fliers" with quick turnaround time. Where the Falcon 9 booster must be transported back from the landing spot to the factory for maintenance and prepping for the next flight, this step has either been simplified or eliminated altogether for the Starship system.

5.4.2 Key Systems of the Starship Launch System

The Starship system has a number of added capabilities. The Starship upper stage has features not present in other typical upper stages. It is designed to land on Earth using propulsion instead of gliding or using parachutes. We will focus on systems related to launch and return.

5.4.2.1 Super Heavy Booster

The Super Heavy's job is to deliver the Starship near the Kármán line and then return to a designated landing site. The key components of the Super Heavy resemble those of Falcon 9. The most notable deviations include the use of hot-gas RCSs and the lack of landing legs.

a) Raptor engine: SpaceX developed the raptor engine with the long-term plan to reach Mars. The raptor engine runs on liquid methane and liquid oxygen (also known as methanox) in anticipation of the lack of fossil fuels on Mars. Carbon dioxide makes up for over 95% of the Mars thin atmosphere. Methane (and water) can be obtained by chemically combining carbon dioxide with hydrogen. Hydrogen can be obtained from dissociating water on Mars. Additionally, methane is abundantly available on Earth, so there is an economic factor for frequent flights.
   The raptor engine has gone through three generations of development. As of this writing (up to eight integrated flight tests), Raptor-2 engines are used in the test flights. It is reported that Raptor-3 will be introduced in the middle of 2025 in upcoming flight tests. Raptor-2 weighs 1,600 kg with a throttling range of 40–100%. The thrust-to-weight ratio is 141. Because the raptor engine runs on a different propellant, comparison with the Merlin engine is not proper. The three generations are shown in Figure 5.24 and shown in Table 5.5.

The sea-level and vacuum versions of Raptor-2 are shown in Figure 5.25. The nozzle of the vacuum version is significantly larger, as explained in Chapter 2.

Figure 5.24 Three generations of the raptor sea-level engine are placed side by side. The general trend is that the plumbing and wiring become progressively less complex and therefore the engine weighs less. *Source:* Ref. [30] / SpaceX.

Table 5.5 Comparison of the key specifications of the three generations of the raptor engine. The mass is getting lighter, while the thrust goes higher or remains the same. Therefore, the thrust-to-weight ratio has improved substantially.

	Raptor-1	Raptor-2	Raptor-3
Mass (kg)	2,000	1,600	1,400
Thrust (tons)	185	230	269
Chamber pressure (bar)	250	300	350
Specific impulse (sec)	330	327	327
Engine TWR	93	144	192

Source: Ref. [31] / Next Big Future.

The general trend of the raptor development is to attain a) a higher thrust-to-weight ratio (TWR) and b) a reduction in the plumbing complexity. TWR is gained by reducing the engine weight while maintaining the thrust at about the same level.

b) Grid fins: Four steel grid fins are located at the top of the booster and are used in the same way as in Falcon 9. They are arranged in a 120-60-120-60 pattern

Figure 5.25 Size comparison between the raptor sea-level (left) and raptor vacuum (right) engines. *Source:* [32] / SpaceX.

around the vehicle. The fins are much larger and weigh 3 tons each. They can support the weight of the Super Heavy and are non-folding, which eliminates the motor system to do the unfolding.

c) RCS: Hot gas reaction control systems are installed on both the Super Heavy and the Starship. They serve the same function as the cold gas thrusters on Falcon 9. Cold gas RCSs do not produce sufficient thrust to move the Super Heavy or the Starship due to their much heavier weights. A hot gas RCS uses methanox routed from the fuel tanks and operates like a small rocket engine to generate thrust.

d) Body material: Both the Super Heavy and the Starship are made of stainless steel, contrary to the conventional practice of using aluminum alloys or lightweight composites. The advantages of steel include a) low cost and readily available, b) easy to machine, and c) tolerance of high temperatures. These advantages allow SpaceX to rapidly modify its designs, run tests, and churn out more test vehicles.

When the Super Heavy descends through the atmosphere, the engine end could experience temperatures as high as 900°C. Steel can tolerate that temperature with its melting point around 1,400°C. Some protective covers may be placed to protect the raptors from the heat produced by the plasma during reentry.

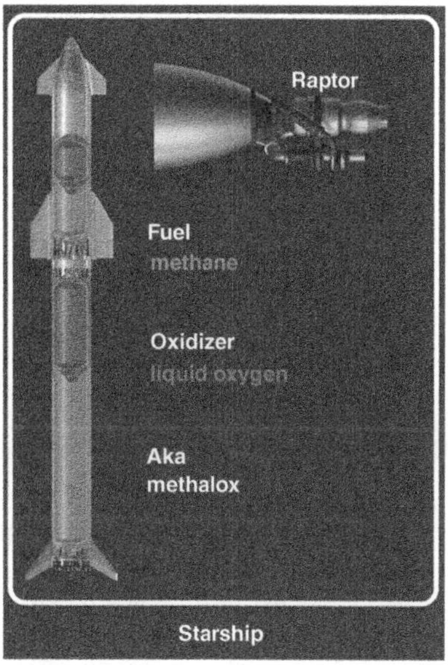

Figure 5.26 Location of the fuel tanks for LO2 and liquid methane in both the booster and the Starship. The LO2 tanks are generally larger and located at the bottom of the craft. The Starship has two "header tanks." *Source:* Ref. [33] / Everyday Astronaut.

e) Fuel tanks: Both the Super Heavy and the Starship run on methanox with the oxygen-to-methane ratio at 3.6 : 1. Figure 5.26 illustrates the locations of the tanks for these propellants. The main LO2 tank is located near the rocket engines of both vehicles.
f) Landing legs: The Super Heavy has no landing legs. Given the weight of the Super Heavy, the legs would have to be very heavy as well. Eliminating the legs reduces the weight and system complexity.

5.4.3 Starship Spacecraft

An orbiting spacecraft in LEO has a speed of nominally 8 km/s. Returning the spacecraft to Earth must dissipate a significant portion of this speed. Unlike the Space Shuttle that glides and lands with wheels on runways, a Starship could land on unpaved grounds off-world. Hence, the Ship must land vertically. A reusable upper stage that can land at designated platforms (instead of splash-down) must be equipped with propulsive capabilities not found in traditional upper stages. The Ship's enormous weight and payload capacity also require multiple engines to work in both the ascent and return portion of its spaceflight. The Ship has the following subsystems related to flight operation.

a) Raptor engines: The Starship has both sea-level and vacuum-level engines on board. All the engines are turned on during the Ship's ascent to LEO after stage separation. The vacuum-level engines come in for high orbits and deep space. The spacecraft uses sea-level engines to execute propulsive maneuvers to land.

The current configuration has six engines: three vacuum engines on the exterior and three sea-level engines on the inside.

b) RCSs: Like those in the Super Heavy, these thrusters are used in coarse attitude adjustment during spaceflight and atmospheric reentry.

c) Fins (flaps): Instead of grid fins, the Starship has two pairs of motor-driven fins located in the forward and the aft sections, as shown in Figure 5.27. Flaps are aerobrakes and require an atmosphere to operate in. The two pairs together are also for steering and attitude adjustment.

At the beginning of the atmospheric entry, the Starship's flaps are fully extended like a skydiver's motion, as shown in Figure 5.28. The "belly" side facing the atmosphere head-on is the windward side, and the opposite side is the leeward side. The Starship is angled to maximize the exposure of its windward side to the atmosphere during descent to bleed off speed.

When the Starship is close to the ground and ready to land, the aft flaps fold in and the three sea-level engines fire to flip the vehicle from horizontal to vertical with the engine-end facing the ground, as shown in Figure 5.29. Landing burn is applied to bring the vehicle to land vertically.

Figure 5.27 One side of the Starship upper stage is covered with heat shield tiles on the backside (known as the windward side) of this image. The front, or leeward, side has no heat tiles. The windward side faces the atmosphere during re-entry. Two pairs of flaps are in the front and aft. *Source:* Ref. [34]. DimaLopatin1999 / Wikimedia Commons / CC BY-SA 4.0.

Figure 5.28 Image describing the Starship re-entry. It maintains almost horizontal until the last moment when the spacecraft flips vertically to land. The spacecraft's two pairs of flaps are extended during descent similar to skydiving. *Source:* Ref. [33] / Everyday Astronaut.

Figure 5.29 In this 2021 flight test, the Starship rose to a 10 km altitude (without the Super Heavy) and returned to the ground. It fires three engines when it is ready to land. The spacecraft flips from horizontal to vertical by folding in the aft flaps while firing three engines. Then, the Starship fires two engines to land. *Source:* Ref. [35] / Everyday Astronaut.

Hence, flaps play critical roles in maneuvering and landing operations. The Starship has no grid fins, so the flaps are the precision control of attitude and steering in the presence of some atmosphere.

d) Header tanks: Header tanks hold small volumes of propellants as reserves. The header tank of liquid oxygen is placed at the Starship's nosecone. The header tank of liquid methane is placed in the vehicle's mid-section between the main methane tank and the oxygen tank. They are shown in Figure 5.30.
Fuels in the two header tanks are reserved for landing. When returning and landing on Earth, for example, the Starship's main fuel tanks could be nearly empty. The vehicle is in a near-horizontal position before flipping vertically, so that the propellants rest on the windward side of the tanks. Drawing propellants from the full header tanks will ensure smooth engine operation for the flipping and then the landing process.
Placing the liquid oxygen header tank at the nosecone has the additional advantage of keeping the spacecraft's center of gravity closer to the spacecraft's midsection since the six raptor engines are heavy. Doing so helps the vertical flipping before landing.

e) Heat shield tiles: A planet's atmosphere, no matter how thin, provides aerobraking to dissipate some of a spacecraft's speed in the process of reentry and landing. Such braking generates plenty of heat and plasma due to the spacecraft's hypersonic speed tearing into the atmosphere.
A Starship entering the Earth's atmosphere from orbit must dissipate much higher ΔV than a Super Heavy booster. Some hotspots on the windward side of the spacecraft are expected to be near or even exceed steel's melting temperatures. Hence, one solution is to use heat shields.

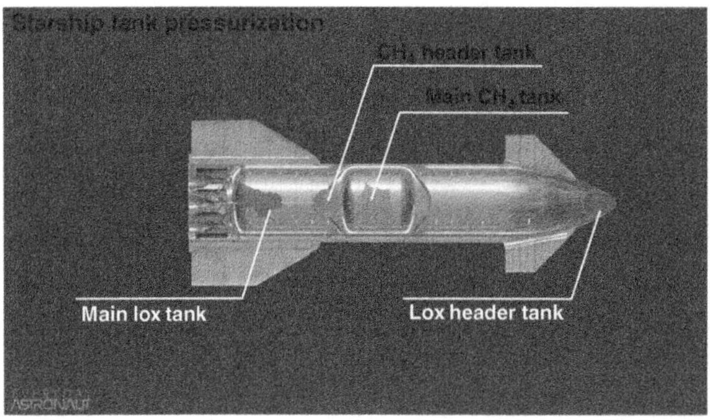

Figure 5.30 Locations of propellant tanks in the Starship. The main tanks are in the midsection and low section. A liquid oxygen header tank is placed at the nosecone and the liquid methane header tank in the midsection. *Source:* Ref. [33] / Everyday Astronaut.

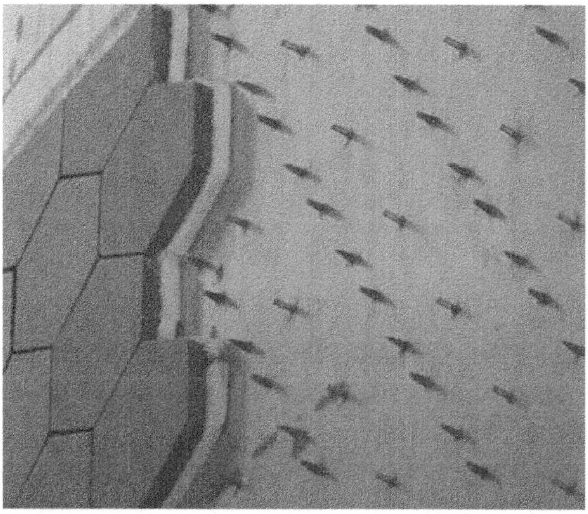

Figure 5.31 Heat shields are mounted on the Starship windward side via the studs. The white foam is a heat insulator, and the black tiles tolerate the high temperatures during reentry. *Source:* Ref. [36] / SpaceX.

Figure 5.32 The windward side (left half) of the Starship is covered with heat tiles, while the leeward side (right half) is bare metal. *Source:* Ref. [37] / NASA / Public domain.

The Starships heat shields are hexagonal tiles made from special foams, as shown in Figure 5.31. Approximately 20,000 tiles are mounted on the spacecraft's windward side and the flaps. The leeward side needs no heat shields. This is illustrated in Figure 5.32.

Heat shield tiles possess low heat conductivity, high specific heat capacity, and a high melting point at around 1,370°C. In other words, the tiles can absorb and store plenty of heat energy and not melt easily.

f) Legs: The Starship has six legs as some variants are for landing on other celestial bodies. It is unclear if the Ship will only land at the Mechazilla like the Super Heavy and not at a nearby landing zone. The legs are normally retracted and deployed when the Ship is ready to land. They are also designed for auto-leveling on uneven ground. See Figure 5.33.

Figure 5.33 The Starship upper stage has six landing legs that are normally retracted. The legs are deployed when landing on a flat surface. The legs are self-adjusting in case the Starship lands on an unpaved area. *Source:* Ref. [38] / SpaceX.

5.4.3.1 Starship Variants

The Starship is intended to be a workhorse for passengers and cargos going to LEO as well as deep space travel. There are five variants of the Ship: Crew, cargo, fuel tanker, lunar, and deep space. Each variant is tailored to serve a dedicated purpose. The five Starship variants are shown in Figure 5.34. SpaceX's plan is for the crew and cargo variants to steadily replace Falcon 9 with shuttle between Earth and LEO. The Tanker variants will also be launched frequently. The crew, cargo, and tanker variants have flaps and heat protection for routine re-entry to Earth.

Flaps and heat shields are, therefore, variant dependent.

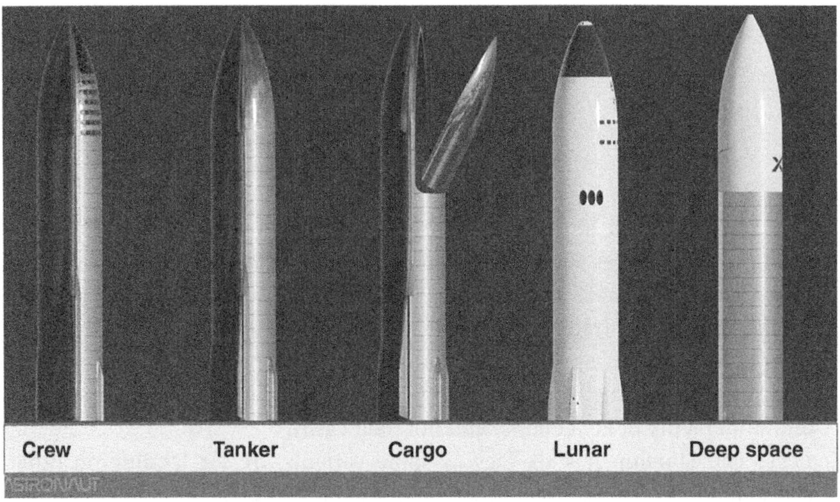

Figure 5.34 There are five variants of the Starship. They serve different purposes. They are crew, tanker, cargo, lunar, and deep space. The crew, cargo, and tanker are expected to be frequent fliers to the LEO. *Source:* Ref. [33] / Everyday Astronaut.

On-orbit refueling is a critical capability of the space transportation infrastructure. Missions to LEO and return do not need refueling. But missions to higher orbits and beyond benefit from refueling because the payload capacity is not sacrificed. Hence, the tankers will be busy transporting fuels from Earth to LEO to refuel other variants.

Crew Variant The crew variant is designed mainly for transporting passengers to different Earth orbits and other celestial bodies such as the Moon and Mars. The current passenger capacity to LEO is 100. It needs flaps and heat shields.

Cargo Variant The cargo variant carries different payloads to Earth orbits or other celestial bodies after refueling. Its current capacity to LEO is 100 tons. The first version has a wide cargo bay of 9 m in diameter and 17 m deep. It can house satellites of different sizes and large space station modules. Whereas the Falcon 9's fairing halves are jettisoned, the cargo bay fairing of the Starship is like a clamshell that opens and closes as illustrated in Figure 5.35.

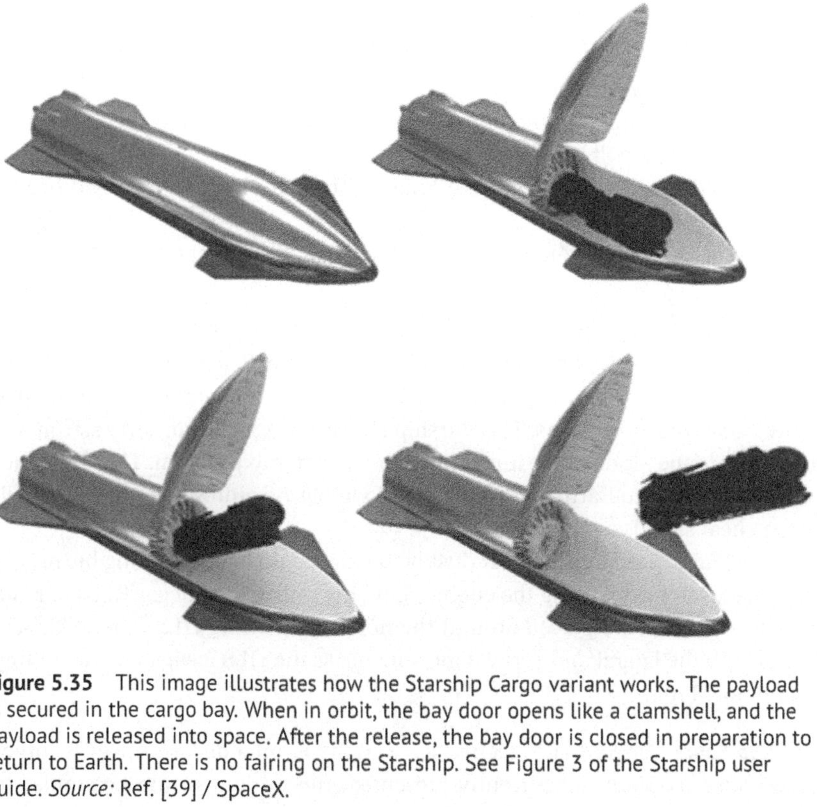

Figure 5.35 This image illustrates how the Starship Cargo variant works. The payload is secured in the cargo bay. When in orbit, the bay door opens like a clamshell, and the payload is released into space. After the release, the bay door is closed in preparation to return to Earth. There is no fairing on the Starship. See Figure 3 of the Starship user guide. *Source:* Ref. [39] / SpaceX.

SpaceX's plan is for the cargo variant to deliver thousands of Starlink satellites to LEO, as will be elaborated in Chapter 6. The cargo variant is expected to fly frequently between Earth and LEO, so flaps and heat shields are needed.

Tanker Variant The tanker's role is like a petroleum truck. They are sent to LEO and possibly other orbits. The tanker's fuel capacity to LEO is 100 tons and may reach 150 tons in the future. The Ship's maximum fuel capacity is 1,200 tons at launch, and plenty of fuel is consumed from stage separation to entering LEO. The ability to refuel will greatly facilitate journeys to higher orbits and beyond. The current estimate is that it will take seven to eight tankers to fill up a regular crew or cargo variant in orbit. Tankers will have flaps and heat shields.

Refueling will be tested in flight after the Starship has completed the integrated flight tests.

Starship Human Landing System (HLS) In 2021, NASA awarded SpaceX US$2.89 billion to develop the Starship HLS as part of the Artemis Program. See Chapter 11.

The Starship HLS's operation has several important tasks in accordance with the Artemis missions:

a) On-orbit refueling in the LEO
b) Transporting cargo to the lunar orbit where the planned space station, Lunar Gateway, will be located
c) Transporting astronauts from the lunar orbit or the Gateway to land on the lunar surface
d) Shuttling astronauts from the lunar surface back to the lunar orbit or the Gateway
e) The first HLS will not return to Earth

The Starship must first demonstrate successful, reliable, and frequent flights to LEO. Second, tanker variants must be placed in LEO to demonstrate a reliable on-orbit refueling operations. The Starship HLS is not assigned to carry astronauts from Earth to the Moon in the initial phase of the Artemis program. The HLS is not required to return to Earth, and since the Moon has no atmosphere, the HLS will have no heat shields and flaps.

To avoid large amounts of lunar dust being kicked up by the landing burns, the HLS's plan is to forego using the engines but only utilize the hot gas RCSs to land. Extra sets of RCSs are placed around the midsection of the HLS. The RCSs will control both the lateral and vertical movements as the HLS navigates to a landing spot on the Moon.

The Deep-Space Variant The deep-space variant is intended to travel to other planets such as Mars. As currently presented, this variant would be similar to the lunar variant with two major exceptions. Since Mars has a thin atmosphere, past Mars probe missions have shown that the thin atmosphere can be leveraged to dissipate the vehicle's kinetic energy to land the spacecraft. Therefore, the deep-space variant will need both heat shields and flaps.

The Starship can carry a sizable team and equipment with different amenities instead of three to four astronauts being cooped up in a small spacecraft for months. Entrepreneurial enthusiasts have put forth various designs online for the long journey to Mars [40–42].

The present exploratory spacecraft going to Mars takes on average of 9 months in the forward journey. However, the return journey can be twice as long due to Mars's different alignment with Earth (see Chapter 13). The criterion for the Starship is establishing the ability to manufacture methanox on Mars. The Ship will also need sufficient propulsion capability to lift off and enter orbit from Mars without the Super Heavy.

5.4.4 Integrated Flight Tests

As of this writing in 2025, SpaceX has been going through a series of integrated flight tests (IFTs) of the Starship system. An IFT is a non-crewed flight that aims to test the full functioning of the booster and the Starship vehicle in orbit. The IFT series evolves as data is collected along the way. Using stainless steel has expedited design changes and fabrication based on data collected from the IFTs.

As of March 2025, there have been eight IFTs since April 2023. The initial phase of the IFTs aimed for the booster landing in the waters of the Gulf of Mexico downrange from launch, while the Starship landed in the ocean after reaching low orbit altitude.

IFT-1, -2, and -3 were not completed due to various issues that occurred soon after launch. Readers are referred to a wealth of information and videos for IFT-1 to IFT-3.

IFT-4 is considered a success in spite of several issues with the Starship while reentering the atmosphere. Both the booster and the Ship successfully splashed down in a controlled fashion in their designated waters.

IFT-5 attracted the most attention as the booster successfully landed at the Mechazilla, demonstrating "booster catch" by the Chopsticks. The Starship landed in the water as planned without the issues observed in IFT-4. IFT-5 will be elaborated below.

IFT-6 took place a month after IFT-5. IFT-6 was also successful although SpaceX chose not to land the booster at the Mechazilla. IFT-7 ended with mixed results. The booster landed at the Mechazilla without any issues. However, the Starship was reported to have been lost while ascending to orbit. As of this writing, the preliminary cause is a potential fuel leak, leading to a vehicle breakup.

The IFT-8 took place on March 3, 2025. The booster demonstrated "picture-perfect" landing at the Mechazilla. The Ship, however, was also lost for reasons of what are believed to be similar to IFT-7.

5.4.5 Integrated Flight Test 5

We will focus on the IFT-5. However, we will use an infographic of IFT-6 by Humanmars.net to illustrate. The planned profile of IFT-6 is the same as that of IFT-5 although a later decision chose not to land the booster. IFT-6's planned flight profile is shown in Figure 5.36. For legibility, we replace the text in Figure 5.36 with numbers. The profile describes the whole flight in several steps.

The steps are as follows: 1) Super Heavy engine start, 2) launch, 3) ascent, 4) most engine cutoff and hot staging, 5) flip maneuver, 6) boost-back burn, 7)

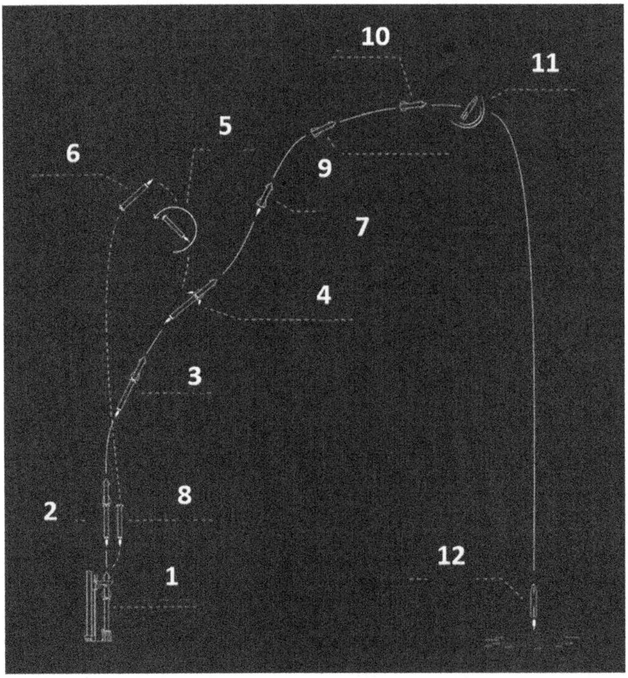

Figure 5.36 The flight profile of the Starship Integrated Flight Test 6 (IFT-6) from the launch to the landing of the booster and the Starship upper stage. The flight plan is to be the same as IFT-5 and calls for the booster to land precisely to the launch tower Mechazilla, and the upper stage lands in the Indian Ocean. However, it was decided that the booster would not return to the tower during flight [43]. The numbers represent: 1) Super Heavy engine start, 2) launch, 3) ascent, 4) most engine cutoff and hot staging, 5) flip maneuver, 6) boost-back burn, 7) Starship ascent, 8) Super Heavy landing burn and catch, 9) Starship engine cutoff, 10) coast phase, 11) Starship entry, and 12) Starship landing burn and splashdown.

Starship ascent, 8) Super Heavy landing burn and catch, 9) Starship engine cutoff, 10) coast phase, 11) Starship entry, and 12) Starship landing burn and splashdown.

IFT-5 planned for the Super Heavy booster to return to the launch site, while the Ship lands in the water as before. According to the online video produced by SpaceX, the flight control called out "Return to Tower" soon after the stage separation. The booster executed its boost-back burn to return to the launch site.

Readers are referred to a video on the SpaceX website [44]. Educational information can be extracted from the telemetry data displayed in this video. Figure 5.37 is a snapshot of the Super Heavy about to be caught by the Chopsticks on the Mechazilla.

Table 5.6 outlines the sequence of events. We also add in a few moments to help us follow the intricacies of the Starship part. For convenience, we convert the speeds from km per hour to km per second. Additionally, two small images in the [44] video show the instantaneous pitching orientations of the Super Heavy and the Starship, which help us follow the orientation of the booster and the Ship.

Figure 5.37 The Super Heavy booster returns to the launch tower in IFT5 and is "caught" by the Chopsticks [45]. *Source:* Steve Jurvetson / Flickr.

Table 5.6 Sequence of events of IFT#5 for both the booster and the Starship.

Time (T+)	Booster event	Super heavy Altitude (km)	Speed (km/s)	Starship event	Altitude (km)	Speed (km/s)
0:00	Liftoff	0	0			
1:00	Max Q (banking from vertical)	8	0.31			
2:37	Most engine cut-off (MECO). Center 3 engines stay on	65	1.44			
2.43	Stage separation, 3 center engines on	70	1.44	Starship 6 engines on		
2:47	Boost-back burn (13 center engines for 1 min)	73	0.49		73	1.47
3:34	10 engines off, only 3 center engines on	94	0.53		101	1.76
3:41	All engines off	95	0.53		106	1.82
3:43	Reaching apogee	95	0.53		106	1.83
4:52		80	0.87		132	2.54
5:08		70	0.97		135	2.75
5:21		60	1.07		138	2.94
5:33		50	1.15		140	3.12
5:43		40	1.19		141	3.29
5:52		30	1.01		142	3.45
6:01		20	0.78		143	3.61
6:07		15	0.53		143	3.73

6 : 12		10	0.38		143	3.83
6 : 19		5	0.32		144	3.98
6 : 27		2	0.06		144	4.14
6 : 31	Landing burn, 13 engines	1			144	4.23
6 : 37	Landing burn, 3 engines	0			144	4.37
6 : 52	Caught by chopsticks	0	0		144	4.75
8 : 00				Turn off 3 engines	147	6.92
8 : 28				SECO	149	7.36
22 : 25				Apogee	212	7.29
25 : 40				Altitude begins to decrease	211	7.28
30 : 32					200	7.30
36 : 32					170	7.34
40 : 48					140	7.37
43 : 24					120	7.40
44 : 00				Preparing for reentry	115	7.41
44 : 32					110	7.41
45 : 44				Sign of the plasma	100	7.42
46 : 55					90	7.43
48 : 10					80	7.41
49 : 58				Peak plasma heating reached	70	7.14
55 : 44					60	5.51

(continued)

Table 5.6 (Continued)

Time (T+)	Booster event	Super heavy Altitude (km)	Speed (km/s)	Starship event	Altitude (km)	Speed (km/s)
59 : 04					50	3.56
1 : 00:40				2nd Max Q	43	2.21
1 : 01 : 02					40	1.88
1 : 02 : 13					30	0.81
1 : 02 : 40				Mach 2	25	0.45
1 : 03 : 04				Transonic	20	0.28
1 : 03 : 28				Subsonic	15	0.21
1 : 04 : 33				Starship horizontal	5	0.12
1 : 05 : 25				Landing burn startup, three engines	0	0.09
1 : 05 : 37				Starship flips vertically	0	0.08
1 : 05 : 42				Soft splashdown	0	0

Referencing Table 5.6, the following observations can be made.

1) All 33 booster engines come on at liftoff
2) Max Q happens at an 8 km altitude. Booster begins pitching away from the vertical.
3) At the 65 km altitude, 30 engines shut down, known as the most engine cutoff (MECO), while three center engines keep running. Hot staging begins (see further below for explanation)
4) At 70 km altitude, Starship ignites six engines, while the booster's three center engines are still running
5) A second later, the Starship separates from the booster
6) RCSs immediately work to turn the booster around with the top end pointing toward the launch site
7) Booster executes boost-back burn with 13 center engines for about 1 minute as the RTLS operation
8) The booster is almost pointing horizontally, while it rises in altitude by inertia. In the meantime, the Starship thrust into orbit, gaining both speed and altitude
9) Booster reaches apogee at 95 km altitude with a speed down to 0.53 km/s, which is all horizontal
10) Booster freefalls with its engine side facing the Earth while picking up speed
11) Booster drops from 60 to 30 km. No reentry burn is called out
12) The booster reaches its maximum speed of 1.2 km/s at 40 km altitude and begins to slow down due to atmospheric drag
13) Booster crossing Mach 1 at 1 km altitude when landing burn starts using 13 center engines
14) The booster approaches Mechazilla and is caught by the Chopsticks
15) The Starship accelerates until it reaches 149 km altitude with a speed of 7.36 km/s when all six engines cut off (SECO)
16) Momentum carries the Starship to its apogee of 212 km altitude, while its speed drops slightly to 7.29 km/s
17) As the Starship drops altitude with horizontal speed, the video camera picks up signs of plasma on the flaps from 103 km altitude onward. At this point, the Starship's body is angled about 60° from the vertical. The evidence of plasma is obvious from here on.
18) Peak heating is called out at 70 km altitude, with its speed coming down to 7.14 km/s
19) (2nd) Max Q is called out at 43 km altitude with a speed of 2.21 km/s. This Max Q is for the Ship returning to the lower atmosphere. The 2nd Max Q can be thought of as the reverse of the 1st Max Q during descent.

20) Between 80 and 40 km altitude, the Starship spends almost 13 minutes to lose (7.41–1.88 =) 5.53 km/s of speed
21) Starship crosses subsonic at 15 km altitude
22) Starship flips to vertical and starts landing burn with three engines at under 1 km altitude (exact height not shown) when speed is 0.08 km/s

The flights of the Super Heavy and Starship from launch to landing are plotted in Figures 5.38 and 5.39 in a fashion similar to Figure 5.10 based on telemetry data extracted from an online video on the SpaceX website [44]. The key events of the booster's RTLS flight are shown in numbers in the figures and explained in the captions.

By all accounts, Test #5 is a complete success without any reported issues. While IFT-4 is a successful flight test overall, IFT-5 goes a major step further by returning the booster to the launch tower and being caught by the Chopsticks with no observable issues. Between IFT-5 and IFT8, "Booster Catch" is successfully demonstrated three times.

A unique feature of the Starship is that the communications and telemetry data go through the Starlink satellites in LEO without experiencing any communication blackout period during reentry.

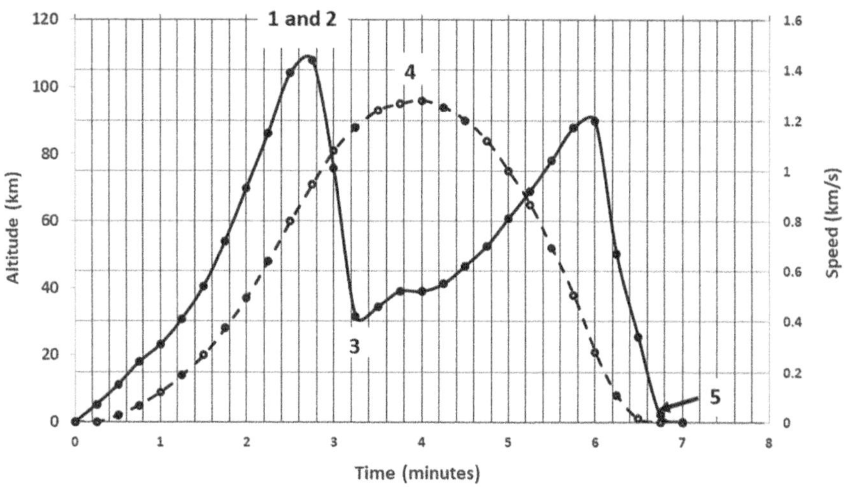

Figure 5.38 Plot of booster height and speed from launch to landing of the Starship IFT-5 based on telemetry data obtained from the SpaceX website. Altitude is represented by the dotted line and the booster's speed by the solid line. The key events for the booster are 1) most engine cutoff (MECO), 2) boost-back burn starts moments later, 3) boost-back burn ends, 4) booster at apogee, and 5) landing burn starts when the booster's altitude is about 1 km. *Source:* Adapted from Ref. [44].

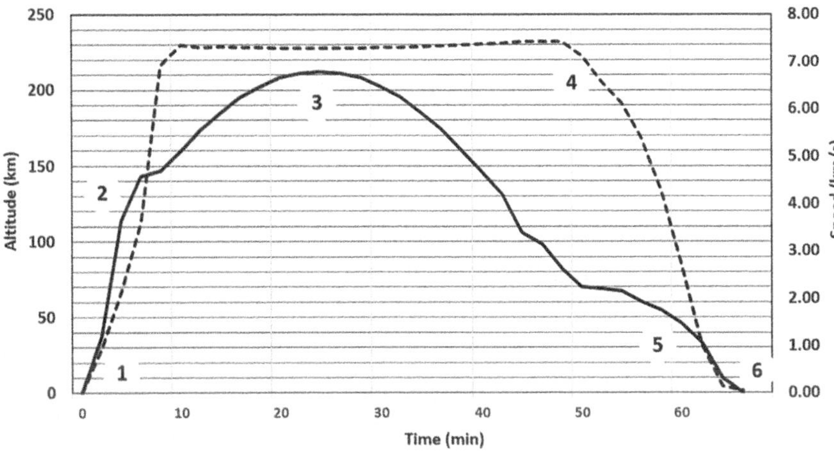

Figure 5.39 Plot of the Starship IFT-5 height and speed from the point of stage separation based on telemetry data obtained from the SpaceX website. Time is referenced to the moment of stage separation. The Starship's altitude is represented by the dotted line and speed by the solid line. The key events for the Starship are as follows: 1) Stage separation and all engines firing, 2) engines partial to total turn off, 3) apogee over 200 km, 4) Starship speed begins to drop significantly upon entry to the upper atmosphere, 5) Starship in lower atmosphere, and 6) water landing at designated location. *Source:* Adapted from Ref. [44].

5.4.5.1 Hot Staging

SpaceX adopted the hot staging technique to replace the regular booster main engine cutoff. Instead of shutting down all booster engines to facilitate stage separation like Falcon 9, three center engines are left running, while the other 30 engines are shut down. The three engines keep a small upward thrust going. The thrust is meant to keep the propellants in the booster at the bottom of the two main propellant tanks. This reduces the chance of the propellants momentarily "free floating" and causes issues when firing up for the boost-back burn.

In hot staging, the Starship ignites six engines a second before being released by the holding mechanism in the booster. The Starship essentially pushes the booster away to free itself. The booster is then turned away to point in the opposite direction of travel via the RCSs. More engines come on making a total of 13 center engines executing the boost-back burn.

To protect the top of the booster from being damaged by the Starship's hot exhaust before the two stages disengage, a temporary hot staging ring with a protective dome is fitted on the booster's top. This metal ring is jettisoned soon after the boost-back burn is completed. One can see the ring falling away from the booster starting at around T + 04 : 50 in the IFT-5 video [44].

In summary, the booster operation works out well as of this writing. The Starship also works out admirably in IFT-5. In IFT-4, visible damage to the flaps by the plasma was observed during descent although no flap failure resulted. The issue with the plasma is not observed in IFT-5 after SpaceX made improvements with the flaps and the heat tiles. Issues related to ship failure in IFT-7 and IFT-8 are being addressed.

5.5 Reusable Rockets in China

China began changing its policy in support of the commercialization of space in the mid-2010s. A number of startups specializing in launch systems and satellites have since been established. As reusable rockets gained popularity in the late 2010s, several of these companies also joined the development endeavors. The reusable rocket technology of commercial aerospace has developed rapidly. LO2/methane and LO2/kerosene are used in most reusable rocket engines.

Research on reusable rocket technology has been conducted by commercial companies and academic institutions. Between 2019 and 2024, reusable rockets were tested in China. Figure 5.40 shows images of reusable rockets produced by several companies.

5.5.1 RLV-T5

RLV-T5 is designed and made by Linkspace. They have developed five prototypes (RLV-T1 ~ T5) and conducted more than 300 test flights. In 2019, the RLV-T5 achieved the first low-altitude flight recovery test.

5.5.2 PALLAS-1

PALLAS-1 is a two-stage, reusable system under development by Galactic Energy. Galactic Energy developed the reusable engine "Welkin" using LO2 and kerosene. In 2022, the first full-system test was conducted successfully.

5.5.3 Nebula

Nebula-1 and Nebula-2 are developed by Deep Blue Aerospace. They designed the kerosene engine "Deep blue thunder-R1" with a throttle range of 30–110%. The 1 km vertical takeoff and landing flight test was achieved in 2022, and the test rocket successfully recovered. The carrying capacity of Nebula-1 and Nebula-2 is 8 tons and 20 tons, respectively. The diameters of Nebula-1 and Nebula-2 are 3.35 and 5 m, respectively.

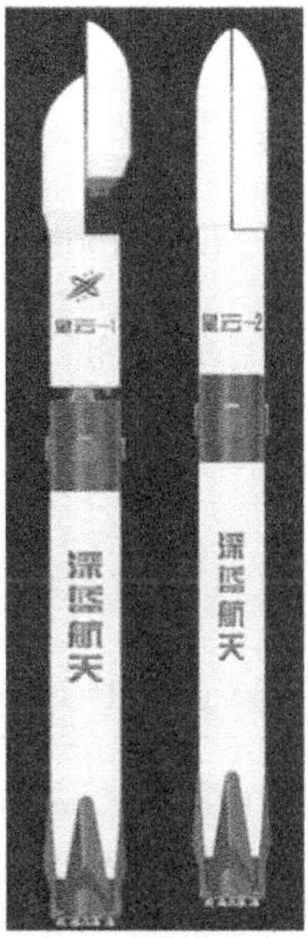

(a) RLV-T5 [46] (b) Nebula-1&2 [47] (c) PALLAS-I [48]

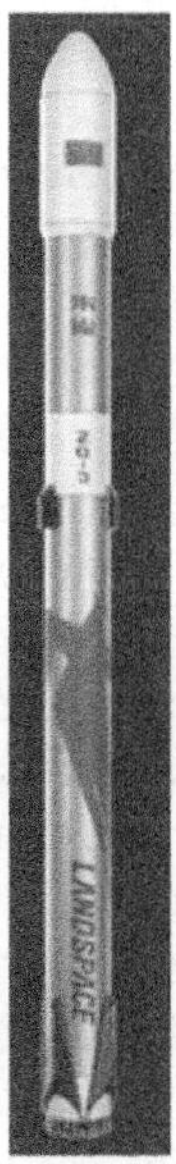

(d) Nebula-1&2 [47] (e) ZQ-2 [49]

Figure 5.40 Images of reusable booster research and products by five space companies in China [46–49]. *Sources:* (a) Ref. [46] / Linkspace; (b, d) Ref. [47] / GALACTIC ENERGY; (c) Ref. [48] / jiangsu Shenlan Aerospace Co., LTD; (e) Ref. [49] / Landspace.

5.5.4 Hyperbola

The "Hyperbola-2" reusable rocket is developed by Beijing Interstellar Glory Space Technology. It is a two-stage liquid launch vehicle with a length of 28 m. The diameters of the first stage and the second stage are 3.35 and 2.25 m, respectively. The payload capacity to the LEO is more than 1.9 tons. The first stage is reusable by means of propulsion. The rocket is powered by a variable-thrust and reusable LO2/methane engine. The "Hyperbola-2" has achieved a full-size booster take-off, landing, and reuse in a flight test in 2023. The booster landed smoothly and precisely at the target site.

5.5.5 ZQ

The "ZQ" series of reusable rockets is designed by LANDSPACE. The length and diameter of the ZQ-3 are 76.6 and 4.5 m, respectively. The booster could be reused up to 20 times. LANDSPACE developed the TQ series engines using LO2 and methane as propellants. The ZQ-3 accomplished a smooth landing after a 60-second flight in 2024 and expects to be flight-capable in 2025.

5.6 Conclusion

Reusable rockets are a major advancement and innovation in the launch industry. Reusability drives down launch costs in terms of dollars per kilogram of cargo. Low launch cost is instrumental in the commercialization of LEO. As of 2024, reusable medium-lift boosters such as the Falcon 9 have achieved a record with over 300 successful launches and recoveries. Other companies are now offering either similar boosters or are in the process of developing them. A reusable booster saves the total hardware cost of a launch system – excluding the payload – by as much as 75%.

The current practice is the upper stage being expended in the low orbit after releasing the payload. The upper stage takes up about 15–20% of the total hardware cost and is the second item on the cost list after the booster. The issue is whether it is economical to design an upper stage that can return from orbit to the ground. It is more economical to have a reusable upper stage with a large payload capacity, such as a heavy-lift system. In other words, it is not economical to have a reusable upper stage that is small and only has a low payload capacity.

Increasing the payload capacity is a strong incentive so that more crew and cargo can be transported in one mission. Commercialization means frequent missions and active participation by the private sector in business applications. High-capacity space vehicles will also serve deep-space missions.

A reusable upper stage combined with a high payload capacity provides a strong incentive for the development of the next-generation launch system. Reusability also means "quick turnaround" time for the rocket stages akin to commercial aviation. As of this writing, this line of development is making rapid progress, as evident from the series of integrated flight tests of the SpaceX Starship system.

A completely reusable launch system with on-orbit refueling will be a key milestone for this decade.

Reading List

Berger E. *Reentry: SpaceX, Elon Musk, and the Reusable Rockets that Launched a Second Space Age*. BenBella Books; 2024.

Hillis PA. *The Future of Reusable Rockets: Falcon 9, Dragon, Neuralink, and Beyond How Elon Musk Changed Space Travel and the Future of Humanity. Independently Published*; 2024.

Scott JZ. *Elon Musk Vs. China's Reusable Rocket: What Happens When the World Imitates Starship*. Independently Published; 2024.

Smibert A. *Reusable Rockets*. Norwood House Press; 2019.

References

1 Pompa F, Padilla R, Thorson M. Anatomy of a SpaceX Launch. USA Today [cited 2025 Feb 17] [online]. Available from: https://www.usatoday.com/pages/interactives/spacex-launch-graphic/.

2 Space X. Falcon 9. Space X [cited 2025 Feb 17] [online]. Available from: https://www.spacex.com/vehicles/falcon-9/.

3 SpaceX. Cited by Timmer J. SpaceX Launches Falcon 9 v1.1, Preps for Reusable Boost Stage. Ars Technica [cited 2025 Feb 17] [online]. Available from: https://arstechnica.com/science/2013/09/spacex-launches-falcon-9-v1-1-preps-for-reusable-boost-stage/.

4 SpaceX. Cited by Marin PD. Lanzamiento del BulgariaSat 1 y segunda reutilizaci'on de un Falcon 9. Eureka [cited 2025 Feb 17] [online]. Available from: https://danielmarin.naukas.com/2017/06/25/lanzamiento-del-bulgariasat-1-y-segunda-reutilizacion-de-un-falcon-9/.

5 SpaceX. X. Cited by Clark S. New Photos Illustrate Progress in SpaceX's Fairing Recovery Attempts. Spaceflight Now [cited 2025 Feb 17] [online]. Available from: https://spaceflightnow.com/2018/06/01/new-photos-illustrate-progress-in-spacexs-fairing-recovery-attempts/.

6 Ralph E. SpaceX Falcon 9 Landing Leg Accidentally Dropped During Retraction Attempt. Teslarati [cited 2025 Feb 17] [online]. Available from: https://www.teslarati.com/spacex-falcon-9-landing-leg-retraction-accident/.
7 The Space Techie. Cold Gas Thrusters: How Does It Work?. The Space Techie [cited 2025 Feb 17] [online]. Available from: https://www.thespacetechie.com/cold-gas-thrusters-how-it-works/.
8 SpaceX. Cited by Evans B. SpaceX Launches Transporter-3 Rideshare Mission, Lands 10x-Flown Booster. America Space [cited 2025 Feb 17] [online]. Available from: https://www.americaspace.com/2022/01/13/spacex-launches-transporter-3-rideshare-mission-lands-10x-flown-booster/.
9 Space X. Transporter-6 Mission. Space X [cited 2025 Feb 17] [online]. Available from: https://www.spacex.com/launches/mission/?missionId=transporter-6-pl.
10 Space X. Space X [cited 2025 Mar 20] [online]. Available from: https://www.spacex.com/mission/.
11 Space X. Telkomsat Merah Putih 2 Mission. Space X [cited 2025 Feb 17] [online]. Available from: https://www.spacex.com/launches/mission/?missionId=merahputih2.
12 Space X. Starlink Mission. Space X [cited 2025 Feb 17] [online]. Available from: https://www.spacex.com/launches/mission/?missionId=sl-6-54.
13 SpaceX. Cited by Ryan J. Meet the SpaceX Falcon Heavy, the World's Most Powerful Rocket. CNET [cited 2025 Feb 17] [online]. Available from: https://www.cnet.com/pictures/meet-the-spacex-falcon-heavy-the-worlds-most-powerful-rocket/.
14 Musk E. SpaceX Flight Profile. X [cited 2025 Feb 17] [online]. Available from: https://x.com/elonmusk/status/960915618865274880?mx=2.
15 SpaceX. Falcon Heavy Test Flight | LZ Landing Drone Footage [video file]. 2020 Apr 09 [cited 2025 Feb 17]. Available from: https://www.youtube.com/watch?v=fbbSCqaeWyA. Cited by Ralph E. SpaceX celebrates historic rocket landings with new 4K footage. Teslarati [cited 2025 Feb 17] [online]. Available from: https://www.teslarati.com/spacex-falcon-heavy-booster-landing-4k-video/.
16 SpaceX Falcon Heavy Launches Classified USSF-67 Space Force Mission, Nails Landings. Video from Space [cited 2025 Feb 17] [online]. Available from: https://www.youtube.com/watch?v=O5NxzcEEg_o.
17 Arabsat-6A Mission. SpaceX [cited 2025 Feb 17] [online]. Available from: https://www.youtube.com/watch?v=TXMGu2d8c8g
18 SpaceX Lands All 3 Falcon Heavy Boosters for the First Time. Galaxy Central [cited 2025 Feb 17] [online]. Available from: https://www.youtube.com/watch?v=sf4qRY3h_eo
19 ViaSat-3 Americas Mission. SpaceX [cited 2025 Feb 17] [online]. Available from: https://www.youtube.com/watch?v=YFbp6PVbJQA

20 Blue Origin. New Shepard [cited 2025 Feb 17] [online]. Available from: https://www.blueorigin.com/new-shepard.

21 Blue Orgin Photoillustration. Cited by Boyle A. Watch Blue Origin Fly Its First Capsule That's Designed to Send People to Space and Back. GeekWire [cited 2025 Feb 17] [online]. Available from: https://www.geekwire.com/2021/watch-blue-origin-fly-first-capsule-thats-designed-send-people-space/.

22 Marshall B. After Unity, New Shepard Gets into Space. DesignSpark [cited 2025 Feb 17] [online]. Available from: https://www.rs-online.com/designspark/after-unity-new-shepard-gets-into-space.

23 Blue Origin. Cited by Pietrobon S. Blue Origin - New Shepard Second Developmental Test Flight and Landing. NASA Space Flight [cited 2025 Feb 17] [online]. Available from: https://forum.nasaspaceflight.com/index.php?topic=38873.260.

24 Blue Origin. Cited by Howell E. New Shepard: Rocket for Space Tourism. Space [cited 2025 Feb 17] [online]. Available from: https://www.space.com/40372-new-shepard-rocket.html.

25 Blue Origin [cited 2025 May 01] [online]. Available from: https://www.blueorigin.com/about-blue.

26 Blue Origin. New Glenn [cited 2025 Feb 17] [online]. Available from: https://www.blueorigin.com/new-glenn.

27 Chylinski R. SpaceX Starship Full Stack [cited 2025 Feb 17] [online]. Available from: https://stock.adobe.com/in/images/spacex-starship-full-stack/490149983.

28 SpaceX. Cited by Clark S. SpaceX Ready for 33-engine Test of Starship's Super Heavy Booster. Spaceflight Now [cited 2025 Feb 17] [online]. Available from: https://spaceflightnow.com/2023/02/08/spacex-ready-for-33-engine-test-firing-of-starships-super-heavy-booster/.

29 Musk E. 3 More Engines for the Meaning of Life. X. Cited by Ali I. Starship to Get 3 More Vacuum Engines for a Total of 42, SpaceX Upgrades OLM at Starbase, Raptor V3 Attains New Milestone. Tesla Oracle [cited 2025 Feb 17] [online]. Available from: https://www.teslaoracle.com/2023/07/13/starship-to-get-3-more-vacuum-engines-for-a-total-of-42-spacex-upgrades-olm-at-starbase-raptor-v3-reaches-new-milestone/.

30 Space X. Raptor 3. X [cited 2025 Mar 20] [online]. Available from: https://x.com/SpaceX/status/1819772716339339664.

31 Wang B. SpaceX Raptor 3 Engines and Beyond. Nextbigfuture [cited 2025 Apr 28] [online]. Available from: https://www.nextbigfuture.com/2023/05/spacex-raptor-3-engines-and-beyond.html.

32 SpaceX. Cited by Ralph E. SpaceX to Build World's Most Advanced Rocket Engine Factory in Central Texas. Teslarati [cited 2025 Feb 17] [online]. Available from: https://www.teslarati.com/spacex-starship-raptor-2-factory-details/.

33 Law A. The Definitive Guide To Starship: Starship vs Falcon 9, What's New and Improved?. Everyday Astronaut [cited 2025 Feb 17] [online]. Available from: https://everydayastronaut.com/definitive-guide-to-starship/.

34 Lopatin D. File:Starship S20.jpg. Wikimedia [cited 2025 Apr 28] [online]. Available from: https://commons.wikimedia.org/wiki/File:Starship_S20.jpg.

35 Cotton E. Starship SN10 — 10 km Test Flight. Everyday Astronaut [cited 2025 Feb 17] [online]. Available from: https://everydayastronaut.com/starship-sn10-km-post-test-flight/.

36 Zapata E. About Starships, and the (Not What You Think) Reusability We Need. Zapata Talks NASA [cited 2025 Feb 17] [online]. Available from: https://zapatatalksnasa.com/2022/06/09/about-starships-and-the-not-what-you-think-reusability-we-need/.

37 Meldner R. From Launch Pad to eBay: SpaceX Starship Heat Shield Debris Surfaces on Marketplace After Explosions. eSeller365 [cited 2025 Feb 17] [online]. Available from: https://eseller365.com/spacex-starship-heat-tiles-surface-ebay/.

38 NASA. Cited by Brown M. SpaceX Starship: New Video Shows Spider-like Legs That Will Land on Mars. Inverse [cited 2025 Feb 17] [online]. Available from: https://www.inverse.com/innovation/spacex-starship-new-video-shows-spider-like-legs-that-will-land-on-mars.

39 X S. Starship Users Guide [cited 2025 Feb 17]. [pdf]. Available from: https://www.spacex.com/media/starship_users_guide_v1.pdf.

40 King P. VR Tour of SpaceX Starship Interior Concept. HumanMars [cited 2025 Feb 17] [online]. Available from: https://www.humanmars.net/search/label/Starship%20interior.

41 Matthew. SpaceX Starship Interior Design – A Glimpse Into Starship Interior. The Aero Blog [cited 2025 Feb 17] [online]. Available from: https://theaeroblog.com/predicting-the-spacex-starship-interior-a-glimpse-into-the-spaceship-interior-of-the-revolutionary-vessel/

42 SpaceX. Cited by SpaceX's Starship Interior Design NEW UPDATE... GREAT SPACEX [cited 2025 Feb 17] [online]. Available from: https://www.youtube.com/watch?v=O5NxzcEEg_o.

43 HumanMars. SpaceX Starship Orbital Flight Test 6 is Scheduled on November 19 [cited 2025 Feb 17] [online]. Available from: https://www.humanmars.net/2024/11/spacex-starship-orbital-flight-test-6.html.

44 X S. Starship's Fifth Flight Test [cited 2025 Feb 17] [online]. Available from: https://www.spacex.com/launches/mission/?missionId=starship-flight-5.

45 SpaceX. Cited by Klotz I. SpaceX Catches Returning Starship Booster With Launch Gantry. Aviation Week Network [cited 2025 Feb 17] [online]. Available from: https://aviationweek.com/space/launch-vehicles-propulsion/spacex-catches-returning-starship-booster-launch-gantry.

46 Linkspace. Cited by JONES A. Private Space Launch Firms in China Race to Orbit. IEEE Spectrum [cited 2025 Feb 17] [online]. Available from: https://spectrum.ieee.org/private-space-launch-firms-in-china-race-to-orbit.

47 Galactic Energy. Galactic Eneregy Successfully Completed the Third Sea Launch Mission of the CERES-1 Rocket [cited 2025 Feb 17] [online]. Available from: https://www.galactic-energy.cn/index.php/En.

48 Jiangsu Shenlan Aerospace. Deep Blue Aerospace [cited 2025 Feb 17] [online]. Available from: http://www.dbaspace.com/indexs.html.

49 LandSpace. ZQ-2 [cited 2025 Feb 17] [online]. Available from: https://www.landspace.com/en/product.html?mao=1.

6

Communications Satellite Constellations

6.1 Introduction

The digital age is in full bloom because of the proliferation of the Internet and mobile devices. Information ubiquity, which is digital information available anytime, anywhere, and affordable, is the goal. Consumers and businesses would panic if they were not somehow "connected." Quality of life and productivity suffer when networks go down even for a few hours.

Information primarily comes in the form of digital data. Furthermore, data drives innovations and productivity improvements across all sectors. From social media to artificial intelligence (AI) to driverless vehicles, large volumes of data are generated, processed, and exchanged all the time.

The current information and communications technology (ICT) infrastructure is good in populated areas as at least 60% of the population worldwide resides in urbanized areas. Areas where the digital divide remains significant include remote areas, maritime areas, low-population communities, emergencies, and disasters.

Installing information infrastructure in remote areas is often hampered by the lack of economic justifications because user density is too low, and the total area with little to no coverage is too large. The access problem is therefore unlikely to be solved by terrestrial solutions. We must look to the space for solutions.

This chapter focuses on large communication satellite constellations in low Earth orbit (LEO). We begin by covering relevant topics with background and then discuss specific constellations.

6.2 Overview of the Satellite Constellations

The advantage of a communication satellite in the geostationary orbit (GEO) is that its footprint can cover nominally one-third of the Earth's surface. The tradeoff

New Space: From Low Earth Orbit to the Moon and Beyond, First Edition.
Edited by Stephen W. Cheung and Chih-Yung Wen.

is signal latency and capacity, that is, the number of users a GEO satellite can serve simultaneously is limited.

Signal latency is the time it takes for a signal to go from the sender to the receiver due to physical distance. Excluding the signal processing time, the time it takes for a signal to reach its receiver by bouncing off a GEO satellite is at least 250 ms (¼ of a second), which is noticeable for voice communication. Hence, a person would need to wait for a nominal delay of over 500 ms to hear back from the other side.

A signal latency of several hundred milliseconds may be tolerated for web browsing. However, it is too slow for new Internet applications such as online videos and remote machine controls such as autonomous driving.

6.2.1 LEO Internet Satellite Constellations

The latency problem can be greatly reduced by placing satellites in low Earth orbit (LEO). If an LEO satellite is placed at a moderate height, such as 600 km, and considering that the users may not be directly below the satellite, the round-trip latency would be about 10 ms. When including the signal processing time at the satellite and the receiving end such as a data center, the round-trip latency would be tens of milliseconds.

LEO communications satellites have several upsides:

1) LEO is the first parking orbit to which all rockets go. The launch cost per kilogram to LEO has been driven down significantly by reusable rockets, and this trend will continue
2) LEO satellites are usually light, weighing between 100 and 1,000 kg versus large GEO satellites weighing several tons
3) The small size and weight mean LEO satellites can be mass produced
4) The low altitude means that the satellite's footprint on Earth is much smaller. The advantage is the ability to serve more users within that small footprint, as well as frequency reuse by other satellites in the constellation

The downside of an LEO satellite is that it moves much faster than the Earth's rotation. An LEO satellite goes around Earth every 90–120 minutes. Hence, a satellite comes within a ground user's view and then disappears in a matter of minutes. Seamless satellite coverage means that several satellites must be constantly within view. A constellation of satellites is therefore needed to fully cover the Earth and provide continuous access like cellular networks. To that end, a satellite constellation can be thought of as a large network of moving cell towers in the sky.

6.2.2 Size of the Constellation

Broadband Internet service is now the primary driving force behind LEO satellite constellations. To achieve continuous Internet coverage from the sky, an LEO constellation necessarily consists of hundreds to thousands of satellites if it is to serve millions of users globally.

The tradeoffs for an LEO constellation are as follows: a) higher altitude has a larger footprint on the ground and requires a smaller number of satellites and b) higher altitude increases signal latency.

6.2.3 Architecture of the Constellation

A satellite is essentially an orbiting relay linking a user with a data source via four radio frequency (RF) carrier beams. The four carriers beams have different frequencies. Imagine a household in Figure 6.1 browsing the net. The user triggers a signal to request information. Such a request is digitally packed in an uplink frequency f1 and is sent via the household's antenna to the satellite above. The satellite receives the signal, decodes it from f1, repacks it with requests from other households in a downlink frequency f2, and sends it to a receiving station on the ground within the line of sight of the satellite.

There are two scenarios with the ground station. The first is that it is connected to a data center nearby terrestrially. The information for all the requests is obtained; the receiving station packs it on an uplink frequency f3 and sends it to the satellite. In this scenario, the ground station is a data gateway. This scenario takes the least time for the household to receive the necessary information.

The second scenario is that no data center is nearby. The ground station has to bounce the requests up to another satellite down the line. The process is repeated until the data is obtained from a data center somewhere and returned to the first ground station. This scenario requires more latency time.

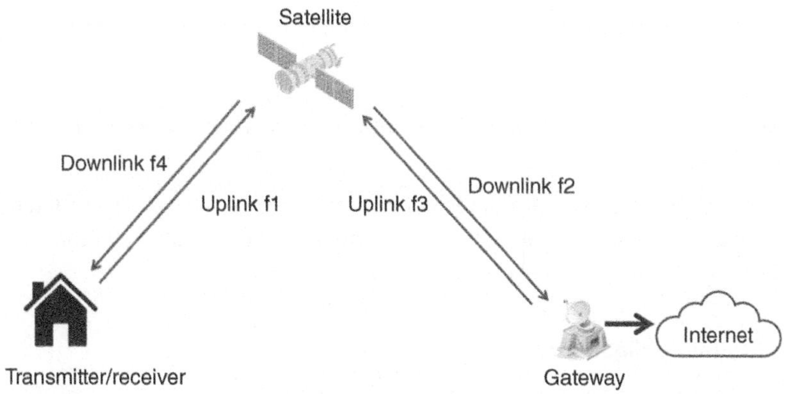

Figure 6.1 A communication satellite links users to a data source such as an Internet data center. Data is usually exchanged by beams of radio frequencies as uplinks and downlinks.

After the first satellite receives the uplink frequency f3 from the ground station, it decodes the data and distributes it to different households below via a downlink frequency f4. The household antenna decodes and feeds the information to the user.

The above process takes tens of milliseconds depending on the altitude of the satellite and the number of bounces to reach a data center and back. The situation has extra complexity in that the satellite in the sky above is in view only for a few minutes. To the user on the ground, the data transfer process must continue seamlessly as long as he is online.

The volume of data sent over the terrestrial and satellite networks keeps growing. We can appreciate this point by the numerous photos and videos we stream, download, and forward to our friends, amounting to tens to hundreds of megabytes a day. Multiplying that by the number of users within the footprint of a satellite, the instantaneous volume of data being uplinked and downlinked is very high.

6.2.3.1 Radio Frequency Bands

A carrier frequency is just a nominal value representing a band of subdivided frequencies. They further fall under a broader band with names such as C-band, Ku-band, and Ka-band. The rule of thumb is that the higher the carrier frequency, the higher the number of subdivided frequencies available to pack more data, that is, bandwidth. The most common frequency bands for satellite data communication are the Ku-band (12–18 GHz) and the Ka-band (26–40 GHz). On the flip side, absorption by rain and water droplets increases with frequency.

Modern electronic products and applications generate large data volumes and require millisecond-level response times. LEO satellite Internet systems should bear the following characteristics:

a) High altitude increases the signal latency. A higher altitude has a larger footprint on the ground and requires a smaller number of satellites
b) Use the highest carrier frequency possible to pack more data
c) A large constellation allows each satellite to serve many users within its footprint.

Point c) is significant in that a large constellation necessitates servicing large user populations to justify the investment. In principle, an LEO constellation can cover the entire Earth and serve everyone from anywhere.

6.2.4 Orbital Configuration

A large constellation is characterized by a series of “shells” at different altitudes and inclinations. We will discuss several features of the constellation.

a) Circular orbits: Satellites of LEO constellations are in circular orbits

b) Altitudes: LEO satellite Internet constellations operate between 500 and 600 km, some are higher near 1,000 km while some choose to operate at a lower altitude around 350 km
c) Inclination and plane: Inclination is the angle of an orbit relative to the equatorial plane. An inclined orbit of 50°, for example, means the satellite traces a band of areas on the globe between 50°N and 50°S. This is illustrated in Figure 6.2
   A plane is a group of satellites sharing the same orbit of specific altitude and inclination
d) Number of satellites per plane: Following (c), each plane consists of a specific number of satellites evenly spaced along the orbit
e) Shell: A shell represents all the planes sharing the same inclination at a designated altitude
f) Number of planes in a shell: Viewed from the north, all the planes in a shell of specific inclination and altitude wrap around Earth like a net. The higher the number of planes, the "denser" is the net
g) Intersatellite link (ISL): Data exchange among satellites of the same constellation via RF or laser devices
h) Angle of elevation on the ground: This is the minimum angle a user's line of sight to a satellite relative to the horizon, ensuring good connectivity. A satellite directly above is 90° angle of elevation

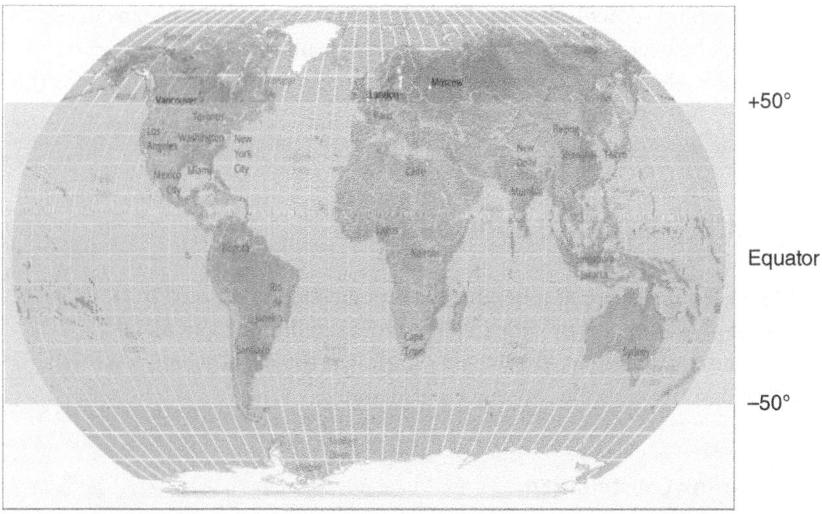

Figure 6.2 A satellite in an inclined orbit covers a band of geographic area. For example, if the inclination illustrated is 50°, the satellite will cover the global geography between 50°N and 50°S latitudes.

A constellation of thousands of satellites is made up of shells of inclination planes at different altitudes. In general, the planes of a given shell are spread evenly wrapping around the globe. Also, all the planes of that shell have an equal number of satellites.

The orbital inclination is chosen in accordance with the intended geographic coverage. From the business perspective, coverage of the most populated latitudes would serve the most users across the globe. Half of the world's population lives between latitudes 20°N and 40°N. China, for example, is from 18°N to 53°N, and the US is from 25.84°N to 49.38°N, while northern countries like Canada and Russia are near 41°N to 82°N.

Hence, orbital inclination from 0° to 55° would service large populations on Earth. A constellation aiming to service highly populated areas needs as many planes as possible surrounding the globe to have enough satellites constantly in the users' view. We should keep this in mind when looking at the orbital inclinations of the constellation configurations.

Beyond 60°N is known as the Arctic region where the population density is low. The region should be serviced by near-polar or high-inclination orbits, usually with government grants or by mandate.

6.2.4.1 Intersatellite Link

Some situations, such as maritime data, require multiple hops among satellites and ground stations to reach a data center. Intersatellite links eliminate the need for multiple uplink-downlink hops. Placing earth stations in the oceans would be impractical. The overall signal latency is improved substantially. The laser has the advantage of higher frequency and can pack more data. The lasers used for this purpose have wavelengths in the near-infrared region.

6.3 Starlink Satellite Constellations

Several constellations are at different stages of development as of 2024. Table 6.1 shows a list of constellations with 1,000 or more satellites proposed.

Among the constellations in Table 6.1, Starlink is the farthest along. We will use Starlink to describe satellite Internet constellations.

6.3.1 Starlink Constellation

Starlink is an Internet service provider (ISP) from space. Its operation is vertically integrated from the design, build, and launch of its thousands of satellites riding on Falcon 9 rockets.

Table 6.1 A list of LEO satellite Internet constellations with 1,000 or more satellites.

Name of the constellation	Country	Number of satellites proposed
Starlink constellation 1	US	12,000
Starlink constellation 2	US	30,000
Project Kuiper	US	3,236
Guo Wang (GW)	China	13,000
Galaxyspace	China	1,000
One-web	UK	6,372

There are two constellations proposed at different time frames. Constellation 1 was announced in 2015. Constellation 1 comprises Phase 1 and 2 with a total of 11,926 satellites. Starlink later received approval for a modified Phase 1. Constellation 2's current plan has 29,988 satellites. The two constellations together add up to almost 42,000 satellites.

Rules by the International Telecom Union (ITU) require that satellites be launched and operational within a prescribed timeline to prevent network operators from hogging allocated frequencies and planes. Starlink has been approved to launch Phase 1 of Constellation 1, as well as part of Constellation 2.

To date, Starlink has two generations of satellites: Gen 1 and Gen 2, also known as versions V1 and V2. Starlink has been progressively launching the two generations of satellites. Each generation also has variations to incorporate new capabilities and applications.

The number of Starlink subscribers has been growing from 2019 to over 3 million in 2024. As of January 2025, there are 6,920 satellites in orbit with 6,882 working [1].

The number of satellites launched is not the same as the number of satellites working. Some satellites are not yet functional or have failed due to various reasons. Starlink satellites take 3–4 months to rise to their operational orbits. Frequent launches mean that tens of satellites are in transit to orbit at any given time.

Table 6.2 outlines the key parameters of the modified Phases 1 and 2 of Constellation 1.

6.3.2 Starlink Constellation 1, Phase 1

The shells of Constellation 1 are numbered from 1 to 8. For example, Shell 1–5 is the 5th shell of Constellation 1.

Table 6.2 The shell number has two digits. The first digit represents Constellation 1, followed by the shell number covering Phases 1 and 2. For example, Shell 1–3 represents the third shell of Constellation 1.

Shell #	Altitude (km)	Inclination (degrees)	# of planes	# of satellites per plane	# of satellites per shell	Subtotal
Modified Phase 1						
Shell 1–1	550	53	72	22	1,584	
Shell 1–2	540	53.2	72	22	1,584	
Shell 1–3	570	70	36	20	720	
Shell 1–4	560	97.6	6	58	348	
Shell 1–5	560	97.6	4	43	172	4,408
Phase 2						
Shell 1–6	355.9	42			2,493	
Shell 1–7	340.8	48			2,478	
Shell 1–8	345.6	53			2,547	7,518
					Total	**11,926**

There are three points to note about the modified Phase 1:

1) The altitudes of the high-latitude shells of the original Phase 1 have been lowered to the 500 km range from 1,100 km
2) Shells 1–1 and 1–2 together have 3,168 satellites with an inclination of around 53°. These satellites account for 72% of Phase 1's total of 4,408
3) A nominal angle of elevation within 40° can be expected by users

There are 72 planes in each of Shells 1–1 and 1–2, and each plane has 22 satellites, making the total number 1,584 units in each shell. Shells 1–1 and 1–2 have high satellite densities and cover the most populated areas on Earth between 53°N and 53°S. The three shells cover higher latitudes and the near-Polar Region in Figure 6.3. These three shells have a smaller number of planes and therefore fewer satellites per shell. Users living in high latitudes, therefore, will see a lower number of Starlink satellites overhead.

Figure 6.3 shows the visual of the modified Phase I with 4,408 satellites. Each shell looks like a net with different thread counts. This visual gives an idea of the satellite distribution. Figure 6.4a and b show two views of all the Starlink satellites in orbit as of January 2025. This satellite map is put up and frequently updated

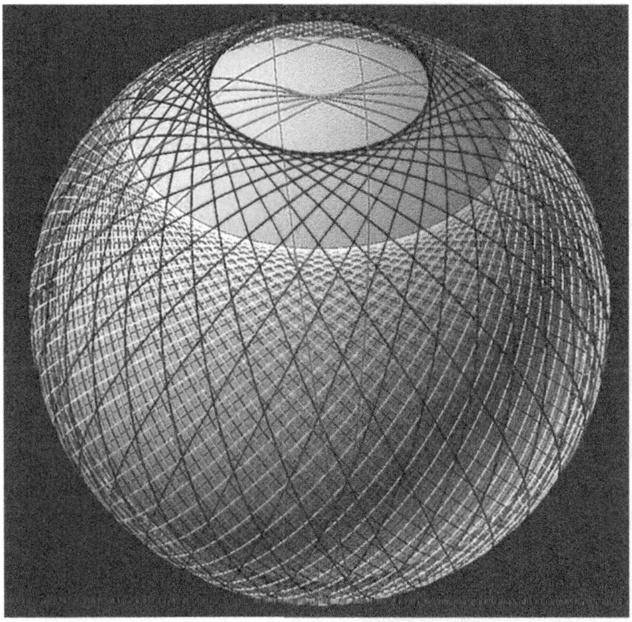

Figure 6.3 Visualization of the orbital planes for 4,408 Starlink Phase 1 satellites. The colors represent different shells. For example, Shells 1–1 and 1–2 represent inclinations 53° and 53.8° and are in the color white. Black and gray represent the high latitude and near-polar orbits [3].

by an entity not affiliated with Starlink [2]. Figure 6.4a is the equatorial view, and Figure 6.6b is a view from a slanted angle from the north. It can be seen that there are far fewer satellites orbiting near-polar orbits.

6.3.3 Starlink Constellation 1, Phase 2

Phase 2 has 7,518 satellites, which accounts for 63% of the total number of satellites in Constellation 1. Table 6.2 shows the configuration of Phase 2. Some details are not yet available. All three shells 1–6, 1–7, and 1–8 are in low altitudes near 350 km as well as inclinations at or below 53°.

Phase 2's configuration shows Starlink's strong tendency to provide coverage to highly populated regions, as evident from the fact that none of the three shells in Phase 2 are for high latitudes. Together with Shells 1–1 and 1–2, a total of 10,686 (or 90% of total in Constellation 1) target populated regions. All the Phase 2 orbits are in the very low earth orbit (VLEO), which is below 400 km.

The advantages of VLEO are a) low signal latency due to its close distance to the ground, b) less launch cost to orbit, and c) the use of higher frequency bands

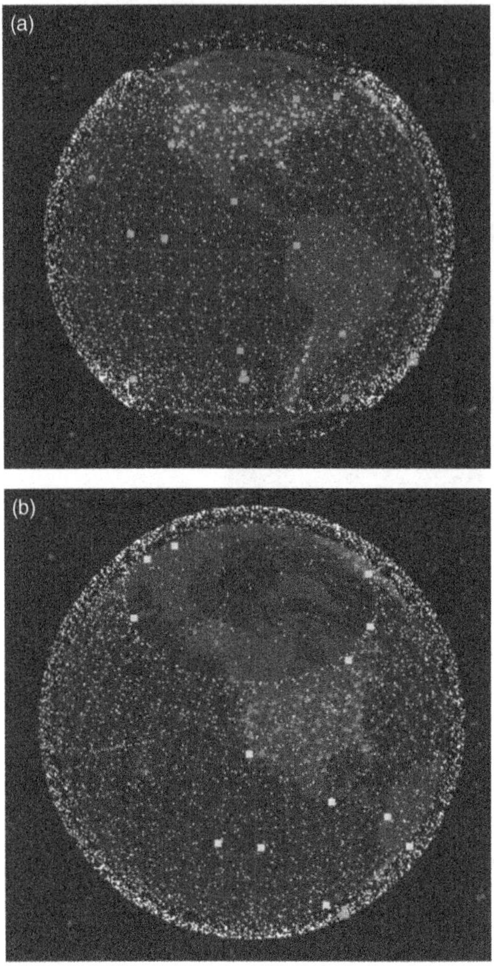

Figure 6.4 Snapshots of the orbiting Starlink satellites as of January 2025. (a) is the side view, and (b) is a slanted view including the North Pole. *Source:* Ref. [4] / Live Starlink Satellite and Coverage Map.

to increase bandwidth but with less atmospheric absorption. The lifespan of satellites in VLEO is shortened accordingly. The disadvantages are a) higher residual drag from the atmosphere and b) higher fuel consumption to compensate for the higher drag.

Phase 2 satellites have added a higher-frequency band with the FCC reportedly granting Starlink approval to launch 7,500 satellites using the E-band (60–90 GHz). The E-band is to be used in the Version 2 satellite series for Phase 2 orbits. Specifically, the satellite-to-ground station downlink is 71.0–76.0 GHz and the ground station-to-satellite uplink is 81.0–86.0 GHz. These two links are for high volumes of data transfer between a satellite and a ground station and not between a satellite and a user.

Starlink claims their satellites will be replaced after nominally 5 years. At that point, the defunct satellite will deorbit to lower altitudes for eventual incineration by the atmosphere. Hence, the expectation for LEO communication satellites is that they will be replaced and upgraded regularly. The demand for replacement satellites and rocket launches will be constant.

6.3.4 Constellation 2

In December 2022, the FCC approved the initial 7,500 satellites for Starlink Constellation 2. Three orbits are granted for altitudes of 525, 530, and 535 km respectively. There are two configurations under consideration. As of July 2024, there is uncertainty about which configuration will be adopted.

6.3.4.1 Constellation 2, Configuration 1

Configuration 1 has nine orbital shells and was planned for deployment by the Starship spacecraft. The shell numbers have not been assigned. The configuration is shown in Table 6.3.

Of the nine orbital shells in Configuration 1, the first 4 are in VLEO from 340 to 360 km, and the rest are in the 500–600 km range. Much emphasis is placed on VLEO, totaling 19,440 satellites or 64.8% of the constellation. This distribution reflects the similarity with Constellation 1. Three of the four VLEO orbits have inclinations at 53° and below, with more satellites per orbital plane (110 each plane). A total of 25,920 satellites (or 86% of the configuration total) service heavily

Table 6.3 Configuration 1 of the Starlink Constellation 2. This configuration plans to utilize the Starship to deliver Gen 2 satellites.

Altitude (km)	Inclination (degrees)	# of planes	# of satellites per plane	# of satellites per shell
340	53	48	110	5,280
345	46	48	110	5,280
350	38	48	110	5,280
360	96.9	30	120	3,600
525	53	28	120	3,360
530	43	28	120	3,360
535	33	28	120	3,360
604	148	12	12	144
614	115.7	18	18	324
			Total	29,988

popular regions at an inclination of 53° and below. The two percentages strongly indicate the constellation's focus on serving highly populated regions and anticipating large volumes of data transfer.

The fourth VLEO orbit is a near-polar orbit at 96.9° with 120 satellites per plane, totaling 3,600. This near-polar VLEO orbit, with another shell at a higher latitude at 115.7° inclination, will increase the coverage of the near-polar regions.

6.3.4.2 Constellation 2, Configuration 2

Configuration 2 is Starlink's alternative in the event the Gen 2 satellites need to be sent to orbit by Falcon 9 due to the delay of the Starship system. Most media reports are staying with Configuration 1. We assume Configuration 1 to be the default plan for Constellation 2 and leave the readers with Configuration 2 as a reference only.

6.3.5 Starlink Satellites

Each version of the Starlink V1 and V2 satellites has several variants. Sixty V0.9 satellites were launched in 2019 on a Falcon 9 rocket, which was the maximum number of satellites that Falcon 9 could carry. Most subsequent launches have been on ridesharing missions. V0.9 satellites were used for initial deployment.

6.3.5.1 Starlink Version 1 Satellites

A V1.0 weighs about 260 kg with dimensions 3.2 × 1.6 m before its solar panels deploy. In May 2021, V1.5 was launched after over 1,100 V1.0s had been sent to orbit. A V1.5 weighs about 300 kg with dimensions 2.3 × 2.7 m. V1.5 is heavier because it has an intersatellite laser link (ISLL) capability.

V1.0 and V1.5 are shown in Figure 6.5. The units look like folded laptop computers. The solar panel unfolds on one side after being released into space. The opposite side facing the Earth has RF antennas of different frequency bands. Figure 6.5 shows an exterior comparison of V1.0 and V1.5.

All Starlink satellites use the Ku and Ka frequency bands for uplink and downlink connecting user terminals, satellites, and ground stations.

Inside the Falcon 9 fairing, the V1.0s and V1.5s are designed to be stacked in a launch rack. Figure 6.6 shows the 51 V1.5s in their first flight. Up to 54 V1.5s can be launched on a Falcon 9 rocket.

The rack is first released by the Falcon 9 upper stage once it reaches around 290 km altitude. Readers can check out this process from online videos [6]. Each satellite is then released consecutively by the rack. The satellites then deploy their solar panels and rise to their designated altitude with onboard ion thrusters. This process normally takes a few months for each satellite to raise its orbit to over 500 km.

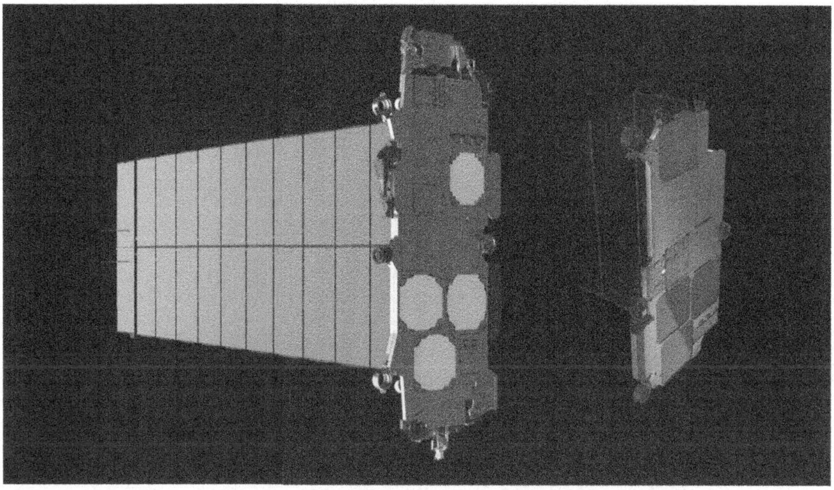

Figure 6.5 Two early versions of the Starlink satellites: V1.0 on the right and V1.5 on the left. *Source:* Ref. [5] / SpaceX.

6.3.5.2 Direct to the Cell Service

In 2022, SpaceX announced its collaboration with T-Mobile, a US mobile operator [8]. The collaboration allows Starlink satellites to include T-Mobile's LTE frequency spectrum. Mobile users can access Starlink in open areas where there is no cellular coverage. Initial usage is for emergency text messages. The service will eventually expand to pictures, voice calls, and data for LTE phones and devices in the world in 2025, pending regulatory approval.

The service is termed Supplemental Coverage from Space (SCS) by the FCC and is a "direct to cell" function between a satellite and a cellular phone. The texting service, according to the SpaceX website, is working in January 2024. Adding the SCS function means adding another transmission/receiving RF system to the satellite for cellular frequencies. SCS capabilities are incorporated in the presently launched V1.5s and some variants of V2 satellites.

6.3.5.3 Starlink Version 2 Satellites

Starlink began introducing Version 2's in 2023. V2 is intended to increase the data capacity of each satellite and the overall bandwidth of the network. This new version aligns with the anticipation of more users worldwide. The original V2 reportedly would weigh close to 2,000 kg and be larger in dimension. It was intended to be launched by the Starship, which has the capacity to launch 200 such units at a time.

Delays in Starship orbital flight tests led Starlink to release smaller V2 variants called V2.0 Mini SCS and V2.0 Mini, both of which can fit in the Falcon 9 fairing.

Figure 6.6 Starlink satellites stacked in a rack inside the Falcon 9 fairing. This image shows 51 V1.5 satellites. *Source:* Ref. [7] / SpaceX.

V2.0 Mini SCS is one with direct-to-cell capability, and V2.0 Mini does not. V1.5 satellites have not yet been discontinued as of this writing.

Table 6.4 lists the known V2 variants mentioned above. V1.5 is included for reference.

To fit the V2 units in Falcon 9, V2.0 Mini is lighter, weighing about 800 kg. The E-band is added to the existing Ku and Ka bands to handle the large data volume linking the satellite and the ground station. The first V2.0 Mini was launched in February 2023 and has been the default model being launched. The highest number of V2.0 Mini's launched on a Falcon 9 is 23. The data capacity of a V2.0 Mini is reported to be 4× more than that of V1.5. Figure 6.7 shows the V2.0 Mini stacked in a rack inside a Falcon 9.

Table 6.4 List of Starlink Version 2 variants. V1.5 is included for reference. SCS: Supplemental Coverage from Space, sometimes called direct-to-cell. The V22.0 Mini and V22.0 Mini SCS are to be launched by Falcon 9. Similar variants are planned for the Starship when the launch system is ready.

Version	Status	Weight (kg)	Length (m) × Width (m)	RF Band	Laser	Thrusters
1.5	Ongoing	300	2.3 × 2.7	Ku + Ka	Yes	Krypton
2.0 Original		1200–2,000	7 × 3.5	Ku + Ka + E	Yes	Kryton
2.0 Mini	Ongoing	800	4.1 × 2.7	Ku + Ka + E	Yes	Argon
2.0 Mini SCS	Ongoing	970	7.4 × 2.7	Ku + Ka + E + directly to cellular	Yes	Argon
2.0 Starship-1	Planned	2,000	6.4 × 2.7	Ku + Ka + E	Yes	Argon
2.0 Starship-2 SCS	Planned	2,000+	10.1 × 2.7	Ku + Ka + E + directly to cellular	Yes	Argon

Figure 6.7 V2 Mini has advanced phase array antennas and E-band frequency, 4× more capacity per satellite, and more bandwidth, as well as new argon thrusters for onboard maneuvering. *Source:* Ref. [9] / SpaceX.

The V2.0 Mini SCS is equipped with direct-to-cell capability. It weighs 970 kg (heavier than the V2.0 Mini by about 170 kg). It is longer in order to include the necessary RF equipment for the cellular spectrum. Starting in January 2024, V2.0 Mini SCS units have been steadily included in Starlink launches together with other V2.0 Mini's. As of September 2024, over 200 V2.0 Mini SCSs are in orbit.

6.3.5.4 Argon Ion Thrusters

The V2.0 Mini is the first Starlink satellite to use argon gas in its ion thruster. Krypton has been the inert gas used in the company's own ion thrusters. Most ion thrusters use xenon gas, but it is the rarest among inert gases. Krypton is less rare but is still more expensive than the more abundant argon. Argon thrusters have 2.4× more thrust. Considering the number of satellites between the two constellations, the decision to switch to argon makes business sense.

6.3.5.5 Intersatellite Laser Links

Optical communication in space is a key technology for intersatellite data transfer. Infrared (IR) wavelength is used in the ISLL. Large-scale deployment of laser links by SpaceX has moved this technology to the forefront.

There are four laser beams on each satellite equipped with an ISLL. Two beams would link the satellite to the satellites in front and behind on the same plane.

Figure 6.8 Starlink satellites deploy their solar panels after being released from the satellite holding fixture. For those satellites equipped with intersatellite laser links, there are four laser beams to optically communicate with four nearby satellites. *Source:* Ref. [11] / SpaceX.

Two beams would link the satellite to the satellites on the adjacent planes to its left and right. Figure 6.8 illustrates this configuration.

Lasers can move huge volumes of data far more than high-frequency RF. Additionally, a laser beam stays narrow over distance compared to RF. A technologist at SpaceX stated in February 2024 that the satellite constellation had over 9,000 lasers at that time and handled over 42 petabytes of data per day. This means that the ISLL has reached maturity and can be used extensively in all the Starlink satellites [10].

6.3.5.6 Future Versions of the Starlink Satellites

Future V2 satellites will be heavier when launched on a Starship. Two Starlink variants are planned, as shown in Table 6.4. They are Starship-1 and Starship-2 SCS. Starship-1 will weigh about 2,000 kg and be significantly longer. Information about the Starship-2 SCS is not available as of this writing. Based on the current dimensions and weight, the Starship vehicle is expected to carry between 50 and 100 Starship-1 satellites.

At the International Astronautical Congress in October 2023 [12], Elon Musk indicated that Version 3 was in the works and could be launched on a Starship sometime in 2024. It was the first time that V3 was known to the public. Little details are known as of this writing.

6.4 OneWeb System

OneWeb is a subsidiary of Eutelsat and is headquartered in London. The company's proposed constellation has 648 satellites in near-polar orbits. OneWeb's primary service is high-speed broadband Internet to enterprises and governments, including fixed and mobile services in rural and isolated areas. OneWeb also targets the maritime market and global coverage. Its advantage is the dense coverage in the high latitude from 50°N and above, including the polar areas.

The first-generation OneWeb satellite weighs about 150 kg. These small satellites are in 12 orbital planes at 86.4° inclination (near polar) and 1,200 km altitude. There are 49 satellites per plane totaling 588 satellites. Over 600 satellites have been launched and are working in orbit. Some are test articles and spare parts.

The OneWeb constellation provides user service in the Ku-band. Links to gateway ground stations are in the Ka-band. The first generation does not have laser links.

The star map in Figure 6.9 shows the near-polar characteristics of the OneWeb orbits. The satellite density is high near the poles.

Figure 6.9 A satellite map showing the satellite distribution of the OneWeb constellation. The constellation configuration consists of all polar orbits. Therefore, the two poles have a higher satellite density than other regions. *Source:* Ref. [13] / Live Starlink Satellite and Coverage Map.

6.5 Kuiper System

Project Kuiper is a subsidiary of Amazon. It was established in 2019 to deploy an LEO satellite Internet constellation to offer broadband services to global users. Kuiper's planned constellation consists of 3,236 satellites operating in three shells at 590, 610, and 630 km. A total of 98 orbital planes are planned, but no additional public information is available.

News reports in late 2023 stated that two prototype satellites were working successfully [14]. As of January 2025, public reports show no change in the working status. Amazon would begin launching satellites by mid-2024. According to the FCC license, Amazon needs to launch and operate half of the 3,236 satellites by July 2026 and the remaining satellites by July 2029.

In 2023, Amazon announced a range of user terminals for different users.

6.6 China GW System

Communication satellite constellations were also proposed by Chinese companies between 2015 and 2020. They had an average of 300–400 satellites in each constellation.

In 2021, the Chinese government consolidated the different proposals and filed with the ITU the GW constellation. GW is often referred to as Guo Wang (national network). The network will be overseen by the state-owned China National Network Group.

The proposed constellation consists of 12,992 satellites. Its configuration is shown in Table 6.5.

From the GW configuration, observations are:

1) 53% of the satellites in number are at high altitudes of 1,145 km
2) Five shells of 10,784 satellites are for 55° inclination or lower, that is, serving populated regions
3) Two shells of 2,208 satellites are at high latitude 60° and near the polar regions
4) High altitudes ensure reasonable elevation angles for users on the ground but add to the signal latency

In December 2024, the media reported that the first batch of 10 GW satellites were launched [15].

Table 6.5 Configuration of China's GW satellite constellation of almost 13,000 satellites. The large number of satellites assigned to low latitudes (less than 55°) suggests that the constellation also aims to serve the populated regions.

Shell #	# of Orbital Planes	# per Plane	Altitude (km)	Inclinatio n (degrees)	Subtotal	Total
I-1	16	30	590	85	480	
I-2	40	50	600	50	2,000	
I-3	60	60	508	55	3,600	6,080
II-1	48	36	1,145	30	1,728	
II-2	48	36	1,145	40	1,728	
II-3	48	36	1,145	50	1,728	
II-4	48	36	1,145	60	1,728	6,912
						12,992

6.7 Business and Economics Angle

The emergence of LEO communication satellite constellations has created unprecedented opportunities and questions. Satellite Internet is a big economic pie, creating new revenue sources in the next two decades. Mega constellations also drive the launch market segment. Like reusable rocket stages, the business practices of the satellite sector have been disrupted. Launching thousands of satellites without reusable rockets would be economically challenging. Hence, the deployment of a mega-constellation hinges on the usage of reusable rockets.

6.7.1 Satellite Ecosystem

Several points can be explored:

1) Mega constellations: The short orbital period at LEO altitudes means that a small satellite constellation is not enough to have good coverage and seamless access, whether it is for servicing the globe or just a country. Starting with tens and then hundreds of satellites in one constellation, now we have mega constellations of thousands of satellites to have coverage intended for millions of users. Only a large user population can justify the capital investment.
2) Small satellites: LEO satellites usually weigh between 100 and 1,000 kg. For communications satellites serving as ISPs, the size and weight of the RF systems are substantial in order to deliver the needed RF power. Satellites of a constellation must be mass produced by thousands. For example, if 1,200 satellites must be launched into orbit each year, the production rate should average 100 units per month or roughly 25 units a week. Whether mass fabrication is to be done in-house or subcontracted out, it is a challenge up and down the supply chain.
3) Unit cost: Mass production usually reduces unit costs. The choice of components and subsystems must bear cost and availability in mind. If the unit cost of a satellite is US$500,000 to fabricate (ignoring other costs such as R&D), a mega constellation of 10,000 satellites would cost US$5 billion on fabrication alone, assuming 100% yield.
4) Frequent launches: It would be unthinkable and uneconomical to launch one satellite at a time. Hence, the number of launches depends on how many satellites can be packed as a large payload from the perspective of weight and volume. A medium- to heavy-lift rocket would be preferred. The number of satellites per payload is in turn determined by the satellite's form factor, weight, system complexity, and the manner of deploying multiple satellites in one launch.

   In the case of Starlink, satellites are often launched on ridesharing missions.

5) Launch cost: The key parameter is the unit weight launch cost, such as USD per kg to LEO. The weight of a Starlink Version 1.5, for example, is 300 kg. Assuming the launch cost is US$2,000 per kg on the Falcon 9, the launch cost per satellite is then 300 × $2,000 = $600,000. For a constellation of 12,000 satellites, the launch cost is US$7.2 billion. The total launch cost would double if the unit-weight launch cost was $4,000 per kg instead.
Conversely, the launch cost could come down to $100 per kg when a high-capacity, reusable upper stage such as the Starship comes online. Savings in launch costs can then be leveraged to build heavier satellites accommodating more capabilities. The SCS function is one such example.
When extrapolating the current trend of Starlink satellites, the unit is actually getting larger and heavier, packing more capabilities.

Combining 3), 4), and 5), a mega constellation must reduce the unit manufacturing cost, have satellites with a form factor that facilitates many units per launch, and use reusable rockets to lower the launch cost.

SpaceX initially stated publicly that Starlink Generation 1 would cost US$10 billion. We can check if the numbers work out. Costs such as R&D, design, production yield, and factory investments are excluded. The estimates below come from reports online and tweets by SpaceX executives. Version 1.0 was available at the time and weighed 260 kg. Manufactured by the company's own production line, the publicly stated unit cost was $250,000, and the number of V1.0's produced per week was 45.

Using US$2,000 per kg as the nominal launch cost on the Falcon 9, the total estimate for 12,000 units of V1.0 = 12,000 units × ($250,000 + $2,000 × 260) = $9.24 billion. This estimate comes close to the company's statement above.

It is the unit cost of each satellite, that is, both the manufacturing and launch costs, that drives the economics of production.

We can also estimate the number of launches needed to send nominally 12,000 satellites to orbit. For V1.0, the factory can produce 45 units a week or 2,340 a year. At this production rate, it would take about 5 years to deliver 12,000 units. Falcon 9 can pack up to 60 V1.0 satellites per launch; and it would take 200 dedicated launches to deliver all the 12,000 units to orbit.

The constellation operator will need to develop its own strategy of launch frequency in order to satisfy the ITU timeline regulation.

As SpaceX continues to put out more advanced and heavier versions of satellites, the three factors – production cost, production rate, and the number of units per launch – change constantly. The Starlink V2 variants are all much heavier due to increases in functionalities to serve more users on the ground and, in the case of Direct-To-Cell, expand the user base. With the Starship, the launch cost per kg is expected to be much lower than Falcon 9's, and an estimate of 100 units of V2's can be packed in the much bigger cargo bay.

LEO Internet satellites should not be expected to have a life expectancy of 10–15 years, like GEO satellites. It is reasonable that Internet satellites will be replaced by more advanced versions in about 5 years. Starlink has stated publicly that its satellites would have a nominal designed life of 5 years and would begin to deorbit after that. The point is that regular replacement of these satellites will keep the design and mass production going.

In summary, LEO mega constellations have disrupted the satellite industry in a big way. It changes the satellite production ecosystem from one unit at a time to mass production of identical units. It greatly drives up the demand for rocket launches toward low cost per unit weight, which in turn speeds the development of reusable rockets and high payload capacity.

6.7.2 Convergence of the Sectors

The 1990s witnessed the convergence of computing, telecom, and broadcasting with the rise of the Internet and the subsequent proliferation of wireless technology. Will the LEO satellite Internet constellation trigger another convergence? This writer offers his own view.

The collaboration between SpaceX and T-Mobile in August 2022 was soon followed by Apple Inc. teaming with Globalstar – a constellation operator – in the same year. This is an early sign of boundary-crossing between the satellite and the terrestrial sector. Instead of going through ground stations, "satellite to cell" is the trend. One can interpret Apple's interest as expanding its cellular phone capability to include direct access to satellites as a supplement to cellular phones. Emergency text messaging would be the first test. From the business perspective, one may question whether the efforts to add the "supplementary coverage from space" capability to some satellites is only for emergency messaging use. Hence, voice and other capabilities are bound to follow.

The second possible boundary crossing is between satellites and automobiles. At this moment, the link between an automobile and the satellite network is global positioning and navigation. It appears to be more natural for applications such as autonomous driving and V2X (vehicle to everything) to connect to the local terrestrial (including cellular) network. However, the terrestrial connection does not work outside coverage.

On the other hand, an electric vehicle is just one dish antenna away from connecting to an LEO satellite network. In doing so, one can envision the convergence of the satellite, terrestrial, and automobile sectors to form a new and much bigger pie. The enormous investment in mega satellite constellations would then be further justified. The vision of seamless integration of space and terrestrial ICT will be realized. How things will play out is too early to speculate. Government policies play a big role in supporting such a change.

6.8 Looking Ahead

Mega satellite constellations comprising thousands of satellites in LEO are at different stages of deployment. When complete, the ICT infrastructure of Earth will have seamless access to broadband and overcome the digital divide not only across communities but also across geographic boundaries, including the oceans. This seamless access is a criterion for new applications such as autonomous vehicles and ships.

The Sixth-Generation (6G) network is expected to be ready for commercial release entering the 2030s. Space is an integral part of 6G, relying on LEO satellites to provide seamless connectivity and global coverage. A single system called "Space-Air-Ground Integrated Network" (SAGSIN) is envisioned in the next cellular network. Hence, the race to build up LEO satellite constellations can be understood.

Satellite Internet is a large piece of the LEO commercialization pie. These constellations are mostly funded and deployed by the private sector. From the economics point of view, the launch cost in terms of $ per kg is dominant in the total cost equation. Low launch cost comes from reusable rocket stages with sufficiently high payload capacity. Payload capacity enables the delivery of large numbers of satellites in one launch.

Reading List

Diego, A.S. (2024). *Starlink's Secret Billion-Dollar Empire: How Elon Musk Is Quietly Rewriting the Rules of Space Business*. Independently Published.

Maarseveen, H.V. (2023). *Starlink: The Quest for Global Connectivity*. Kindle.

Nichols, K. (2025). *Geometry of the Satellite Constellation*. Independently Published.

Zhang, J. and Li, J. (2023). *Laser Inter-Satellite Links Technology*, 1ste. Wiley-IEEE Press.

References

1 McDowell J. Starlink Launch Statistics. Jonathan's Space Pages [cited 2025 Feb 21] [online]. Available from: https://planet4589.org/space/con/star/stats.html.

2 Constellation Starlink - Current Positions. Satellitemap [cited 2025 Feb 21] [online]. Available from: https://satellitemap.space/?constellation=starlink.

3 Lentz D. Evaluating SpaceX's Starlink Push. NASA Space flight [cited 2025 Feb 21] [online]. Available from: https://www.nasaspaceflight.com/2020/06/evaluating-spacexs-starlink-push/.

4 Live Starlink Satellite and Coverage Map. Satellitemap [cited 2025 Feb 21] [online]. Available from: https://satellitemap.space.

5 SpaceX. Cited by Ralph E. SpaceX CEO Elon Musk Reveals Next-generation Starlink Satellite Details. TESLARATI [cited 2025 Feb 21] [online]. Available from: https://www.teslarati.com/spacex-elon-musk-next-gen-starlink-satellite-details/.

6 Pultarova T. Starlink Satellites: Facts, Tracking and Impact on Astronomy. Space [cited 2025 Feb 21] [online]. Available from: https://www.space.com/spacex-starlink-satellites.html.

7 SpaceX. Cited by Ralph E. SpaceX's First West Coast Starlink Launch Orbits 51 New 'Space Laser' Satellites. TESLARATI [cited 2025 Feb 21] [online]. Available from: https://www.teslarati.com/spacex-first-west-coast-starlink-launch-space-lasers/.

8 Reuters. SpaceX and T-Mobile to Announce Plans that Boost Connectivity. Reuters [cited 2025 Feb 21] [online]. Available from: https://www.reuters.com/business/media-telecom/spacex-t-mobile-announce-plans-that-boost-connectivity-2022-08-24/.

9 SpaceX. X. Cited by Ralph E. SpaceX Unveils Next-gen Starlink V2 Mini Satellites Ahead of Monday Launch. TESLARATI [cited 2025 Feb 21] [online]. Available from: https://www.teslarati.com/spacex-unveils-next-gen-starlink-v2-mini-satellites-ahead-of-monday-launch/.

10 Shiller D. IR Lasers Link 9,000 Starlink Satellites And Move 42 Million GB Per Day. LightNOW [cited 2025 Feb 21] [online]. Available from: https://www.lightnowblog.com/2024/02/ir-lasers-link-9000-starlink-satellites-and-move-42-million-gb-per-day/.

11 SpaceX. Cited by Wang B. SpaceX Starship Starlink Gen 3 Will Be Nearly as Wide as the Space Station. Next Big Future [cited 2025 Feb 21] [online]. Available from: https://www.nextbigfuture.com/2024/03/spacex-starship-launched-starlink-gen-3-unfolded-nearly-as-wide-as-the-space-station.html.

12 Foust J. Starship Satellite Launches a Year Away. Space News [cited 2025 Feb 21] [online]. Available from: https://spacenews.com/starship-satellite-launches-a-year-away/.

13 Constellation Oneweb - Current Positions. Satellitemap [cited 2025 Feb 21] [online]. Available from: https://satellitemap.space/?constellation=oneweb.

14 Roulette J. Amazon's Prototype Kuiper Satellites Operating Successfully. Reuters [cited 2025 Feb 21] [online]. Available from: https://www.reuters

.com/technology/amazons-prototype-kuiper-satellites-operate-successfully-2023-11-16/.

15 Daws R. China Launches First Satellites for GuoWang Constellation. Telecoms [cited 2025 Feb 21] [online]. Available from: https://www.telecomstechnews.com/news/china-launches-first-satellites-guowang-constellation/.

7

Earth Observations

7.1 Introduction

Earth observation (EO) by satellites is the second largest application in LEO. The primary purpose of EO is to collect information about the target objects and track their changes over time. EO satellites weigh from several hundred to over 1,000 kg. Educational institutions and startup companies are sending nano-satellites for different EO objectives as well.

In this chapter, the images are digital files. EO satellites collect terabytes of data every day. The traditional workflow of EO starts with a satellite generating raw data from a target and beaming back the cumulated raw data to the ground periodically. The data is stored, processed, and analyzed by resident experts, and the results are distributed to end users. The computing power resides with the resident facilities in this workflow.

The rapid culmination of data volume has rendered the traditional workflow impractical. Advances in data science, Cloud computing, and artificial intelligence (AI) have disrupted traditional practices and improved the efficiency of the work-flow to keep up with the volume of incoming data. AI efficiently sorts out changes and trends. All the above technologies translate to productivity improvement, i.e. reduction of latency from the time of raw data collection to actionable intelligence delivered to stakeholders' hands. As data volume increases, data processing productivity must improve accordingly to keep up with the rate of data accumulation.

Commercial entities are also capitalizing on turning EO data into sector-specific business intelligence. They launch their own satellites often with higher resolution and dedicated coverage to collect and process data faster to serve their commercial needs.

New Space: From Low Earth Orbit to the Moon and Beyond, First Edition.
Edited by Stephen W. Cheung and Chih-Yung Wen.

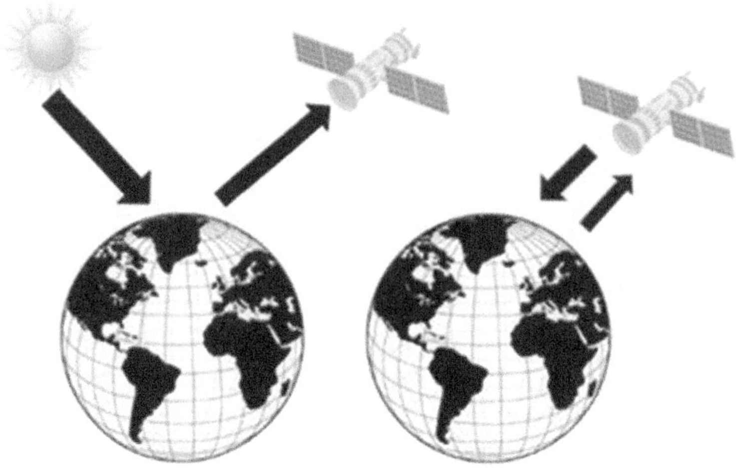

Figure 7.1 Passive versus active earth observations from space. Passive EO examines sunlight reflected from the Earth. Active EO employs a satellite and sends a beam of radar frequency. The signals returned from the target are measured.

7.1.1 Passive Versus Active Imagery

An EO satellite is essentially an orbiting camera taking pictures of the Earth. The camera can be highly sophisticated. Passive imagery means the sensor system just collects the light bounced off from its targets. The source of illumination is the Sun.

Active imagery means that the interrogatory signal comes from the satellite, and its sensor system collects the reflected signal. The two types are illustrated in Figure 7.1.

This chapter covers both passive and active imagery. We focus on optical and radar imaging and explain the terminology and applications with examples.

7.2 Passive Imaging

The Sun generates electromagnetic waves over a broad spectrum of wavelengths, a narrow portion of which is visible to our eyes, but most are invisible.

7.2.1 Electromagnetic Spectrum for Passive Earth Observation

Infrared (IR) and radio waves have longer wavelengths than the visible colors. Optical wavelengths are often expressed in nanometers (nm) or micrometers (μm).

The wavelengths of the visible and IR spectrum are:

- Violet: 380–450 nm
- Blue: 450–495 nm
- Green: 495–570 nm
- Yellow: 570–590 nm
- Orange: 590–620 nm
- Red: 620–750 nm
- Near infrared (780–1,400 nm)
- Shortwave infrared (1,400–2,700 nm)

Most EO applications use visible and IR wavelengths.

7.2.2 Atmospheric Absorption

Figure 7.2 shows the absorption characteristics of the atmosphere over a range of wavelengths. The vertical axis represents the level of opacity from 0% to 100%. There are bands of wavelengths that can pass through the atmosphere with little to no absorption, i.e. low opacity. The Earth's atmosphere filters out most of the wavelengths shorter than violet. The atmosphere is almost transparent to the visible colors but for some scattering.

In active and passive imagery, objects have reflection and absorption characteristics. We are used to seeing objects in their visible colors. Actually, a lot happens outside the visible band, especially on the IR side. Some IR wavelengths are absorbed due to the presence of gases such as water vapor, carbon dioxide, and methane in the atmosphere, and relatively transparent to other IR wavelengths. IR is useful in imaging outside the visible spectrum in interrogating a target's spectral attributes.

Microwaves from 1.5–30 cm wavelength (frequency 1–20 GHz) are used in the active imagery. From space, active illumination is a train of microwave pulses, and the reflected pulses are collected. The collected data is processed, and the corresponding images are reconstructed.

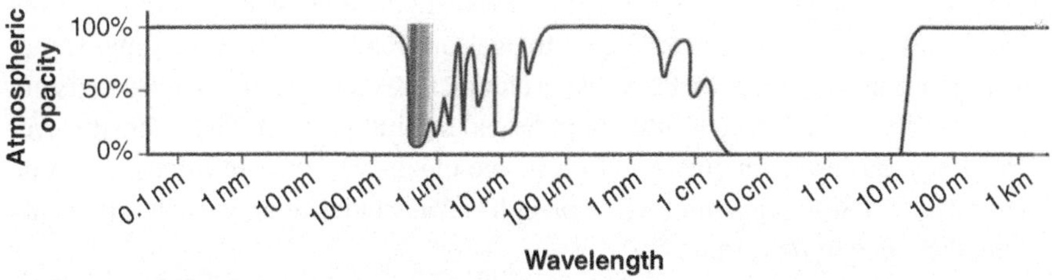

Figure 7.2 Atmospheric absorption (opacity) of a range of wavelengths of electromagnetic waves. 100% opacity means the waves are fully absorbed, and 0% is fully transparent. There are bands of wavelengths, such as visible light, infrared, and radio waves, with small amounts of opacity. See Ref. [1] for color version of this figure. *Source:* Ref. [1] / Edwtie / Wikimedia commons / Public domain.

7.2.3 Terminology

There are several key terms in the field of Earth observation.

7.2.3.1 Sensor

A sensor system is a collection of subsystems that collect and convert reflected photons into raw data. Higher intensity means more photons are collected. Sensors work across a selected range of wavelengths. A camera is a sensor that consists of optics and electronic recorders such as charged-coupled devices (CCDs). Sometimes, a sensor may include a signal source.

7.2.3.2 Reflectance

Reflectance is the reciprocal of the absorption. It is the portion of the radiation reflected from an object. The reflectance of an object ranges from 0% to 100% and is wavelength dependent. An object's color works this way. A green plant, for example, reflects green light and several wavelengths invisible to our eyes.

7.2.4 Resolution

Resolution is a critical characteristic in describing an EO system. It is the level of detail a sensor system can distinguish from a target. There are four types of resolution: radiometric, spatial, spectral, and temporal. The more one wants to know about an object, the finer the resolution one needs, and the bigger the volume of data generated.

7.2.4.1 Radiometric Resolution

Radiometric resolution is the measure of the smallest change in photon counts that a sensor can distinguish. Digital images are made up of a two-dimensional array of pixels on the sensor. Pixel is short for picture elements. Each pixel registers the intensity of the light (number of photons) impinging on it. The intensity level is digitally expressed as bits. An N-bit image means 2^N levels of intensity can be separated. An 8-bit black and white pixel has 256 shades of gray; each represents a small change in light intensity.

For color imagery, there are three basic colors: Red, Green, and Blue (RGB). Other colors come from combining different shades of RGB. The Landsat 8 satellite has a 16-bit radiometric resolution ranging from 0 to 65,535, i.e. over 65 thousand possible shades of light intensities for each of the three basic colors. We can begin to appreciate the volume of data for each snapshot taken.

7.2.4.2 Spatial Resolution

Spatial resolution is a measure of the smallest physical separation between two adjacent objects in an imaging system. A pixel is the finest granularity in a given sensor. A high number of pixels means that each pixel registers more granularity of the target. Today, cellular phones offer millions of pixels in their camera system.

Figure 7.3 shows the same image taken with three different spatial resolutions per pixel: 30 m (left), 100 m (middle), and 300 m (right). Resolution of 30 m, for example, means that two objects less than 30 m apart on the ground cannot be detailed. The image resolution of 300 m is coarser.

The spatial resolution can be improved via a higher number of pixels on a sensor and with better optics. The finer the spatial resolution, the more details can be discerned, but the data volume gets larger. An example is the moderate resolution imaging spectro-radiometer (MODIS) system on board some EO satellites. Some MODIS bands have a spatial resolution of 1 km, while other bands have a spatial resolution of 250 and 500 m.

7.2.4.3 Spectral Resolution

Spectral resolution is a sensor system's ability to discern two adjacent bands of wavelengths. A sensor operates with several bands of wavelength and uses software to reconstruct a picture according to the emphasis on these colors. For example, a consumer camera has three filters (RGB) in the visible color range. The human eye possesses three independent channels to process these three basic colors that form a color image.

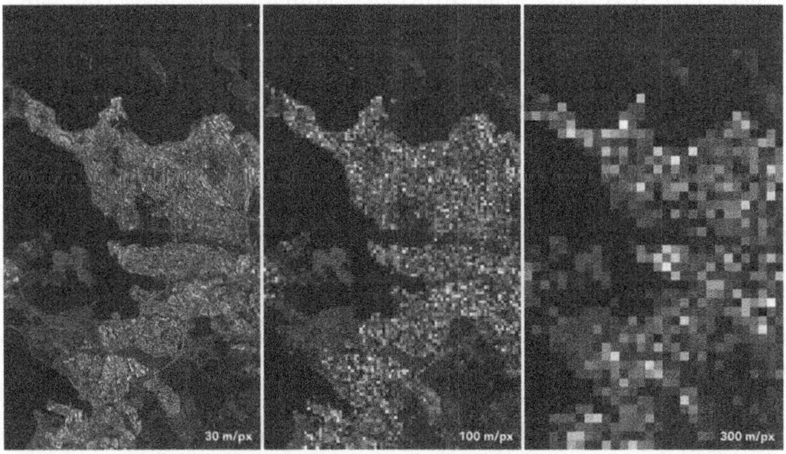

Figure 7.3 Pixel comparison of images of three resolutions by Landsat 8 satellite taken off Reykjavik, Iceland. The spatial resolutions per pixel are 30 m (fine), 100 m (medium), and 300 m (coarse). See Ref. [2] for color version of this figure. *Source:* Ref. [2] / NASA / Public domain.

Table 7.1 There are three imaging types: panchromatic, multispectral, and hyperspectral. Panchromatic means all colors in the visible spectrum.

Imaging type	Number of spectral bands
Panchromatic (all colors in the visible spectrum)	1–2 (typical)
Visible and near infrared (NIR)	3–4 (Red, green, blue, and NIR)
Multispectral	5–36
Hyperspectral	37 and to hundreds

Source: Adapted from Ref. [3].

An imaging system is often described by the number of spectral bands it is capable of sensing. The USGS (United States Geological Survey) provides the guidelines in Table 7.1. Panchromatic means that all colors in the visible spectrum are received and processed. Multispectral and hyperspectral go to IR wavelengths and are our interests in this chapter.

7.2.4.4 Temporal Resolution

Temporal resolution is the amount of time it takes for a satellite to revisit the same location. In EO, this duration is determined by the orbital characteristics of the satellite.

A LEO satellite completes one orbit in about 90–120 minutes, so the satellite should return often. But the Earth's rotation comes into play. Therefore, it will take a longer time for a satellite to return to the same ground track, especially at the same time of the day as is required in some applications. Geostationary satellites have near-perfect temporal resolution. EO satellites are usually in near-polar LEO orbits with temporal resolutions that vary from 1 to 16 days, depending on their altitudes and orbital shape.

For example, the MODIS sensors aboard NASA's Terra and Aqua satellites are both in sun-synchronous (near-polar) orbits. Terra is at 705 km with the ground track repeating every 16 days, while Aqua is at 705 km and repeats every 14.5 days. This bi-monthly ground track repetition is useful to track changes in plant growth, for example.

As a quick summary, it would be ideal to pack good radiometric, spatial, spectral, and temporal resolution all in one sensor system. Reality is that trade-offs must be made in accordance with mission objectives. Daily weather requires high temporal resolution. For agriculture, spectral and spatial resolution are preferred at weekly or monthly intervals.

7.2.5 Spectral Characteristics

It is not enough to photograph the Earth with visible colors. Matters on Earth, from atmospheric gases, plants, water, and soils to man-made objects such as buildings, all reflect wavelengths in and outside the visible spectrum when illuminated by sunlight. One must interrogate the target by deliberately looking for such wavelengths.

Hence, an optical sensor system is designed to discern the selected wavelength bands. Common ways are for the reflected light to be directed to a prism or diffraction gratings to be split into multiple wavelengths of interest, and their corresponding intensities to be recorded separately.

EO images encompass the basic colors in the visible band and the key IR bands. The IR covers a wide range of wavelengths. It is therefore customary to subdivide the IR into broad and fine sub-bands.

For example, the color blue is for atmosphere and deep-water imaging and can reach depths of 50 m in clear water. Green is for vegetation and deep-water structures as well as up to 30 m in clear water. Red is used to detect man-made objects, soil, vegetation, and up to 9-m water.

Right next to red is the Red Edge (670–760) and then the much broader near-IR (750–1,500 nm). Both are important when looking at vegetation. Beyond the NIR is the shortwave IR (1,400–3,000 nm), mid-IR (3,000–5,600 nm), and far-IR (5,600 nm–1 mm) are useful in vegetation, soil moisture, and fire.

7.2.6 False Colors

A natural or true color composite to our eyes is in reality a combined image of three basic colors (RGB) with their respective intensities. This is how a camera takes pictures.

False colors are often used in image analysis and reconstruction to highlight the special attributes of a target whose wavelengths are invisible. For example, false colors are used to represent the IR or a specific band inside the IR. These false colors are often defined in the composite image.

7.3 Multispectral and Hyperspectral Imaging

Most passive EO satellites are designed for multispectral imagery, which goes beyond RGB and extends into the infrared region. These sensors typically detect 5 to more than 30 spectral bands.

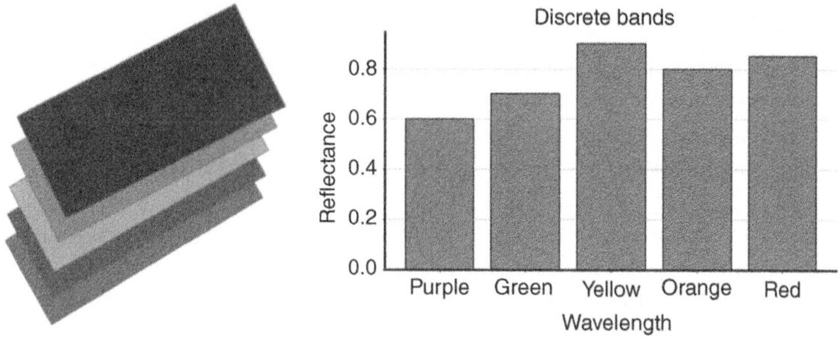

Figure 7.4 This example is for multispectral scanning with five monochromatic wavelengths (purple, green, yellow, orange, and red). The entire display comprises five monochromatic images. Each pixel picks up the intensity of each wavelength, and their reflectance values are plotted on the right. *Source:* With permission of ZJ Zhang.

Figure 7.4 illustrates the multispectral imaging with five spectral bands. This means that each snapshot taken of a target is actually a stack of five monochromatic images of five colors. As an illustration, we assume all the five colors are visible. If the reflectance of each band is taken on a single pixel, a plot is formed on the right side of Figure 7.6.

Each pixel carries a subset of five reflectance values for the five wavelengths. The reflectance value of each wavelength is, in turn, characterized by its radiometric resolution (number of bits). If there are one million pixels in the sensor, then there will be one million of these subsets of reflectance data.

For most of the multispectral cameras on the market, the spectral range is between 400 and 900 nm, i.e. from visible to NIR. Their spatial resolution varies from band to band, i.e. not every band has the same width and therefore its spatial resolution varies. For reference, we list the spectral bands and their associated spatial resolution from the Sentinel-2 satellites below.

7.3.1 Sentinel 2

Sentinel-2A and 2B are a pair of multispectral satellites developed by the ESA and were launched in 2015 to a sun-synchronous orbit of 786 km at 98.62° inclination. They are 180° apart. The pair's mission is to monitor changes in the land surface conditions, including vegetation. Its spatial coverage is between latitudes 56°S and 82.8°N. Each satellite's revisit time is 10 days but 5 days as a pair.

Table 7.2 shows Sentinel-2A's 13 spectral bands and their corresponding bandwidths and spatial resolutions. Bands 5 to 12 have narrow wavelength bands (width 10–20 nm) in the NIR and shortwave IR regions, so the sensor can pick out these specific bands with good spatial resolutions. Readers interested in the applications of the 13 bands are referred to the Sentinelhub website [4].

Table 7.2 The Sentinel-2 EO satellite has 13 multispectral bands ranging from visible colors to beyond short-wave infrared (SWIR). The spatial resolution of each band is 10–20 m.

Sentinel-2A Bands	Wavelengths (nm)	Spatial resolution (m)
Band 1 – Coastal aerosol	433–453	60
Band 2 – Blue	457–523	10
Band 3 – Green	542–578	10
Band 4 – Red	650–681	10
Band 5 – Vegetation Red Edge	698–713	20
Band 6 – Vegetation Red Edge	733–748	20
Band 7 – Vegetation Red Edge	773–793	20
Band 8 – NIR, shorelines, and biomass	789–895	10
Band 8A – Vegetation classification	858–879	20
Band 9 – Water vapor	935–955	60
Band 10 – SWIR, Cirrus Cloud Detection	1,360–1,390	60
Band 11 – SWIR1 moisture in soil and vegetation	1,565–1,655	20
Band 12 – SWIR2 moisture in soil and vegetation	2,100–2,280	20

7.3.2 Landsat Series

The Landsat Program is a series of satellites as a joint effort between NASA and USGS to collect multispectral data of Earth's resources and environment. Currently, Landsat 8 and Landsat 9 are operating. Landsat 8, for example, follows a sun-synchronous orbit at an altitude of 705 km with a temporal resolution of 16 days. Both Landsat 8 and Landsat 9 have 11 bands.

Next in line is the Landsat Next constellation to be launched in 2030. The constellation will have three satellites, 120° apart in the same orbit, 653 km at 98°. Landsat Next will each have 26 bands. Both the spatial and spectral resolutions are expected to be better than those of Landsat 8 and 9. Figure 7.5 shows the graphical comparisons of the bands of Landsat 8 and 9 and Landsat Next with various usage of the bands provided by the USGS. For legibility, we replaced the fine print on the image with numbers and provided the information in Tables 7.3 and 7.4.

7.3.3 Multispectral Images

Below are examples showing how an image would appear under different combinations of the wavelength bands of Landsat 8 and 9. A three-digit number

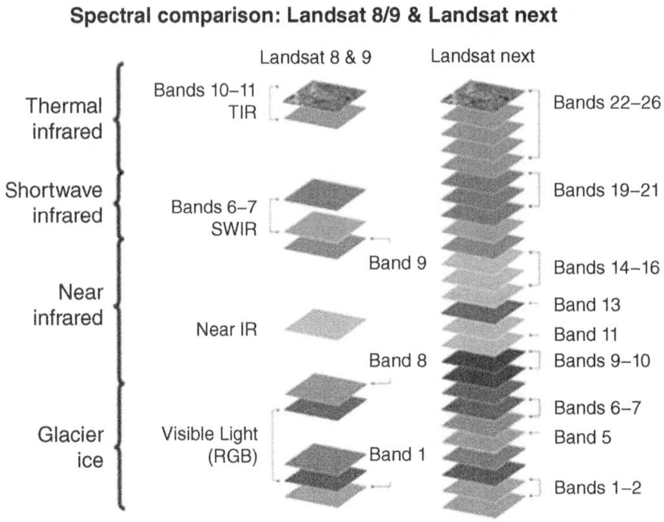

Figure 7.5 Landsat Next is a new generation multispectral EO satellite to be launched in 2030. Landsat 8 and Landsat 9 have been in operation. Landsat Next has more spectral bands. Information about the bands is in Tables 7.3 and 7.4. See Refs. [5, 6] for color version of this figure. *Source:* Adapted from Refs. [5, 6].

Table 7.3 Band designations of Landsat 8 and 9 in Figure 7.5.

	Landsat 8 and 9
Band 1	Imaging shallow water and coral and tracking dust/smoke
Band 8	Panchromatic (15 m resolution)
Band 6–7	SWIR vegetation (type, biomass, health), soil, and burned areas
Band 9	Cirrus cloud detection (atmospheric correction)
Bands 10–11	Surface temperature, crop water use (evapotranspiration), cloud detection, and active fire volcanos

commonly represents three spectral bands selected for special purposes. Each of the selected bands is depicted by a false color Red, Green, and Blue, respectively. We encourage the readers to visit the references and see all the images in color.

a) Agriculture in northern Australia with Landsat 7: Figure 7.6A is taken in natural color and coincides with the false colors. Using the Landsat bands, the three-digit number (432) means that the leftmost digit 4 is Visible Red and is also represented as false color Red, middle digit 3 is Visible Green represented as false color Green, and the rightmost digit 2 is Visible Blue represented by

Table 7.4 Band designations of Landsat Next in Figure 7.5.

	Landsat Next
Bands 1–2	Improved aerosol and inland/coastal water quality
Band 5	Vegetation health
Band 6–7	Water quality, harmful algae blooms
Bands 9–10	Leaf area, chlorophyll, and early plant stress
Band 11	Vegetation indices
Band 13	Water vapor, for improved atmospheric corrections
Bands 14–16	Snow cover and ice melt (liquid water)
Bands 19–21	Soil quality and crop management
Bands 22–26	Surface temperature, crop water use, mineral and surface comp mapping, cloud detection, active fire/volcanoes

false color Blue. The landscape shows obvious iron oxide while the soil and rocks are in blue and gray. Vegetation and water are not evident.
Figure 7.6B shows the same image using channels (752), which means the false color Red is SWIR2 (Band 7), the false color Green is near-infrared (Band 5), and the false color Blue is Visible Blue. This channel combination is often used in geology and shows the rocks in purple and blue colors.
The false colors add visibility to otherwise invisible SWIR 2 and NIR.

b) Vegetation focus: Figure 7.7 shows the channel combination (543) of the rural agricultural town of Chowchilla in California. Band 5 (NIR) is in the false color Red. Band 4 (Visible Red) is in false color Green. Furthermore, Band 3 (Visible Green) is in the false color Blue. The highlight is to show the strong reflectance of NIR by vegetation in false color Red. Note that the original visible colors, such as visible Red and visible Green, are now represented by different false colors.

7.3.4 Hyperspectral Imaging

Hyperspectral imagery expands the number of wavelength bands further. Hyperspectral goes from the visible spectrum out to 15,000 nm in the IR spectrum to short-wave infrared (SWIR). The width of each band is much narrower than that of the multispectral distinct bands. Hyperspectral has tens to hundreds of contiguous narrow bands across the visible and the IR spectrum. The wavelength coverage now appears continuous. This is illustrated in Figure 7.8.

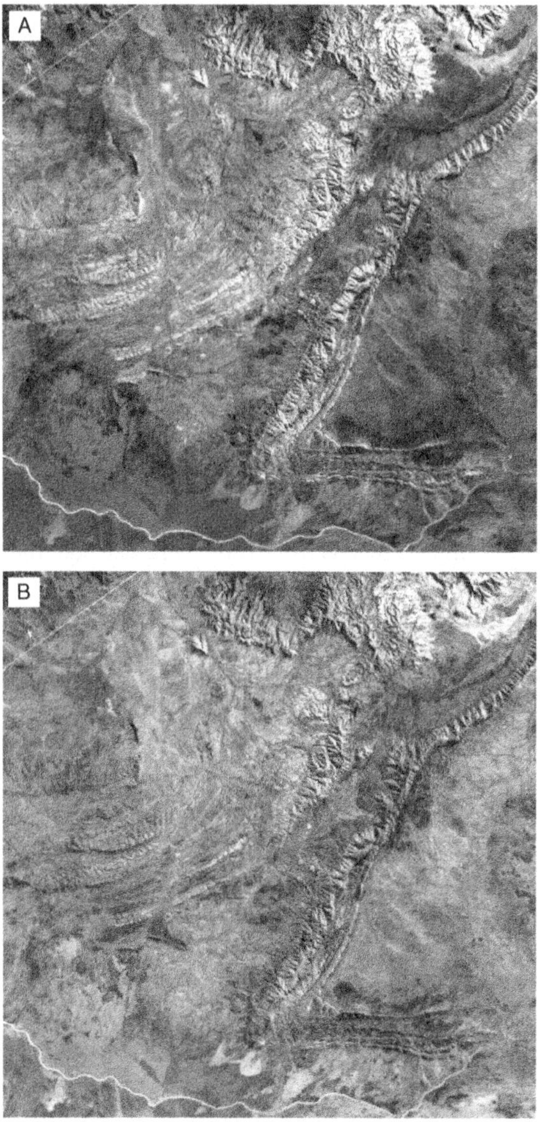

Figure 7.6 (A) A satellite image of the ground with color bands (432), that is, with Band 4 (Visible Red), Band 3 (Visible Green), and Band 2 (Visible Blue) of Landsat 7. The bands happen to coincide with the natural colors. As a composite, Band 4 is represented by the false color Red, Band 3 by the false color Green, and Band 2 by the false color Blue [7]. (B) The same satellite image of Figure 7.6A using a different combination of channels (752), where Band 7 is short-wave IR, Band 5 is NIR, and Band 2 is Visible Blue. In that order, the composite of the image is in false color Red, false color Green, and false color Blue, respectively. See Ref. [7] for color version of this figure. *Source:* Ref. [7] / Micromine.

The plot on the right of Figure 7.8 shows the reflectance values across wavelength bands in one pixel. The full hyperspectral picture is the sum of the plots over the pixels in the sensor, which is just one snapshot at a given moment. The volume of data cumulated over an orbital period is many times larger than that of the multispectral data. A multidimensional data cube, also known as a hypercube, is formed.

Figure 7.7 Image taken by Landsat 8 highlighting vegetation with NIR in Chowchilla, California. The channels are (543). Channel 5 is the NIR in the false color Red, 4 is the natural color red in the false color Green, and 3 is the natural color Green in the false color Blue. See Ref. [8] for color version of this figure. *Source:* Ref. [8] / Esri.

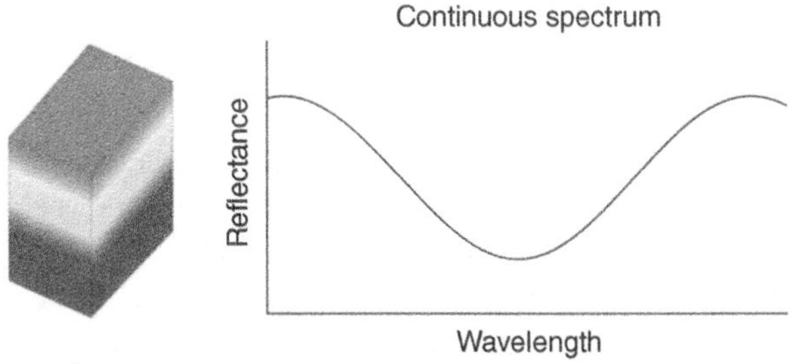

Figure 7.8 Similar to Figure 7.4, the hyperspectral imagery consists of many wavelengths from the visible across to the infrared. Each pixel detects the reflectance values of the many wavelengths and is plotted on the right. The entire image is the sum of all these values of the pixels.

The data continuum in a hypercube allows analysis in ways that cannot be done with distinct bands. Cutting and slicing across different layers of data is one example of the new possibilities. One advantage is the comparison of data of the same target over a period or over select wavelengths to look for changes and trends. Hyperspectral sensing has been expanded from mineral identification to include forestry, agriculture, mineralogy, environment and surveillance, and water and air quality.

All objects have unique spectral fingerprints in hyperspectral imaging. A digital library of the hyperspectral spectrum can be compiled to sort out different

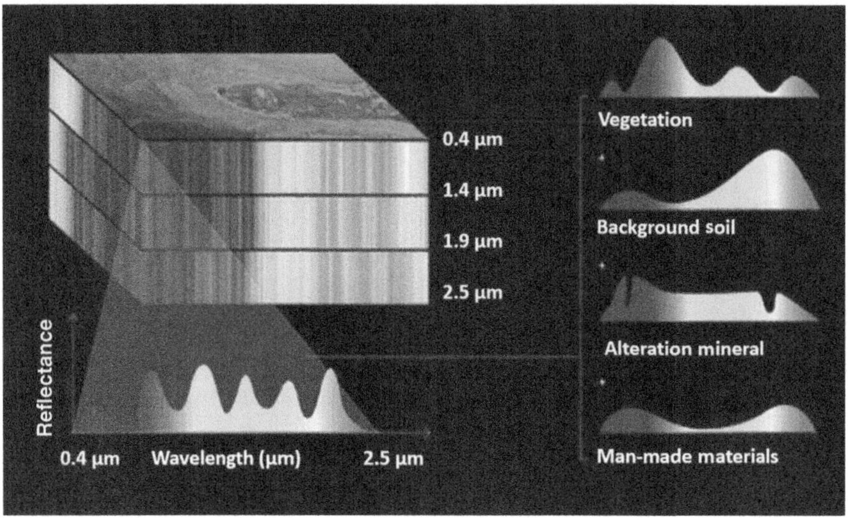

Figure 7.9 Illustration of the hyperspectral scans of vegetation, soil, and man-made materials. Each pixel scans over several wavelength bands from the visible to the infrared. The wavelength unit is μm (1,000 nm). See Ref. [9] for color version of this figure. *Source:* Ref. [9] / Pixxel.

materials. With hyperspectral, plants no longer look alike. Multispectral scanning cannot produce such fingerprints reliably due to the insufficient number of wavelength bands to produce precise fingerprints. The hyperspectral spectrum reveals more details not possible with simple RGB or multispectral methods.

Figure 7.9 shows the hyperspectral fingerprints of four reference materials on the Earth's surface: soil, vegetation, mineral, and man-made materials over a spectrum from 400 nm (0.4 μm) to 2,500 nm (2.5 μm). Different materials can be identified and sorted by computing power.

This enormous amount of data and the rate of accumulation make it impossible for traditional data storage in individual organizations and analysis to be handled manually. Now, users can access the data via the Cloud, and AI is a common tool for this data-intensive work. Researchers and business analysts no longer need to have large data storage and computing power on their own premises to serve their business verticals.

Figure 7.10 shows a base image overlaid by a hyperspectral image displayed in a small window by the company Pixxel in 2022 [10]. The company's hyperspectral instrument on their satellite has 150 bands. The background in the image is RGB. A smaller window can be moved around to different locations of interest to show the corresponding hyperspectral perspective. False colors are used in these hyperspectral windows. As elaborated by the caption of these photos, the hyperspectral

Figure 7.10 Lithological mapping of minerals in the Permian Basin, USA. The photo's regular presentation is in natural colors. A square window can show the view using the hyperspectral capability in the software. See Ref. [10] for color version of this figure. *Source:* Ref. [10] / Pixxel.

images show more material signatures. This kind of analysis cannot be performed with multispectral images.

7.4 Precision Agriculture

Precision agriculture discerns the health status of vegetation at precise locations and monitors the progress of crop growth. This is achieved by a digital system integrating satellite data with other ground data and delivers information that suits the farming trade. A farm operator needs accurate positioning data so that he can take corrective action.

Agriculture is one of the major applications of EO with a trillion-dollar market worldwide. The gross production value in the agriculture market is projected to be US$4.59 trillion in 2024. Food production is a major part of the world economy. Fluctuations in outputs in major food-producing nations greatly affect the world's supply and economic stability.

Figure 7.11 illustrates the global perspective of stability in agricultural production. The Agricultural Market Information System (AMIS) is a satellite-based platform to "enhance food market transparency," according to the organization's website [11]. AMIS is under the United Nations. AMIS members are G20 (Group of Twenty) members, major exporting and importing countries of agricultural commodities.

AMIS assesses global food supplies with a focus on four major crops: wheat, maize, rice, and soybeans. It provides information to policymakers to monitor

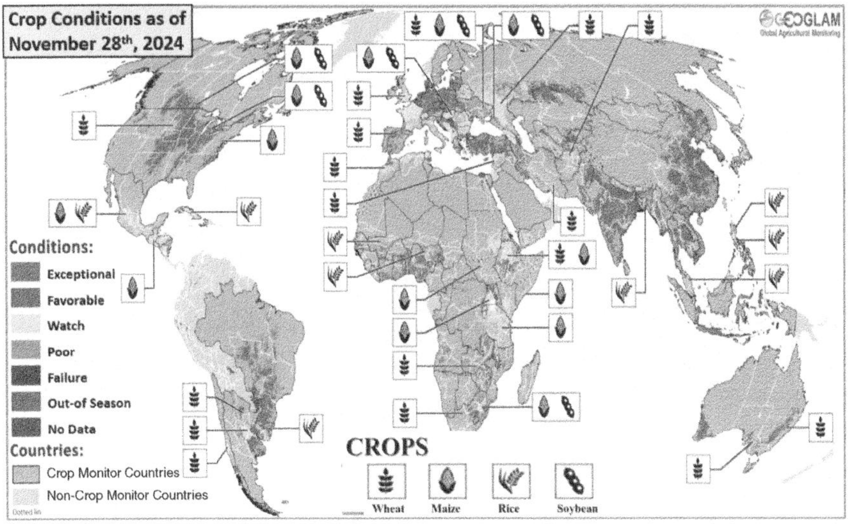

Figure 7.11 The Agricultural Market Information System (AMIS) uses satellite data to track the health of four major crops worldwide. The platform aims to provide market transparency. See Ref. [12] for color version of this figure. *Source:* Ref. [12] / GEOGLAM.

global food market stability and impending uncertainty. Figure 7.11 is a snapshot taken in November 2024. The graphics and related data are updated regularly. The color codes in Figure 7.11 show the crop conditions as explained on the left side of the image.

Agriculture has entered a full digital age. Digital images from satellites and aerial vehicles, coupled with ground sensors, form an information system that elevates farm productivity to a precision level. Computing power analyzes the vast volume of data from the sky, while local sensors measure parameters such as temperature, irrigation, and soil chemistry. Farmers now know the conditions of their crops down to meter-level precision.

The US Department of Agriculture in 2024 states that the average farm size in the US is about 450 acres (1 acre = 4,000 square meters), while large farms average 3,000 acres. Technology helps to manage crop and vegetation health as well as soil conditions down to precise locations in these large farms.

Factors affecting crop and vegetation health include temperature, water, diseases, nutrition, and the chemical conditions of the soil. Environmental factors such as rainfall and temperature are external and meteorological. Diseases, nutrition, and water level are local and within the farm operator's control. Crop conditions also change over the growing season.

Satellites offer wide perspectives to track changes during the season. The spatial resolution of the satellite determines precise locations where the crop is doing well or poorly. As the spatial resolution improves to several meters, coupled with the global positioning overlay, farm operators are able to pinpoint their actions.

Commercial firms have been key in bridging the gap between specialized satellite data and farm operators. This level of service is best carried out commercially. These firms are steadily forming an ecosystem to support the integration of satellite data, ground sensors, and communications.

7.4.1 Plant Health

Plant health is commonly manifested in its "greenness." Greenness comes from the presence of chlorophyll. Chlorophyll in a healthy plant absorbs blue and red light in the visible spectrum, while water absorbs several wavelengths in the IR spectrum. Mesophyll cells in leaves capture red light to energize photosynthesis. A healthy plant also reflects green, thereby appearing green. Therefore, the reflectance from the colors Blue and Red is low and high for Green.

Figure 7.12 shows the reflectance of vegetation, soil, and water across the visible and IR wavelengths. An interesting characteristic of vegetation is that its reflectance goes from absorptive Red to reflective at the beginning of the NIR. This transition is known as the Red edge.

When viewed across the visible and infrared spectrum, we see that materials display their own reflectance signatures. Different crops have different signatures,

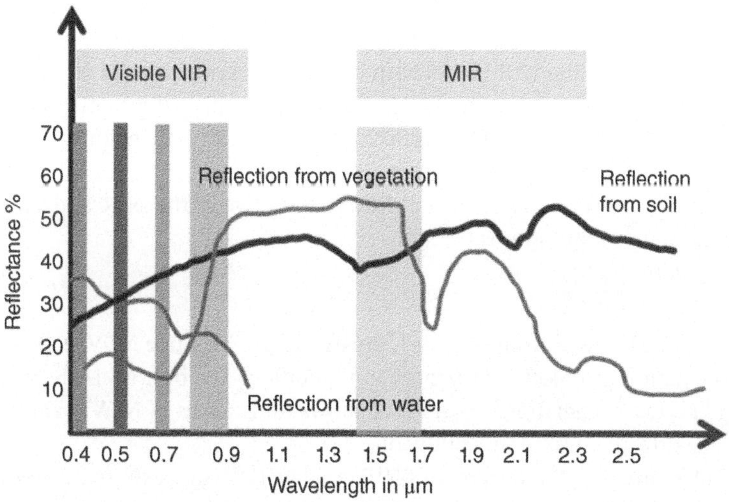

Figure 7.12 Reflectance properties of the three essential materials: plants, water, and soil. Wavelength coverage goes from visible to the middle of the infrared spectrum.

but the general characteristics remain. Reflectance dips in the middle IR due to water absorption in the plant. Such signatures come in handy when hyperspectral imaging is used. Water and soil have different reflectance characteristics, as shown in Figure 7.12. Water reflects blue the most and is the reason water appears blue. Soil's reflectance trends with wavelength increase and reflects IR in general.

7.4.2 Vegetation Indexes

Indexes are numerical values that objectively describe the health status of green plants. Different indexes, but not all, are usually based on the reflectance values of certain wavelengths. More specialized indexes indicate special attributes. We should emphasize that no single index can fully represent the status of plant health.

Several common indexes are introduced here with a focus on NDVI:

1) Normalized difference vegetation index (NDVI)
2) Green normalized difference vegetation index (GNDVI)
3) Red edge normalized difference vegetation index (RENDVI)
4) Moisture stress index (MSI)

The first three indexes bear the same concept of "normalized difference vegetation index," the wavelengths are different to reflect different emphasis.

7.4.3 Normalized Difference Vegetation Index

NDVI is the most used index for green vegetation. Since healthy vegetation absorbs blue and red while reflecting NIR plants with active photosynthesis can be rated by comparing the reflectance of red and NIR. Figure 7.13 depicts the healthy and unhealthy cases as far as the red and NIR reflectance are concerned. Healthy plants can be seen to absorb more red.

NDVI is a ratio. The numerator is the difference between the NIR and red reflectance, and the denominator is the sum.

$$\mathrm{NDVI} = \frac{\mathrm{NIR} - \mathrm{RED}}{\mathrm{NIR} + \mathrm{RED}} \tag{7.1}$$

The value of NDVI ranges from -1 to $+1$, crossing zero. Negative NDVI represents dead vegetation or objects not related to vegetation. The best NVDI is $+1$, representing "perfect" vegetation. Figure 7.14 shows four ranges of NDVI values representing healthy plants and unhealthy ones.

Healthy plants have a high reflectance of NIR and low reflectance for red, so the NDVI is high. Unhealthy vegetation contains less chlorophyll and reflects more red so the numerator becomes small. NDVI crosses zero when NIR equals red with

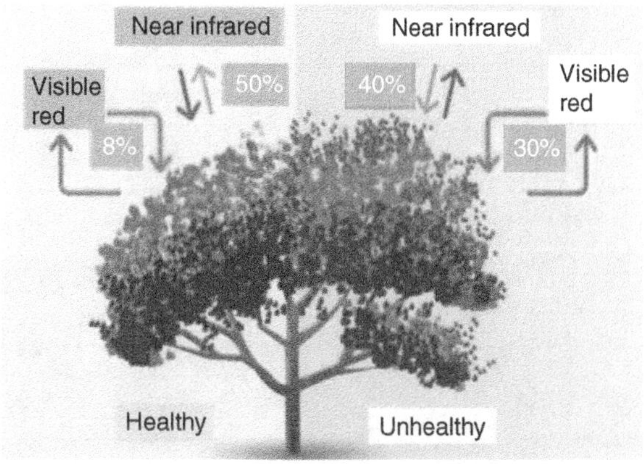

Figure 7.13 The absorption and reflectance values of a healthy plant are compared to those of an unhealthy one. Healthy plants absorb red and reflect near-infrared, whereas the unhealthy plant reflects red and reflects less NIR. See Ref. [13] for color version of this figure. *Source:* Ref. [13] / EOS Data Analytics, Inc.

Figure 7.14 Range of NDVI values of plants. Dead plants are 0 and below. The NVDI values of very healthy plants are between 0.66 and 1.0, while those of unhealthy plants are between 0 and 0.33.

dead plants or non-vegetated matter such as water, snow, ice, rock, and bare soil. NDVI also varies with seasonal changes in climate. Figure 7.15 shows the NDVI for different crops among nearby plots of land. The difference could be due to the operators' attention to different factors.

NDVI only uses NIR and ignores wavelengths beyond. The ratio is simple but does not provide enough indications of other attributes to plant health. NVDI is affected by the environment, such as clouds and soil colors, and can be incorrect in situations with dense vegetation.

7.4.4 Other Indexes

We will briefly describe the other indexes mentioned above.

a) Green normalized difference vegetation index (GNDVI): This index is similar to NDVI. The Red reflectance is replaced by Green in the NDVI equation, thereby

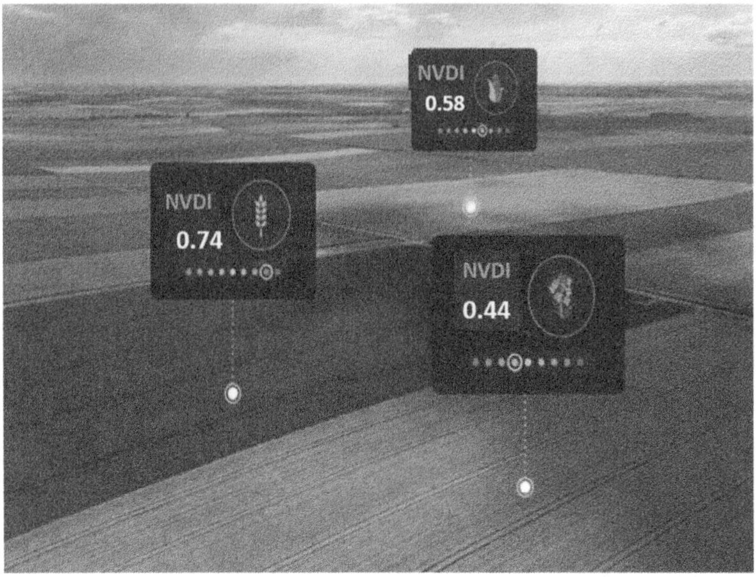

Figure 7.15 NDVI values of different plots of farmland representing the health status. *Source:* Ref. [14] / EOS Data Analytics, Inc.

placing emphasis on chlorophyll.

$$\text{GNDVI} = \frac{\text{NIR} - \text{GREEN}}{\text{NIR} + \text{GREEN}} \tag{7.2}$$

The range of GNDVI is also from −1 to 1.

b) Red edge normalized difference vegetation index (RENDVI): This index replaces the reflectance of Red by that of the Red Edge in the NDVI equation. The Red Edge reflectance changes rapidly. Any stress within the vegetation will show up here. Hence, this index is for detecting vegetation stress. RENDVI is useful for the middle to late growing season. It also ranges from −1 to +1.
c) Moisture stress index (MSI): MSI is the ratio between the NIR and SWIR. These wavelength bands fall outside the visible spectrum, but the water in plants is sensitive to them. MSI indicates the change in water content in the leaves. Lack of water leads to moisture stress in the vegetation. High MSI values indicate greater water stress and low water content. The value of this index ranges from 0 to more than 3. The common range for green vegetation is 0.4 to 2.

In summary, NDVI is a good general-purpose indicator of plant health, but more specifics need to come from other indexes.

7.5 Other Applications of Passive Earth Observations

Passive EO has many applications other than agriculture. This section describes several applications with examples.

7.5.1 Large Project Management

Mega construction projects take up large plots of land, often measured in square kilometers. Construction progress is usually monitored by aerial vehicles or satellites. The Line in Saudi Arabia is an example.

Announced in 2021, The Line (Saudi Arabia Neom Megacity) is a US$500 billion megacity project. The Line is to be a 170-km straight line city in the Tabuk province of Saudi Arabia. Aiming to house 9 million residents, The Line is 170 km long, 200 m wide, and lined with high rises. The megacity aims to be "zero-carbon" and car-free. Underground trains and air taxis will be the means of transportation. Figure 7.16 is a satellite image of the project in progress. The long and thick line represents The Line while support facilities are nearby. Given the megacity's dimensions and requirement of straightness, satellite imagery is a very suitable tool to monitor progress.

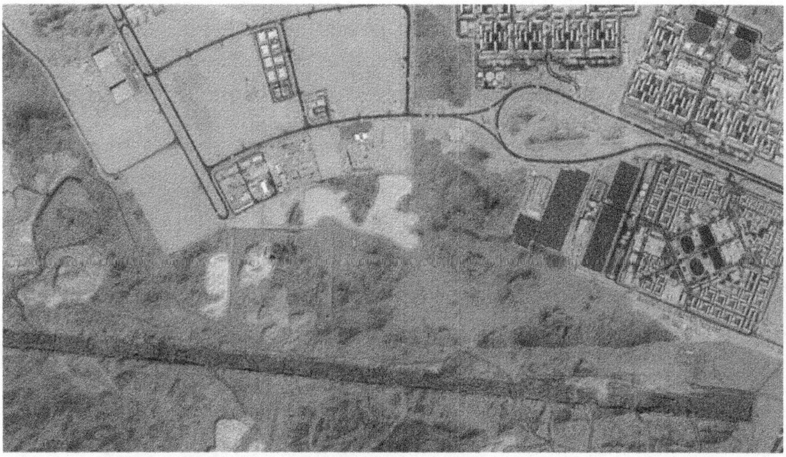

Figure 7.16 The Line project is highlighted in dark gray as viewed from space. The black dots represent excavators around the line. Support facilities are shown in the upper half of the photo. Light gray dots near the top of the photo represent construction vehicles, and the black rectangles are arrays of solar panels. See Ref. [15] for color version of this figure. *Source:* Ref. [15] / Soar.Earth Ltd.

7.5.2 Disaster Monitoring

Satellite and aerial imaging are critical after major natural disasters strike. Examples include floods, earthquakes, storms and hurricanes, volcanoes, tsunamis, tornadoes, wildfires, and major oil spills. Aftermaths of disasters are often accompanied by the tremendous destruction of surface infrastructure such as roads and utilities, as well as property and environmental damage.

Images from satellites and aerial vehicles often provide unobstructed views of the extent of the "ground zero" area and near real-time developments. Immediate assessments from "the eye in the sky" of most disasters assist response teams and humanitarian relief with the needed guidance, such as access routes. Government agencies and insurance companies can also estimate the extent of damage and respond accordingly.

We show satellite images of several events as examples of but without detailed elaboration.

a) Wildfires

In Figure 7.17, a total of 1,140 square km was burned, and over 1,000 structures were destroyed in the 2017 Thomas Fire in Ventura County, California. The fire lasted for 40 days. The cause was believed to be due to high winds causing power lines to come into contact.

Wildfires occur in all countries every year. They could be caused by nature (e.g. lightning strikes), human carelessness, powerline accidents, and sometimes by control-burns. The average number of wildfires in the US is around 55,000 a year. Climate change is believed to have exacerbated the number and frequency of wildfires and large fires.

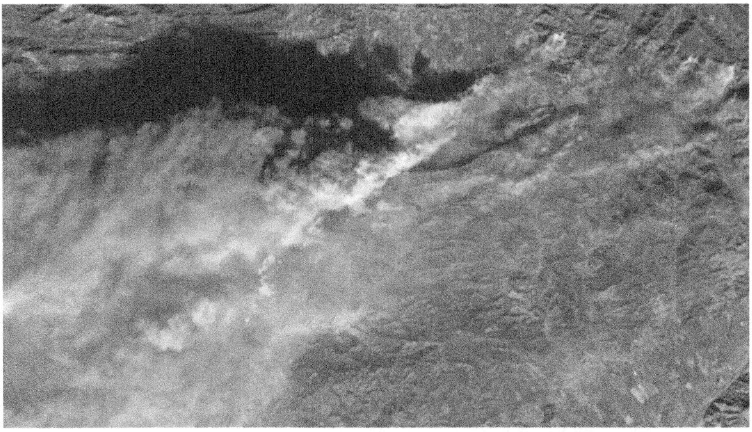

Figure 7.17 Satellite image of the big fire in Ventura County, California, December 2017. See Ref. [16] for color version of this figure. *Source:* Ref. [16] / NASA / Public domain.

b) Hurricanes and Floods

Mega storms and hurricanes occur frequently and incur tremendous property damage and loss of lives. Torrential rains cause floods and mudslides. Today, weather satellites can accurately project the trajectories of impending storms and hurricanes and their intensities by category. Hence, early warning and community preparation have reduced the extent of damage.

Damage assessment often compares "before and after" images of the impacted areas. In 2017, Category 5 hurricane Irma caused tremendous damage to the Caribbeans. Figure 7.18 shows the "before and after" photos of the British Virgin Islands. The "after" photo shows obvious browning of all the islands due to vegetation being wiped away by strong winds.

c) Earthquakes

Major earthquakes cause tremendous casualties and property loss. Great earthquakes rank among the top natural disasters owing to the extent of damage and human casualties. There is no accurate way to predict earthquakes.

Examples of major earthquakes include:

a) The Kahramanmaraş Turkey-Syria earthquake (magnitude 7.8) in 2023, with over 55,000 people killed.

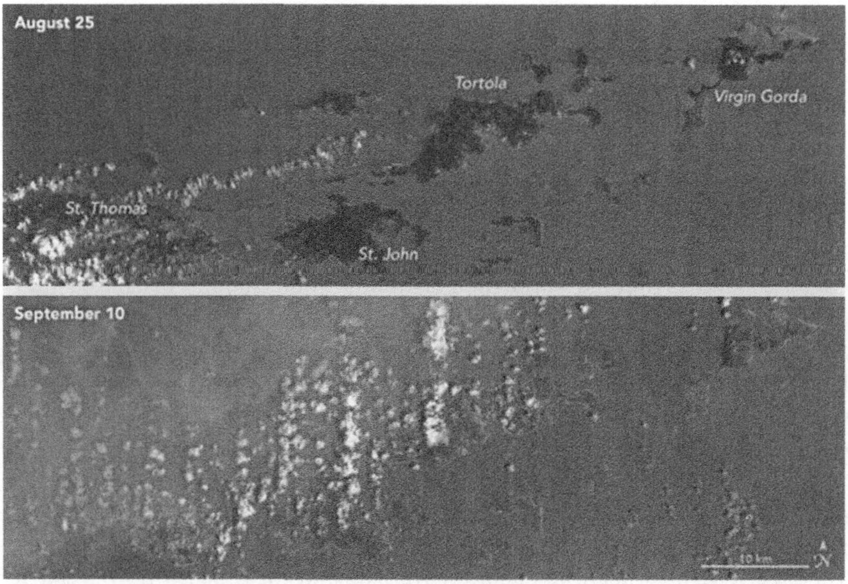

Figure 7.18 Photo comparison of the British Virgin Islands before and after Category 5 Irma in December 2017. See Ref. [17] for color version of this figure. *Source:* Ref. [17] / NASA / Public domain.

Figure 7.19 A large fishing boat was washed ashore by the tsunami. See Ref. [18] for color version of this figure. *Source:* Ref. [18] ChiefHira / Wikimedia Commons / CC BY-SA 3.0.

b) Morocco earthquake (magnitude 6.8) in 2023, with over 2,960 deaths and 5,674 injured.
c) Tohoku earthquake and tsunami (magnitude 9.0) in 2011. The casualty was over 16,000 deaths, and many thousands injured or missing. The tsunami waves reached 10 km inland, damaging bridges, roads, and buildings. See Figure 7.19.

7.5.3 Minerals

Minerals are critical resources for a nation's industrial foundation. Although minerals are found all over the world, their distribution and amounts of deposits vary. There are more than 4,000 natural minerals on Earth. Their unique chemical characteristics are manifested by their spectral signatures when illuminated by sunlight. These spectral signatures go beyond the color appearances in the visible spectrum. Multispectral and hyperspectral images provide data for analysis by experts in mineralogy.

EO satellites help identify large-scale distribution of minerals to expedite prospecting and provide accurate locations of mining sites. Figure 7.20 shows two images of the Mountain Pass Mine located in the Mojave Desert, California. The area is rich in deposits of rare earth elements. The image on the left is a natural color composite. The spectral bands of the image on the right are enhanced to show the iron minerals.

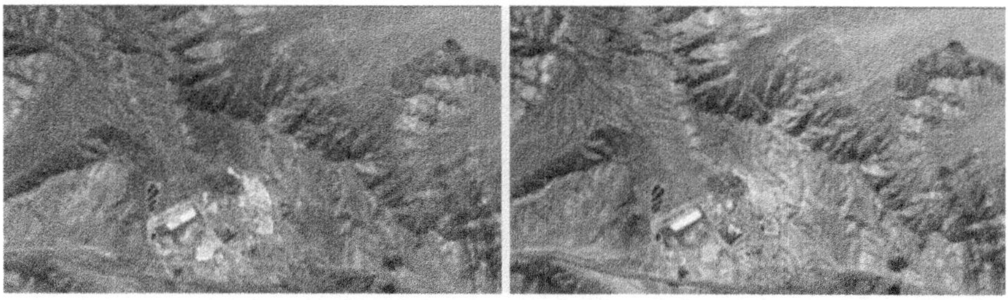

Figure 7.20 Images of the Mountain Pass Mine. The image on the left is in natural color and the image on the right has a false color representing iron minerals. See Ref. [19] for color version of this figure. *Source:* Ref. [19] / Exploration Mapping Group, Inc.

7.5.4 Climate

The global climate is the Earth's physical systems at work. No single parameter can completely represent the environment. There are many parameters that impact our climate, which in turn affect the entire environment. The EO satellite is a tool to measure some of these key parameters or the results of their actions. We will show satellite images representing the parameters of our global environment with two cases for illustration.

EO imagery is the best tool to measure different parameters of the oceans.

a) Ocean surface salinity: All the water on Earth circulates through the water cycle. The oceans hold 97% of the water and supply almost all the precipitation in the world. The remaining 3% of the water is found in rivers and lakes, underground water, ice, and glaciers.
   Salinity is a measure of the salt content (mainly sodium chloride). Sea water salinity is the ratio of salt in grams to 1 liter of water (about 1,000 g) expressed as parts per thousand (ppt). Salinity ranges between 33 and 37 ppt and averages 35 ppt. The ocean surface salinity changes with seasons and varies with regions. The salinity at the equator is lower than that at the poles due to plenty of rainfall.
   Climate change-induced temperature rise is the result of an increase in evaporation from the oceans, which drives up the ocean surface salinity. Such an increase results in more extreme weather events such as torrential precipitation. The ocean surface salinity helps us understand the water cycle and changes over the globe.
   Figure 7.21 shows a snapshot of a running composite of sea surface salinity in the world via satellites from NASA's "Sea Surface Salinity from Space" program. The color codes represent the salinity values in ppt.

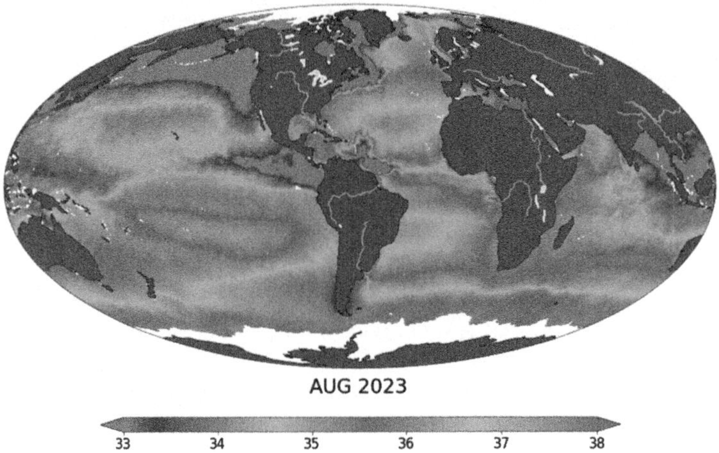

Figure 7.21 Global sea surface salinity in August 2023, shown in this map, is "optimum interpolation salinity" based on combined data from different satellites. The Salinity of 35 ppt (parts per thousand) is the average. See Ref. [20] for color version of this figure. *Source:* Ref. [20] / NASA / Public Domain.

b) Coral reefs and ocean surface temperature: Coral reefs are often called "rainforest of the sea" as they provide the habitats and biodiversity that marine life lives on. Approximately 25% of ocean fish depend on healthy coral reefs.
The health of coral reefs greatly depends on the water temperature. Corals live in water between 20°C and 29°C. They will become stressed and vulnerable to diseases when living in water that is too warm for more than a month. This is known as coral bleaching. If the stress is not promptly removed, the coral will become less likely to reproduce and susceptible to diseases, leading to death.
Climate change leads to an increase in ocean temperatures. The US National Oceanic and Atmospheric Administration operates the Coral Reef Watch (CRW), which reports the global sea surface temperature daily. CRW offers an alert map with different colors and risk levels. Figure 7.22 is an example of the sea surface temperature map in August 2024.

Other pollutions: Satellites can be used to detect other pollution and climate effects from space. They include ocean pollution such as the accumulation of trash, release of toxic waste, reduction in ice covers in the poles, and sudden bursts of emission of greenhouse gases.

7.6 Synthetic Aperture Radar (SAR)

Synthetic aperture radars have their own signal source to illuminate the target. SAR antennas are mounted on the nadir side of the satellites. The source generates

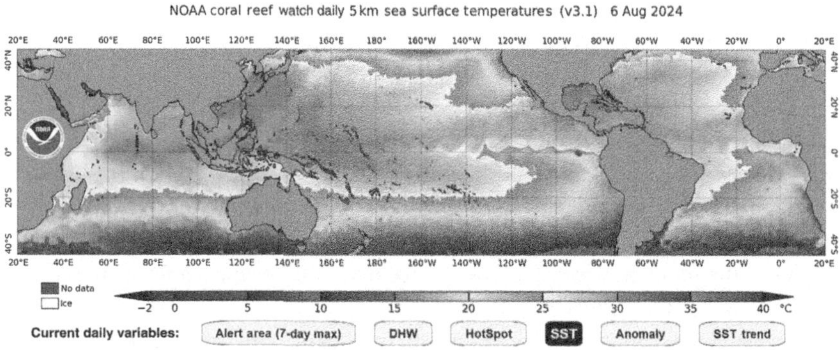

Figure 7.22 This map shows the sea surface temperatures. Different colors represent different temperatures. See Ref. [21] for color version of this figure. *Source:* Ref. [21] / NOAA / Public Domain.

pulsed microwaves at gigahertz frequency. These frequencies can pass through the atmosphere with little to no absorption even in the presence of water vapor and clouds. SAR works 24 hours a day and under almost all-weather conditions. This is an advantage over optical satellites that need sunlight as the source.

In optical EO, sunlight is composed of all kinds of wavelengths. In SAR, the microwave source has a single wavelength. Therefore, synchronization, that is, timing, between the transmitted and received signals is feasible.

A SAR satellite antenna is a long strip, typically several meters long. The antenna length is lined up with the flight path known as the azimuth. SARs are designed to look sideways. The reflected signals from the side contain more perspectives as the area being scanned is much larger than looking straight down. The reflected microwave pulses are collected, and a two-dimensional picture of the scanned region is constructed from the data.

7.6.1 Microwaves Interaction with Matter

The interaction between microwaves and matter is between the wave's electric field (E-field) and the object's electrons. The dielectric constant measures the extent of this interaction. Perfect conductors have an infinite dielectric constant, which means that electrons move freely under the influence of the incoming E-field. The external E-field is reflected away from the conductor's surface with no penetration. On the other hand, the electrons in a perfect insulator are tightly bound in the atoms and do not react to an external electric field. The dielectric constant of a perfect insulator is 1 (for true vacuum).

Most materials are neither perfect conductors nor perfect insulators. Their dielectric constants are finite in value. Incoming microwaves partially go through and partially get reflected. Wet soils, for example, have higher dielectric constants and reflect microwaves more than dry soils. Radar signals can therefore penetrate deeper in dry soils and less in moisture-loaded farmland.

7.6.1.1 Scattering

Reflection and scattering from a surface are related to its roughness. Reflection only occurs with a perfectly smooth surface. Scattering occurs when the surface has roughness and irregularities. Roughness is a relative term. It is common to express the dimension of the jaggedness of a surface feature relative to the wavelength λ of the incident wave. Thus, a surface may appear smooth for a long wavelength but rough for a short wavelength.

There are three scenarios that can happen to a beam of microwaves from a satellite (denoted Object A in Figure 7.23) hitting a surface at an oblique angle.

1) The surface is smooth: the jaggedness h is less than λ/32. The beam is reflected in a direction away from the incident beam. A lake is an example. Hence, the SAR satellite pinging the lake's surface receives nothing back, and the smooth surface would appear black when such an image is reconstructed from the reflected signals.
2) The surface is slightly rough: h is between λ/32 and λ/2. Reflection still happens from the surface. The reflected signal is the same as (1) but is also accompanied by wavelets spread around the point of contact. The wavelets are known as scattering.
3) The surface is very rough: h is larger than the λ/2. Scattering now dominates, and the scattered beam goes in all directions.

Scattered signals that bounce back toward the source are termed backscattering. The three scenarios in Figure 7.23 show reflection and scattering in a single

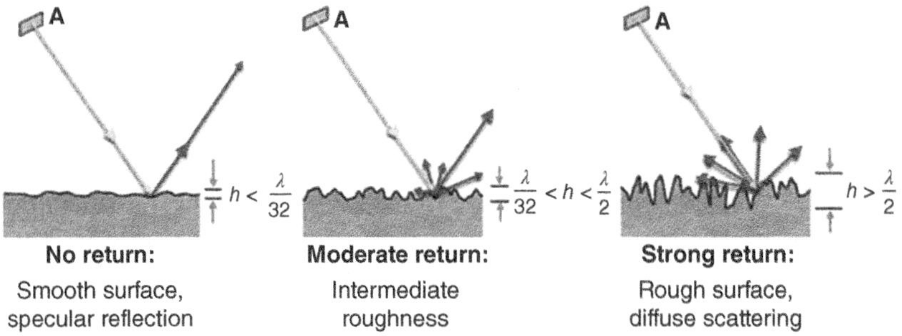

Figure 7.23 A quantitative indication of reflection and scattering from surfaces of different roughness h. The effects are wavelength dependent. When the surface roughness h is smaller than λ/32 of the incident wavelength, the reflection would be dominant. If *h* is larger than λ/2 of wavelength, scattering will be dominant. See Ref. [22] for color version of this figure. *Source:* Ref. [22] / NASA / Public Domain.

bounce – off the primary point of contact. Multiple bounces can happen. A double bounce means that the scattered signals hit another object nearby and result in a second reflection or further scattering. For example, buildings in a city are near each other. Relative to a single bounce, signals after a double bounce and multi-bounces returning to the satellite are progressively weaker with different time delays. Figure 7.24 illustrates the backscattering phenomenon of the single and double bounces.

Backscattering depends on the wavelength of the incident beam. A beam of incident microwaves can discern height variations on the target surface in the same order as the wavelength. As the SAR beam scans the surface while the satellite moves forward from above, a wealth of time-delayed backscattered signals is collected from the surface. Note that electromagnetic waves travel much faster than the moving satellite, so the satellite position appears "frozen in time" relative to the signal propagation. A two-dimensional profile of the target surface can now be constructed via computing algorithms.

Microwaves with long wavelengths penetrate nonconductive materials more than those with short wavelengths. This is illustrated in Figure 7.25. Thus, long-wavelength microwaves give a coarse indication of the object's granularity they rebound from but with deeper penetration. Short-wavelength microwaves result in finer granularity but with a shallower penetration depth.

In general, high-frequency microwaves do not penetrate objects easily, thereby only reflecting off surface features. Hence, high-frequency bands such as the Ka, K, and Ku bands are rarely used in SAR. The X-band at about 10 GHz has several uses but only stops at interrogating the top layer of a target. Long wavelengths,

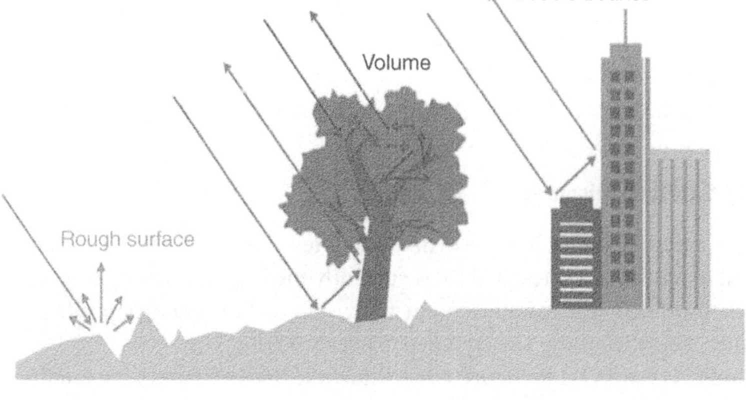

Figure 7.24 Illustration of the single bounce (left) and double bounce (right). The diagram in the middle shows the penetration ability of some RF signals. See Ref. [23] for color version of this figure. *Source:* Ref. [23] / NASA / Public Domain.

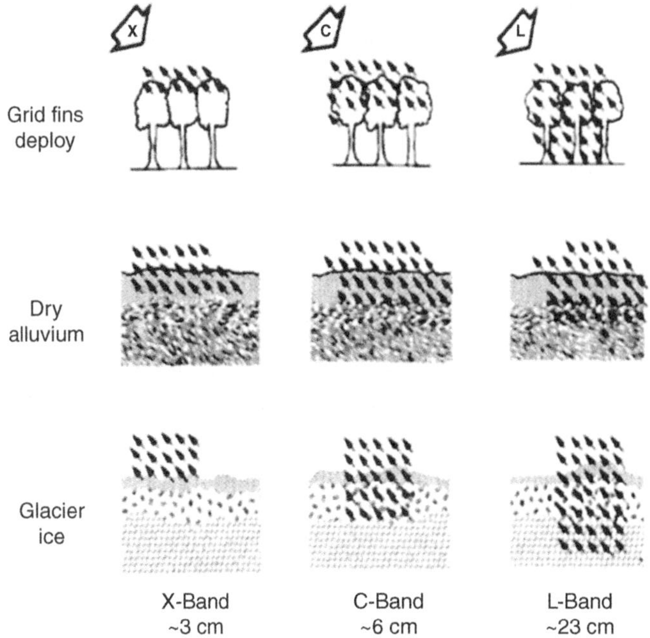

Figure 7.25 Penetration depths in different media of SAR frequencies from low (L-band) to high (X-band). *Source:* Ref. [22] / NASA / Public Domain.

such as the C-band, are used in applications where penetration is needed, such as in the leaves and branches in a forest, as depicted in Table 7.5.

Wavelength is also a factor when it comes to penetrating other land cover types such as soil and ice. This capability of subsurface penetration has enabled archeologists to use long-wavelength SAR to look for ancient sites that have been hidden by soil, sand, or vegetation [24].

Table 7.5 lists the frequency bands and the corresponding SAR applications.

7.6.2 SAR Satellite Missions

The recent trend is the constellations of SAR satellites. These constellations range from several to tens of satellites. ICEYE is currently the largest constellation with 18 satellites in orbit. These satellites are usually sent to sun-synchronous orbits. SAR satellites weigh from the 100 kg range to over 1,000 kg.

We will use RADARSAT-1 in Figure 7.26 to illustrate the SAR part of the satellite. The RADARSAT-1 of the Canadian Space Agency weighs 2,750 kg and operates in a sun synchronous orbit at an altitude of about 800 km and an inclination of 98.6°. Launched in 1995, RADARSAT-1 was decommissioned in 2013 after operating for

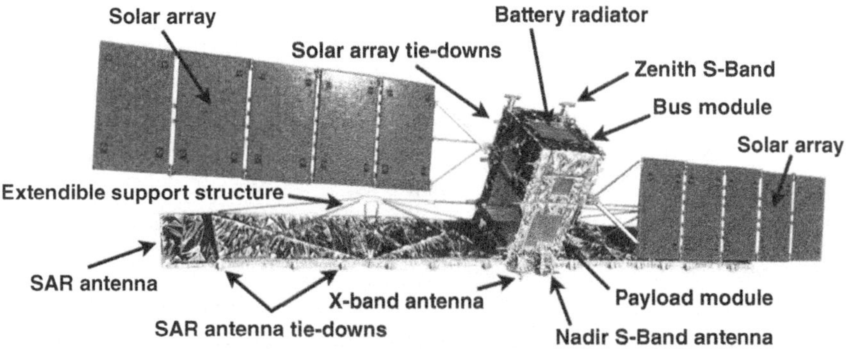

Figure 7.26 Functional diagram of Canada's RADARSAT-1. The SAR antenna is the central feature and is facing the Earth. There are different antennas for communications and telemetry. See Ref. [25] for color version of this figure. *Source:* Ref. [25] / Canadian Space Agency.

Table 7.5 Typical applications of several common frequency bands in SAR imaging. Note that high-frequency bands such as the X-band have high spatial resolution. A low RF frequency offers more penetration.

Band	Frequency (GHz)	Typical Application
X	8–12	High-resolution SAR (urban monitoring, ice and snow, little penetration into vegetation cover; fast coherence decay in vegetated areas)
C	4–8	SAR workhorse (global mapping, change detection, monitoring of areas with low to moderate penetration, higher coherence); ice, ocean maritime navigation
S	2–4	Agriculture monitoring
L	1–2	Medium-resolution SAR (geophysical monitoring, biomass and vegetation mapping, high penetration; InSAR)

Source: Ref. [23] / NASA / Public Domain.

17 years – 12 years beyond its 5-year design life. RADARSAT-2 was launched in 2007 and is still operational.

Figure 7.25 shows a labeled diagram of RADARSAT-1. The SAR antenna is at the nadir side facing the Earth. The SAR antenna is 15×1.5 m in size and operates at 5.3 GHz (C-band). The X-band and S-band antennas in the figure are for communications and telemetry with ground commands.

Table 7.6 provides a list of SAR satellite missions and operators. Table 7.7 lists more details of four SAR satellites listed in Table 7.6. The image of the Synspective satellite is shown in Figure 7.27.

Table 7.6 A list of sample SAR missions: their country, RF band, and spatial resolution.

Company Name/Satellite Mission	Country	Band	Best Resolution
AirBus (TerraSAR)	Germany	X-Band	72 cm
e-Geos (CosmoSkyMed)	Italy	X-Band	1 m
Capella Space	USA	X-Band	50 cm
MDA (RCM)	Canada	C-Band	3 m
ICEYE	Finland	X-Band	50 cm
Umbra Space	USA	X-Band	16 cm
Synspective	Japan	X-Band	1 m
iQPS	Japan	X-Band	75 cm
Spacety	China	C-Band	3 m
China Siwei Survey and Mapping Technology Co. Ltd	China	L-Band	50 cm

Source: Adapted from Ref. [26].

Table 7.7 More details for four SAR satellite constellations in Table 7.5. Their orbital altitude, weight, and number of satellites per constellation are shown.

	ICEYE	Synspective	China Siwei	e-Geos
Orbit	Sun-synchronous orbit (97.7°)	Sun-synchronous orbit (97°)	Sun-synchronous orbit (97°)	Sun-synchronous orbit (97.9°)
Altitude (km)	560–580 km	560	530	619
Proposed number of satellites	40+	30	8	5 (so far)
Weight of the satellite (kg)	85	100	560	1,700

7.6.3 Scanning Modes

As an SAR satellite orbits Earth, it illuminates targets on either side of the flight path. Perpendicular to the flightpath on the ground is called the range. The SAR sends a fan beam that scans along the range direction on the ground. The look angle is the angle between the nadir and the beam. A typical look angle is between 20° and 60°.

Figure 7.27 Artist impression of an SAR satellite by Synspective of Japan. See Ref. [27] for color version of this figure. *Source:* Ref. [27] / Synspective Inc.

The transmitted power is packed within the swath of the fan beam. A narrow fan beam results in better spatial resolution as more RF power is available to paint the target, that is, the beam intensity is higher than that of a wide beam. Hence, a large swath yields poorer spatial resolution.

There are multiple modes in which a SAR beam can operate and can be switched on or changed electronically by command. The RADARSAT-2, for example, can operate in different modes. The largest swath area has the lowest spatial resolution (100 m), while the "very high resolution" mode only illuminates a small area on the ground and has the best resolution (3 m) [28].

The SAR satellite images of the two cities are shown in Figures 7.28 and 7.29 to demonstrate the level of granularity.

7.6.4 Interference SAR (InSAR)

Interference occurs between two superimposed waves of identical wavelengths but differ by some phase angle. In optics, such interference manifests itself as fringes or concentric rings. The relative phase angle between the two waves is between 0° and 360°, i.e. within one wavelength.

7.6.4.1 InSAR Mapping

SAR data are echoes collected from backscattered signals off the objects in the scanned area. The echoes come from different distances of travel from objects to the satellite.

Figure 7.28 A SAR image of the Port of Alexandria, Egypt, taken by ICEYE. See Ref. [29] for color version of this figure. *Source:* Ref. [29] / ICEYE.

Figure 7.29 A TerraSAR-X image acquired in 2021 showing Washington, DC, including the Washington Monument, the White House, and the Lincoln Memoria. *Source:* Ref. [30] / Airbus U.S. Space & Defense, Inc.

Short for Interferometric SAR, InSAR is a powerful tool. When two sets of SAR data of the same area but collected at different times are compared, the differences in the distances of the signal travel can be considered as the differences in the phase angles between the two superimposed waves. Such differences manifest as fringes expressed in false colors. Note, however, that in InSAR, we are comparing data collected at two different times using algorithms but not actual waves. Thus, InSAR is mathematical manipulation and not a physical one.

When more sets of SAR data of the same target area taken at different times are compared, the time variation of "surface displacements" can be obtained. If the wavelength from a SAR system is several centimeters, an interference pattern can detect displacement changes in the same order. If shorter wavelengths are used, smaller displacements can be measured.

InSAR is applied to examine changes in the landscape due to ground movements or displacements. These movements may occur abruptly or over a long period. Abrupt changes can be due to tectonic events such as earthquakes and volcanoes. Examples of slow changes include land subsidence because of oil and gas extraction, excessive underground water pumping, ice and glacier movement, and major underground works such as tunnels.

7.6.4.2 InSAR images

The InSAR image of the volcanic action of the Kilauea Volcano, Hawaii, is shown in Figure 7.30. The Kilauea Volcano on the Island of Hawaii is active. The ground around it also goes through various degrees of deformation. InSAR is the perfect tool to record the time variation of the ground movement. The interference pattern shown in Figure 7.30 is data from April 11, 2015, to May 22, 2015. The pattern represents about 10 cm of ground deformation, and the rings are fringes representing a change in the distance from the satellite to the ground as recorded by an InSAR satellite at different times. The fringes represent land motion (uplift or subsidence). Each circle is equal to half of the SAR's wavelength, that is, in the centimeter range.

7.7 Looking Ahead

Earth observations are the second-largest segment of satellites in LEO by number. Government agencies are sending EO satellites in the 1,000 kg range, such as the Landsat series. Smaller satellites and nanosatellites with narrower objectives by commercial enterprises are gaining popularity.

The popularization of nanosatellites such as CubeSats in the past two decades has led to the launch of over 2,000 nanosatellites, most of which are by non-government entities, according to the Nanosats Database [32]. Most of these

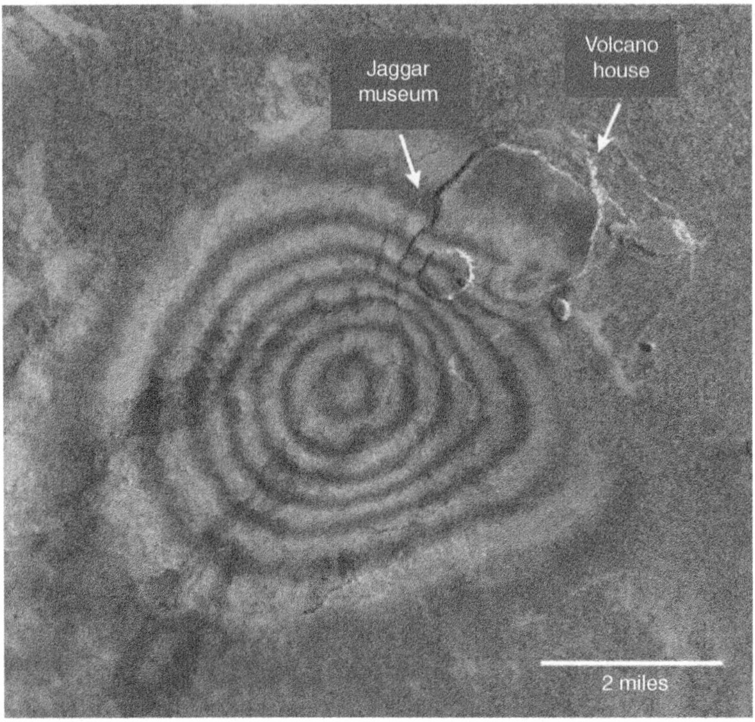

Figure 7.30 An InSAR image of the volcanic activity of the Kilauea Volcano in Hawaii. SAR data over one month in 2015 are compared and computed. See Ref. [31] for color version of this figure. *Source:* Ref. [31] / United States Geological Survey / Public domain. The fringes represent deformation the size of several centimeters – about 10-cm ground deformation. The SAR wavelength is about 3 cm.

satellites are designed to monitor phenomena on Earth's surface and atmosphere. Currently, nanosatellites are often used as "pathfinders" to test out new instruments before implementing them on larger platforms. Constellations of nanosatellites are also gaining popularity due to their relatively low cost of development and launch.

7.7.1 AI in EO Satellites

There are more optical EO satellites in orbit than SARs. Optical sensors are limited by sunlight availability and local weather, such as clouds and rain. A good portion of the raw images turns out to be compromised by the weather conditions. The application of AI to sort out usable images is increasing.

Figure 7.31 Image of the PhiSat-1 nanosatellite by the Universitat Politècnica de Catalunya of Spain in 2020. On board is a commercial microprocessor trained to sort out images within the satellite before beaming back to Earth. See Ref. [33] for color version of this figure. *Source:* Ref. [33] / NanoSat Lab / Polytechnic University of Catalonia (UPC).

In 2020, the nanosatellite PhiSat-1 was launched to a sun-synchronous orbit of 530 km altitude. It was equipped with an AI-enabled commercial processor. PhiSat-1 was originally proposed by the Universitat Politècnica de Catalunya of Spain and funded by the ESA, as shown in Figure 7.31. Prior to launch, the algorithms on board PhiSat-1 had been "trained" with past EO data. The processor sorted out unobstructed images to beam back to Earth. As of December 2023, PhiSat-1 is still operational.

With AI embedded in EO satellites, productivity from imaging by the satellite to workflow on the ground will improve accordingly. A clear benefit is conserving RF power in beaming only prescreened data sets back to Earth.

In 2024, ESA launched PhiSat-2, which is equipped with multispectral and AI processing. It is a 6U satellite launched to the sun-synchronous orbit capable of real-time AI processing of imaging data [34].

AI also improves productivity with storage and computation on the ground. Improvement in end-to-end productivity is critical considering the high growth in data volume. The term "latency" describes the time from the images being taken to the actionable information delivered to decision makers. This latency has been pushed down from days and weeks to hours and minutes, i.e. near real time.

7.8 Conclusion

Earth observation is an important application of satellites. Placing EO satellites in LEO offers various advantages. The number of EO satellites in LEO exceeded

1,100 in 2023. Both government agencies and the private sector are launching EO satellites regularly to service sectors including government, education, commerce, and climate sciences.

EO satellites are data generators producing enormous volumes of information every day. Managing such an ever-increasing volume of data has shifted to the use of advanced hardware and software, such as AI to expedite the workflow.

EO images, when integrated with information such as global positioning and geospatial systems, have high commercial value, forming a new digital platform that businesses can use innovatively. Such a platform aims to transcend proprietary systems that business users can just pay to use the platform. In other words, data from EO satellites is migrating toward some form of convergence with other satellite applications and simplifying the human-machine interface for end users.

The field of earth observations is expanding and is made possible by the low cost of access to space, small satellites, and advances in sensors and instruments. The more one looks, the closer one wants to look and learn more.

Reading List

Alcaraz-Segura, D., Bella, C.M.D., and Straschnoy, J.V. (2013). *Earth Observation of Ecosystem Services (Earth Observation of Global Changes)*, 1ste. CRC Press.

Chen, K.S. (2020). *Principles of Synthetic Aperture Radar Imaging (Signal and Image Processing of Earth Observations)*, 1ste. CRC Press.

Committee on NASA-NOAA Transition from Research to Operations, National Research Council (2013). *Satellite Observations of the Earth's Environment: Accelerating the Transition of Research to Operations*. National Academy of Sciences.

Lamine, S., Srivastava, P.K., and Kayad, A. (2023). *Remote Sensing in Precision Agriculture: Transforming Scientific Advancement into Innovation (Earth Observation)*, 1ste. Academic Press.

Lein, J.K. (2011). *Environmental Sensing: Analytical Techniques for Earth Observation*, 2012e. Springer.

Pandey, P.C., Srivastava, P.K., and Balzter, H. (2020). *Hyperspectral Remote Sensing: Theory and Applications (Earth Observation)*, 1ste. Elsevier.

Rani, M., Seenipandi, K., and Rehman, S. (2020). *Remote Sensing of Ocean and Coastal Environments (Earth Observation)*, 1ste. Elsevier.

Singh, A.K. and Tiwari, S. (2022). *Atmospheric Remote Sensing: Principles and Applications (Earth Observation)*, 1ste. Elsevier.

References

1 NASA. Stylized Graphical Representation of Which Electromagnetic Wavelength Regions are Blocked by Earth's Atmosphere. Wikimedia [cited 2025 Feb 21] [online]. Available from: https://commons.wikimedia.org/wiki/File:Atmospheric_electromagnetic_transmittance_or_opacity.jpg.

2 Remote Sensing. NASA Earthdata [cited 2025 Feb 21] [online]. Available from: https://www.earthdata.nasa.gov/learn/earth-observation-data-basics/remote-sensing.

3 Hyperspectral Imaging from Space. GISGeography [cited 2025 Feb 21] [online]. Available from: https://gisgeography.com/hyperspectral-imaging/.

4 Sentinel-2 Bands. Sentinel Hub Custom Scripts [cited 2025 Feb 21] [online]. Available from: https://custom-scripts.sentinel-hub.com/custom-scripts/sentinel-2/bands/.

5 Landsat Next. NASA. Cited by Landsat Missions. Landsat Next. U.S. Geological Survey [cited 2025 Feb 21] [online]. Available from: https://www.usgs.gov/landsat-missions/landsat-next.

6 Landsat Next. NASA. Cited by Communications and Publishing. Spectral Comparison of Landsat 8–9 and Landsat Next. U.S. Geological Survey [cited 2025 Feb 21].[online]. Available from: https://www.usgs.gov/media/images/spectral-comparison-landsat-8-9-and-landsat-next.

7 Bilki F. Satellite Imagery in Mineral Exploration: Part 1. Micromine Global [cited 2025 Feb 21] [online]. Available from: https://global.micromine.com/satellite-imagery-in-mineral-exploration-part-1/.

8 Landsat 8 Imagery Available for Online Users. ArcNews [cited 2025 Feb 21] [online]. Available from: https://www.esri.com/about/newsroom/arcnews/landsat-8-imagery-available-for-online-users/.

9 Hyperspectral Imagery. Pixxel [cited 2025 Feb 21] [online]. Available from: https://www.pixxel.space/hyperspectral-imagery.

10 The World's Highest Resolution Hyperspectral Imagery. Pixxel [cited 2025 Feb 21] [online]. Available from: https://www.pixxel.space.

11 Home Page. Agricultural Market Information System (AMIS) [cited 2025 Feb 21] [online]. Available from: https://www.amis-outlook.org/home.

12 Global Crop Monitor. GEOGLAM Crop Monitor [cited 2025 Feb 21] [online]. Available from: https://www.cropmonitor.org/global-crop-monitor-202406.

13 Cherlinka V. NDVI FAQ: All You Need To Know About Index. EOS Data Analytics [cited 2025 Feb 21] [online]. Available from: https://eos.com/blog/ndvi-faq-all-you-need-to-know-about-ndvi/.

14 Satellite Crop Monitoring Software for Farm Advisors. EOS Data Analytics [cited 2025 Feb 21] [online]. Available from: https://eos.com/products/crop-monitoring/agriculture-consulting/.

15 Harris M. These Exclusive Satellite Images Show that Saudi Arabia's Scifi Megacity is Well Underway. MIT Technology Review [cited 2025 Feb 21] [online]. Available from: https://www.technologyreview.com/2022/12/09/1064544/satellite-images-line-megacity-google/.

16 Smoke and Fire in Southern California. NASA; [cited 2025 Feb 21] [online]. Available from: https://earthobservatory.nasa.gov/images/91379/smoke-and-fire-in-southern-california. Cited by Clifford C. NASA shared pictures of the California wildfires from space — and the images are unreal. Consumer News and Business Channel [cited 2025 Feb 21] [online]. Available from: https://www.cnbc.com/2017/12/08/nasas-pictures-of-the-california-wildfires-from-space.html.

17 Joshua Stevens. Cited by Brito C. Before & After Satellite Photos of Hurricane Irma's Destruction. CBS News [cited 2025 Feb 21] [online]. Available from: https://www.cbsnews.com/pictures/before-and-after-satellite-photos-hurricane-irma/4/.

18 ChiefHira. Shishiori-Karakuwa Station after Tsunami. Wikimedia [cited 2025 Feb 21] [online]. Available from: https://en.m.wikipedia.org/wiki/File:Shishiori-Karakuwa_Station_after_Tsunami.JPG.

19 Exploration Mapping Group, Inc. Cited by Baumann J. Mineral Exploration from Space. ArcNews [cited 2025 Feb 21] [online]. Available from: https://www.esri.com/about/newsroom/arcwatch/mineral-exploration-in-the-hyperspectral-zone/.

20 Aquarius/SMAP Optimum Interpolation Maps: Sea Surface Salinity. NASA Salinity [cited 2025 Feb 21] [online]. Available from: https://salinity.oceansciences.org/oi-salinity.htm#123.

21 Daily Global 5km Satellite Sea Surface Temperature (a.k.a. CoralTemp). Coral Reef Watch [cited 2025 Feb 21] [online]. Available from: https://coralreefwatch.noaa.gov/product/5km/index_5km_sst.php.

22 Flores-Anderson A, Herndon K, Thapa R, Cherrington E. The SAR Handbook: Comprehensive Methodologies for Forest Monitoring and Biomass Estimation; 2019.

23 Synthetic Aperture Radar (SAR). Earth Data [cited 2025 Feb 21] [online]. Available from: https://www.earthdata.nasa.gov/learn/earth-observation-data-basics/sar.

24 Secrets Beneath the Sand. Earth Observatory [cited 2025 Feb 21] [online]. Available from: https://earthobservatory.nasa.gov/images/90847/secrets-beneath-the-sand.

25 Components and Specifications. Canadian Space Agency [cited 2025 Feb 21] [online]. Available from: https://www.asc-csa.gc.ca/eng/satellites/radarsat1/components.asp.

26 SAR Satellite Companies. Synthetic Aperture Radar (SAR) [cited 2025 Feb 21] [online]. Available from: https://syntheticapertureradar.com/sar-satellite-missions/.

27 Satellite. Synspective [cited 2025 Feb 21] [online]. Available from: https://synspective.com/satellite/.

28 The RADARSAT Constellation. Canadian Space Agency [cited 2025 Feb 21] [online]. Available from: https://www.asc-csa.gc.ca/eng/multimedia/search/image/4251#longdesc.

29 ICEYE. Earth Online. European Space Agency [cited 2025 Feb 21] [online]. Available from: https://earth.esa.int/eogateway/missions/iceye.

30 Airbus Defense and Space GEO Inc. U.S. 2021. Cited by Airbus U.S. Synthetic Aperture Radar (SAR) Data Evaluation Completed. Earth Data [cited 2025 Feb 21] [online]. Available from: https://www.earthdata.nasa.gov/news/airbus-u.s.-synthetic-aperture-radar-sar-data-evaluation-completed.

31 Hawaiian Volcano Observatory. Volcano Watch — Rainbows on the Ground: How Satellite Radar Helps Us Monitor Volcano Deformation. U.S. Geological Survey [cited 2025 Feb 21] [online]. Available from: https://www.usgs.gov/observatories/hvo/news/volcano-watch-rainbows-ground-how-satellite-radar-helps-us-monitor-volcano.

32 World's Largest Database of Nanosatellites, Over 4400 Nanosats and CubeSats. Nanosats Database [cited 2025 Feb 21] [online]. Available from: https://www.nanosats.eu/#figures.

33 FSSCat. NanoSat Lab. Universitat Politècnica de Catalunya [cited 2025 Feb 21] [online]. Available from: https://nanosatlab.upc.edu/en/missions-and-projects/fsscat. Cited by PhiSat-1 & -2 Nanosatellite Mission. eoPortal [cited 2025 Feb 21] [online]. Available from: https://www.eoportal.org/satellite-missions/phisat-1#%CF%86-sat-1-mission.

34 New Satellite Demonstrates the Power of AI for Earth Observation. European Space Agency [cited 2025 Feb 21] [online]. Available from: https://www.esa.int/Applications/Observing_the_Earth/Phsat-2/New_satellite_demonstrates_the_power_of_AI_for_Earth_observation.

8

Space Solar Power

8.1 Introduction

This chapter focuses on space solar power. It is the interception of solar energy in space. The working principles and relevant technologies will be discussed.

Two important numbers are stated upfront:

1) The solar power density in the space around Earth is 1.37 kW/m^2. This means that an area of 1 square km intercepts 1.37 GW of solar power. This value is based on the Earths distance from the Sun and is constant
2) 1 GW power plant in the power grid can power over 800,000 typical American households

Space solar power (SSP) aims to capitalize on the power density given in 1) modern civilization cannot function without electric power. Society's economic well-being depends on the supply of affordable energy to sustain industrialization and our personal standard of living. Projecting forward to 2050 and beyond, global demand for energy will continue to grow at a high rate due to population increase, urbanization, and the industrialization of more countries.

Over 80% of global energy consumption comes from fossil fuels, which include oil, natural gas, and coal. The rest comes from nuclear and renewable sources, such as solar, wind, hydrogen fuel cells, hydro, and others. Many countries are investing in large solar farms and renewable energy platforms, some more aggressively than others.

Renewable energy is motivated by 1) the depletion of fossil fuels or at least getting very difficult to extract and 2) international push to reduce greenhouse gas emission.

There are two practical issues to consider in renewable energy:

1) Capacity factor: This factor is the percentage of time, in which an energy generator is available on a day. Terrestrial solar, for example, varies with time of day

New Space: From Low Earth Orbit to the Moon and Beyond, First Edition.
Edited by Stephen W. Cheung and Chih-Yung Wen.

and is unavailable at night. Say solar energy of a particular region is available only for 8 hours a day, then the capacity factor is 8/24 or 33%.

2) Infrastructure buildup: Mature economies have built up their energy delivery infrastructures mainly around fossil fuels. Such infrastructure includes transportation, material processing, pipelines, power plants, and retail gasoline stations. In the case of electricity generation, power grids are the norm and have been extensively built in industrialized nations. Currently, renewable energies mostly serve as supplements to power grids. As the market share of renewable energy grows, new infrastructure will be added. The price per unit of power consumption will be affected accordingly. These issues will all take time and extensive government investment to resolve them.

   Space solar power has a 100% capacity factor except on rare occasions. Governments must play a critical role in full-scale-deployment of space solar to replace fossil fuels.

8.1.1 Early Space Solar Power Proposals

Space solar power generated by an orbiting satellite was first proposed by Peter Glaser in the 1970s. Studies at the time were mostly about placing mid-sized satellites and directing power beams to other orbiting satellites and space stations as well as underserved areas on Earth.

Early designs entail two key points:

a) Utilization of high-power microwave tubes used in wireless power transmission
b) A team of astronauts working in high orbit to assemble a large structure

Today, microwave tubes in point a) would be too heavy due to the weight of these tubes and the high voltage equipment to power them, not to mention these tubes' conversion efficiency was not high. Solid-state devices would do much better today. Point b) calls for extravehicular activities (EVA) that still have not been done in orbits beyond the LEO. Assembling all modules in high orbit is a recognized challenge but is worth undertaking. Recent thinking is to use autonomous and semiautonomous machines to assemble the SPS satellite in high orbit.

8.2 Solar Power

Solar power is the use of photovoltaic (PV) materials to convert light to electricity. The PV device is a semiconductor device. A solar panel is made of PV cells.

The conversion efficiency is a primary performance indicator of a PV cell and panel. It is the percentage of incoming solar energy being converted into electric energy. Currently, the efficiency of commercial PV panels is 15-20%, some reaching as high as 25%. The efficiency is PV material dependent. PV cells in R&D laboratories are reported to be nearly 50%.

Next, we elaborate on end-to-end efficiency as it is the overall performance indicator of the entire SPS system.

8.2.1 Terrestrial Solar Power

For terrestrial solar power, a significant amount of sunlight energy is lost to the atmosphere before it reaches the the PV panels on the ground. Earth's reflection (known as albedo), atmospheric absorption, day/night cycle, geographic locations, climate, and weather conditions all reduce the solar intensity by large factors. The solar intensity on the ground can be reduced to less than 10% of that in space. The overall reduction in usable sunlight to convert to solar power terrestrially is significant.

8.2.2 Space Solar Power

In essence, space solar power works by going through a series of power conversions from sunlight collected in space to electric power at the grid on the ground. Most proposals aim to net ~1 GW at the grid. Using 1.37 kW/m^2, one square kilometer of solar collector would lead to 1.37 GW worth of solar power collected in space. Inefficiencies in multiple power conversion stages bring down the power consumption at the grid substantially. Hence, the solar collector must be four to five times larger in collection area to result in 1 GW at the grid.

Figure 8.1 shows the block diagram outlining the stage-by-stage power conversion of an SSP system. The first segment occurs in space and the second segment

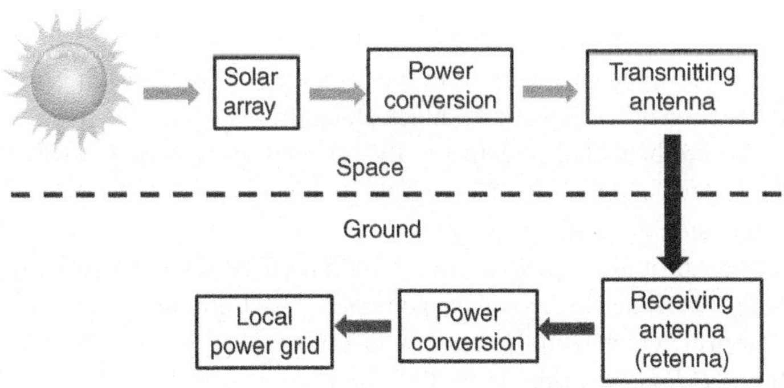

Figure 8.1 Flow chart of a space solar power system. The solar arrays convert solar power into electric power and then microwave power to be transmitted to the ground. The received power on the ground is converted back into electric power.

on the ground. The power is transmitted wirelessly to a receiving antenna on the ground. The SSP satellite is placed in a geostationary orbit to collect sunlight continuously.

For elaboration, we assume microwaves to be the wireless medium, and the goal is 1 GW at the grid. Solar power is collected by PV panel arrays on a big satellite. The solar panel area is several square kilometers. The sequence of conversion starts with solar power to direct currents from the PV cells and then to microwaves. The microwave power is beamed back to the ground via a circular transmitting antenna on the satellite. The beam of transmitted microwaves broadens to a larger circular area on the ground. The receiving antenna elements, collectively known as the Retenna, convert the microwave power back to electric current and is then fed to the grid. There are subsystems along the blocks in Figure 8.1 to condition the output of each conversion block.

The key advantage of SSP is its 24×7-hour acquisition in GEO. Terrestrial solar is hampered by different capacity factors. But they both share the same efficiency challenge in PV cells. Therefore, improvement of PV efficiency and commercialization of such new materials will benefit both space and terrestrial solar energy.

An SSP satellite has enormous size and weight and costs tens of billions of dollars to realize. Launching such a heavy satellite into geostationary orbit takes up a large portion of the project's budget. Only about 10% of the total project cost is allocated to building the receiving segment on the ground.

None of the blocks in Figure 8.1 looks difficult as the basic technologies are all there. The weight, size, and power level of the satellite as well as the orbital altitude make the whole idea super expensive and challenging.

The 24×7-hour advantage eliminates the capacity factor due to diurnal cycles and other factors. To the community SSP is charged to service on the ground, the satellite looks like a stationary power plant that continuously feeds power wirelessly.

End-to-end efficiency is the overall efficiency of the SSP system. It measures from the front end (solar power collection) to the back end (power grid). It is the product of the conversion efficiency of each block along the way. The unconverted power mostly turns into heat. This means the PV cells and panels are hot when illuminated by high intensity sunlight. The thermal management of PV cells is an important design consideration of space solar power.

Most published SSP proposals quote their end-to-end efficiency to be between 15% and 20%. Using 20%, it means we need to generate 5 GW from the solar panels to have 1 GW at the grid, which would need about 4 km^2 of solar panels on the SSP satellite to produce 1 GW at the grid.

We will use one published proposal to illustrate the conversion efficiencies and other key parameters in Section 8.3. Below we will first look at several key subsystems of an SSP system. The goal is assumed to have 1 GW at the grid.

8.2.3 Space Solar Power Subsystems

An SSP system is made of a number of key subsystems working together in Figure 8.1. Most of the subsystems are on board the SSP satellite. The subsystems on the ground are relatively straightforward.

8.2.3.1 Photovoltaic Panels

Published proposals have always assumed the use of PV cells to convert sunlight to electric power. The most popular PV cell material is silicon. The push on PV cell research has been in several directions:

a) Improving conversion efficiency with advanced PV materials
b) Thermal management
c) Performance at high temperatures
d) Performance under radiation

As terrestrial solar gains market share, commercial PV cells have seen their price drop and efficiency increase. Advanced PV materials such as gallium arsenide reportedly offer 40% efficiency, but silicon is still widely used in the market (more than 90% of its market share).

With the PV conversion efficiency of the 20% range now and toward 40+% in research labs, 1 square meter of a PV panel can get hot owing to several hundred watts of continuous heat generation from unconverted solar energy. For 4 km^2 of PV cells collecting 5 GW of sunlight, More than 50% of that power would turn into heat borne by the PV panels.

8.2.3.2 Microwave Power Transmitter

Tens of thousands of solid-state amplifiers now replace the microwave tubes, reducing the weight and equipment complexity in the original SSP design. Current proposals use arrays of solid-state microwave amplifiers, each of which can output tens to hundreds of watts of microwave power.

The transmitting antennas in the current proposals are phase array antennas in which the phase of each solid-state amplifier can be controlled electronically. Such phase variations effectively create an electronically steerable beam of microwave power that points to one or multiple locations on the ground. This function allows for a larger footprint on the ground or servicing multiple communities on demand.

Two frequencies have been chosen for microwave power transmission. They are 2.45 and 5.80 GHz, with 2.45 GHz appearing more frequently in publications. These two frequencies are the least absorbed by the atmosphere and are outside the communication bands. The size of the power transmitter on the satellite using 2.45 GHz is typically about 1 km in diameter; it is smaller for 5.80 GHz.

The rule of thumb for electromagnetic wave transmission is the higher the frequency, the smaller the microwave devices and the less power each device can handle. But the transmitted beam is not as spread out at higher frequency.

The large transmitter area can serve several purposes:

1) A wide transmitter produces a wide beam. The purpose is for the peak power density within the beam to be sufficiently below the safety limit along the transmission path and on the ground. Simplistically, the power density follows a Gaussian distribution curve, with the center being the maximum.
2) Accommodate thousands of solid-state amplifiers and their respective transmitting antennas.

8.2.3.3 Receiving Antenna (Retenna)

A microwave beam slowly broadens to large cross-sections after it leaves the transmitting antenna. The spread is frequency dependent. Hence, a 1-km diameter 2.45-GHz beam leaving the transmitting antenna can spread out to nearly 10 km in diameter on the ground nominally 36,000 km away. A 5.8-GHz beam would spread out less. For an SSP satellite directly over the Equator, a circular beam out of the transmitter creates a larger circular footprint on the ground directly below. This footprint would turn elliptical if projected to locations of nonequatorial latitudes. The higher the latitude, the more elliptical the footprint becomes.

On the ground, the receiving antenna is a collection of dipole antennas spread out inside a large circle (or ellipse depending on latitude). Often times, the antenna dipoles are slightly elevated, so the pasture can double as farmland.

Safety considerations refer to the microwave power intensity that would not cause harm to humans, animals, and plants on the ground as well as aviation. The safety standard set for microwave power is 5 mW/cm^2 or 50 W/m^2.

If a 1-GW microwave beam is evenly spread over a circle of 10 km diameter without loss along the way, the intensity would reach about 10 W/m^2, which is below the 50 W/m^2 limit. Using Gaussian distribution with standard deviation σ, 68% of the 1-GW microwave power would be concentrated within a circle of diameter 2σ, while circles of diameters 4σ and 6σ would encompass 95% and 99.7%, respectively. If we take the diameter of the retenna, viz. 10 km, to be 6σ, then σ is 1.7 km, which means that the retenna area contains 99.7% of the microwave power. The microwave power in the inner circle of a 1.7 km radius would have 0.68 GW, which works out to be 35 W/m^2 and is within the safety limit.

8.3 Multi-Rotary Joints System

The Multi-Rotary Joints (MRJ) system is proposed by a team from the China Academy of Space Technology in 2015. The publication tabulates the numerical

Figure 8.2 Artist depiction of China's multi-rotary joints (MRJ) proposed SSP system. The solar panels are mounted along a frame. The disk at the center is a microwave transmitter facing Earth. *Source:* [2] / Ohio University.

values of the proposed SSP system that would serve our purpose. Interested readers are also referred to online videos on this proposal and other designs (e.g. [1]).

Figures 8.2 and 8.3 show the concept of the MRJ system and the dimensions of the proposed satellite. Systems information is contained in Tables 8.1 and 8.2.

The center structure is a 1-km-diameter microwave transmitting antenna facing the Earth. On each side stands a structure holding up 25 subarrays of solar panels. The satellite is 11.8 km long end to end. The area of each PV subarray is $600 \times 200\,\text{m} = 0.12\,\text{km}^2$. The total solar panel area is therefore $0.12 \times 50 = 6\,\text{km}^2$.

The total solar power collected is $1.37\,\text{kW/m}^2 \times 6\,\text{km}^2 = 8.2\,\text{GW}$.

The end-to-end system efficiency from solar panel to grid is 13.8%, as shown at the bottom of Table 8.1. This means the power at the grid is $8.22\,\text{GW} \times 0.138 = 1.13\,\text{GW}$.

The end-to-end efficiency of 13.8% is based on a PV cell efficiency of 40% (Table 8.1). The proposal assumes the use of advanced PV cell materials. Including other efficiencies such as "gap of solar cells" and "angle of sunlight," the

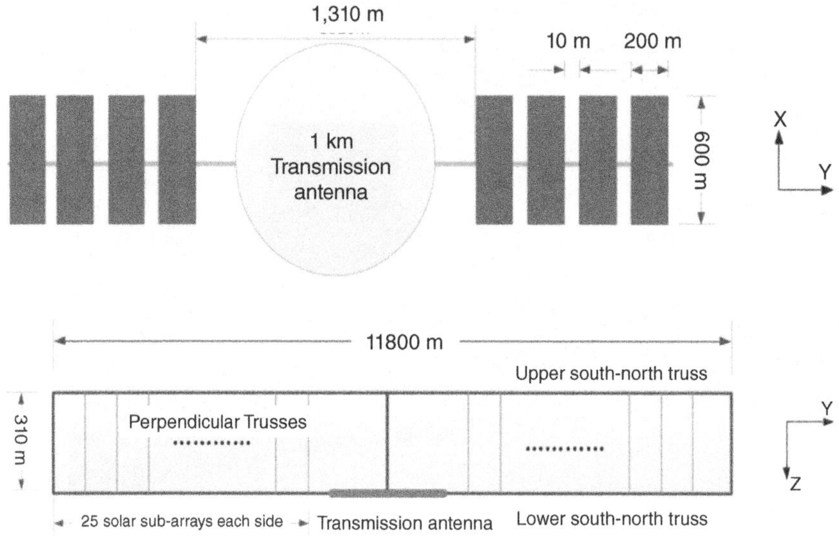

Figure 8.3 Dimensions of the multi-rotary joints solar power satellite. The end-to-end length is 11.8 km. Each of the solar panels is 600 × 200 m in area. The transmission antenna is 1 km in diameter. Multi-rotary joints SPS. *Source:* Ref. [2] / Ohio University Press.

overall efficiency of the PV block is reduced to 29%. We will return to these numbers in the thermal management discussion.

The following selected values listed in Tables 8.1 and 8.2 are worth exploring further.

1) The total mass of the proposed satellite is 10,000 metric tons (or 10 million kg)
   a) Solar panel-related mass = 2,000 tons (20% of total)
   b) Solar power management = 2,500 tons (25% of total)
   c) Microwave power management and transmission = 4,000 tons (40% of total)
   d) Supporting structure = 1,200 tons (12% of total)
   e) Other items = 200 tons (2% of total)
2) Many SSP satellite designs have a total weight from several thousand metric tons to 20,000 metric tons. The obvious impact is the total launch cost. Hundreds of trips would be needed if delivered by the SpaceX Starship and even more if by Falcon Heavy to high orbit. The payload capacity on the Starship to GEO depends on on-orbit refueling.
3) The proposal assumes a launch cost of US$1,000 per kg utilizing SpaceX reusable rockets. The total launch cost for 10 million kg would be $10 billion. The cost per kg to GEO is a volatile variable in the cost equation.

Table 8.1 Conversion efficiency values of the function blocks of the MRJ system.

Factors	Efficiency	System efficiency
Solar energy collection and conversion (0.29)		
Solar cell	0.40	0.4
Error of Sun-pointing	0.99	0.396
Gap of solar cells	0.85	0.336
Angle of sunlight	0.958	0.322
Space environment effect	0.9	0.290
Power transmission and management (0.854)		
Voltage conversion in solar array	0.95	0.276
Transmission	0.95	0.262
Voltage conversion in antenna	0.95	0.249
Consumed by service devices	0.999	0.248
Microwave power conversion and emitting (0.833)		
Microwave generator	0.85	0.211
Microwave regulation	0.98	0.207
Microwave power transmission		
Microwave transmission	0.90	0.186
Microwave power receiving and conversion (0.765)		
Receiving antenna	0.9	0.168
Rectifier circuits	0.85	0.143
Electric power regulation (0.97)		
Electric power collection	0.98	0.140
Voltage conversion	0.99	0.138

Source: Ref. [2] / Ohio University Press.

4) The solar subarrays occupy most of the satellite's size and, together with power management subsystem, take up 45% of the total weight. The microwave portion including the transmitting antenna occupies another 40% of the total weight.
5) The supporting structure of the whole satellite occupies 12% of the total weight. Structure materials should be light and strong. Under microgravity conditions in GEO, the subsystems that carry large heat loads, such as the PV panels, expand and flex due to thermal effects.

Table 8.2 Specifications of key subsystems of the MRJ system [2].

SPS system	Orbit	GEO
	Delivered power	~1 GW
	Life time	30 years
	Efficiency	~13%
	Total mass	~10,000 t
Solar energy collection and conversion	Solar cell	Thin-film GaAs
	Efficiency	~40%
	Area of solar array	~6 km2
	Output power	~2.4GW
	Voltage of solar array modules	~500 V
	Mass	~2,000 t
Microwave power transmission	Frequency of microwave	5.8 GHz
	Efficiency	~5.4%
	Diameter of transmitting antenna	1,000 m
	Number of antenna modules	128,000
	Transmitting power of an antenna module	12.5 kw
	Mass	4,000 t
	Diameter of receiving antenna	5 km
Power transmission and management	Style	Mix of distributed and centralized
	Voltage of main cable	20 kV
	Voltage of solar sub-arrays	5,000 V
	Number of rotary joints	100
	Mass	2,500 t
Structure	Module	Deployed truss
	Mass	1,200 t
Attitude and orbit control	Thrusters	1 N electric thruster
	Mass	100 t
Others	Mass of thermal management	150 t
	Mass of ISRM	50 t
Operation mode		Continuous transmission

Source: Ref. [2] / Ohio University Press.

6) The system utilizes rotary joints to connect the trusses to form the satellite bus. Different proposals offer unique designs.
7) Solar power management: The electricity from the PV panels needs to be conditioned by the power management system before being directed to produce microwave power and support all onboard electronics.
8) Transmitting antenna modules: These modules are solid-state microwave amplifiers. The efficiency at the transmitter stage is 90% (under the Microwave Power Transmission segment in Table 8.1). However, the cumulative efficiency at the microwave output (relative to the solar collection front-end) decreased to 18.6%. The microwave power to be transmitted is 18.6% of the 8.22 GW collected power or 1.53 GW. There are 128,000 solid-state modules according to Table 8.2. Therefore, each module produces an average of 12 kW of microwave power. Compared with microwave tubes, solid-state amplifiers are more efficient. Solid-state devices can also be mass produced.
9) The high heat load in the whole SSP satellite system requires the key components and devices to perform continuously at high temperatures. This is a challenging requirement for all devices.
10) The SSP satellite needs to make two regular adjustments. The first is the attitude of the PV collector structure, so the solar panels always face sunlight normally (90°) to have maximum exposure. The second adjustment is due to a slight misalignment of the center of gravity of the satellite as a result of slight imbalance in symmetry. This would cause the satellite's attitude to drift and the microwave beam to move.

8.3.1 Thermal Management

The range of conversion efficiency of PV cells is between 20% and 50% and reportedly higher from research papers. There are other inefficiencies in the solar collection system. This suggests on average 50–80% of the incoming solar power could turn into heat within the PV subsystem. From Table 8.1, one can estimate that the power left in the solar collection block is about 30% of the incident power, i.e. there is $8.22\,\text{GW} \times 0.7 = 5.7\,\text{GW}$ of unconverted power most in the form of heat in the solar collection block. This amount of continuous heat will drive up the operating temperature of the PV cells, which will in turn affect their conversion efficiency. The other subsystems also generate heat, but the PV panels are the most impactful.

Nearly all the conversion efficiencies in the electricity-to-microwave conversion function are in the high 90%. Hence, the amount of heat generated there is not as big a contributor.

The temperature at GEO swings between −150°C and 150°C. Since the SSP is designed to have the solar panels point at the Sun continuously to generate continuous power, the satellite runs very hot constantly.

Thermal management is a big issue in SSP. The space environment does not allow cooling by convection (air circulation). Liquid cooling can be part of the design but would add to the weight calculations. Eventually, the large amount of continuously generated heat must go somewhere. The "somewhere" is one or more heat reservoirs whose role is to collect the heat and then discharge it into space.

According to thermodynamics, a heat reservoir needs to be massive with the material property to absorb and retain heat. Radiation is the only way to emit the stored heat into space. The amount of thermal radiation is proportional to the fourth power of the absolute temperature of the radiating body. Absolute temperature is "273° + temperature in Celsius." This means that the hotter the radiating body, the more heat it will radiate away.

Electronic devices usually operate less efficiently at high temperatures. Hence, devices such as PV cells, solid-state microwave amplifiers, and all the processing equipment need to be designed to perform at high temperatures without sacrificing their efficiency and life expectancy.

Summarizing the above, the heat reservoirs add weight to the SSP system and run very hot. The implication of the conversion efficiency of PV cells is clear. High PV efficiency would not only reduce the total area of the solar panels but also the amount of heat generated. The weight of the solar portion and heat management would then be reduced accordingly. In turn, the weight reduction also leads to a lower launch cost of the satellite.

8.4 Space Solar Power Using Optical Reflectors

Instead of deploying solar panels over a large area to collect solar power directly, another type of SSP system uses optical reflectors as the first stop. The reflectors direct sunlight toward a PV cell bank. The area of the PV cells is presumably reduced although the power intensity of the PV cells increases accordingly. Hence, there is a focusing action by the reflectors. An increase in power density means the PV panels will receive higher solar influx and operate at even higher temperatures within a smaller area.

The reflectors add an extra step to the system in the process of converting solar energy to electricity. The weight of the (reflectors + solar) combination must be favorable compared to the "direct collect" design. Modern designs using optical reflectors also count on many optical modules that can be mass fabricated and assembled. The modules also allow their orientations to be electronically tuned to maximize reflection.

8.4.1 Solar Power Satellite via Arbitrarily Large PHase Array (SPS-ALPHA)

SPS-ALPHA is proposed by J. Mankins under a study funded by NASA in 2011. There are several design versions that work in similar ways. The two versions are shown in Figures 8.4 and 8.5. Each weighs around 8,000 tons.

There are three main points about SPS-ALPHA:

1) Constructed from modules
2) Optical reflectors in modular form
3) A sandwich structure with PV cells on one side receiving reflected sunlight and RF transmitters on the side facing the Earth. The sandwich structure is also assembled in modular form.

The sandwich structure in 3) lets the PV cells to collect more sunlight than their physical size allows. Modern fabrication techniques allow the PV cells and transmitter modules to be in one piece.

The SPS-ALPHA has eight basic families of modular elements. The sheer size of an SPS satellite means that each module needs to be mass produced in large numbers, often in the order of thousands. The eight families include:

a) Optical reflector modules
b) Solar power-generating modules (PV units)

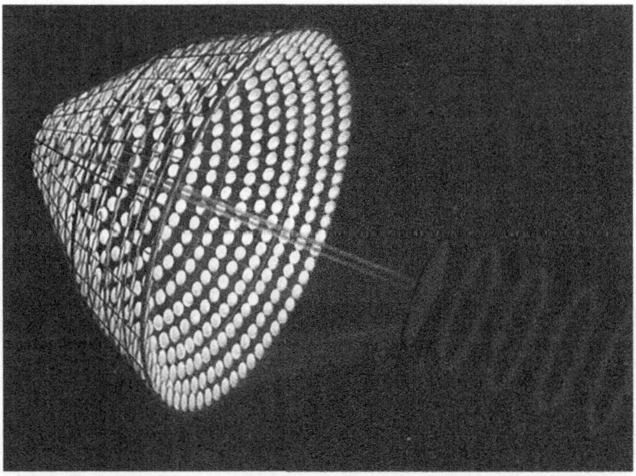

Figure 8.4 Design of a space solar power satellite using optical reflectors to collect sunlight. Optical reflectors are placed in a half-ellipsoid shape in this 2011 version. *Source:* Ref. [3] / American Institute of Aeronautics and Astronautics.

Figure 8.5 Optical reflectors are placed in a sigmoid shape in this 2012 version. *Source:* Ref. [4] / American Institute of Aeronautics and Astronautics.

c) Wireless power transmission modules
d) HexFrame structural modules
e) Robot modules
f) Interconnects
g) Propulsion/attitude control modules

The SPS functionality comes from a), b), and c) above.

An SPS-ALPHA begins with a large, rigid platform serving as a scaffold – described as primary structure assembly (PSA) in Mankin's report. The PSA consists of two segments: a solar reflector assembly (SRA) and a primary array assembly (PAA). The two segments are connected by a truss. Large numbers of different modules will be assembled on the scaffold and interconnected in space.

The SRA is where the optical reflectors are mounted. The shape of this platform takes on several viable geometries, as shown in Figures 8.4 and 8.5. Figure 8.4 is a half-ellipsoid shape (2011 version). Figure 8.5 is a sigmoid shape (2022 version). The difference between the two versions is primarily the integrated shape of the reflectors. Unlike the conventional designs, which deploy PV cells linearly to maximize sunlight collection, the SPS-ALPHA can be thought of a different geometry but with the same collection area.

The PAA consists of primary power/transmitter array (PPTA) modules. They are arrays of solar power-generating (SPG) modules of PV cells on the side facing the reflectors. Wireless power transmission modules are on the side facing the Earth. The power output from each PPTA module is immediately converted into microwaves.

The eight modules are explained as follows.

8.4.1.1 Reflectors Module

Reflectors are mirrors – usually made by depositing thin-filmed coating of aluminum on Kapton to save weight – on a nonrotating structure forming a module. These modules are estimated to be 75-100 kg each, and 4,000-5,000 modules will be needed. Each structure is anchored to the platform by interconnects identified as f) above. The module is referred to as a heliostat, which is a device with a reflective surface that can be moved individually by a control mechanism to always face the light source. The individually adjusted modules ensure that sunlight is always optimally collected and redirected to the solar power-generating modules.

8.4.1.2 Solar Power-Generating Module

The SPG module has a side made of six triangular-shaped PV cells forming a hexagonal piece known as HexBus. The module receives the reflected sunlight from the SRA. These modules are estimated to weigh 15-20 kg each. About 200,000–300,000 modules will be needed.

8.4.1.3 Wireless Power Transmission module (WPT)

The WPT module is on the other side of the HexBus. It consists of solid-state devices and transmitting elements in each module. The circuit sandwiched between the SPG and WPT modules converts solar power into electricity and then immediately microwaves and transmits it via the transmitting elements. The phase angle of each transmitting element is controlled, so the whole transmitting system functions like a phase-array antenna. The WPT modules weigh 50 kg each. The number of WPTs is the same as that of SPG modules, that is, 200,000–300,000.

8.4.1.4 HexFrame Structure Module

These modules are mechanical beams that support the reflector modules and connect to the SPG side. These beams weigh 50 kg each, and about 5,000 units will be needed.

8.4.1.5 Robotic Module

These modules are robotic arms that do various assembly and actuation in space. These units weigh about 10 kg each, and about 5,000 units are required.

8.4.1.6 Propulsion/attitude Control Module

Spread around the satellite platform, these modules provide the propulsion needed for station keeping, navigation, and guidance.

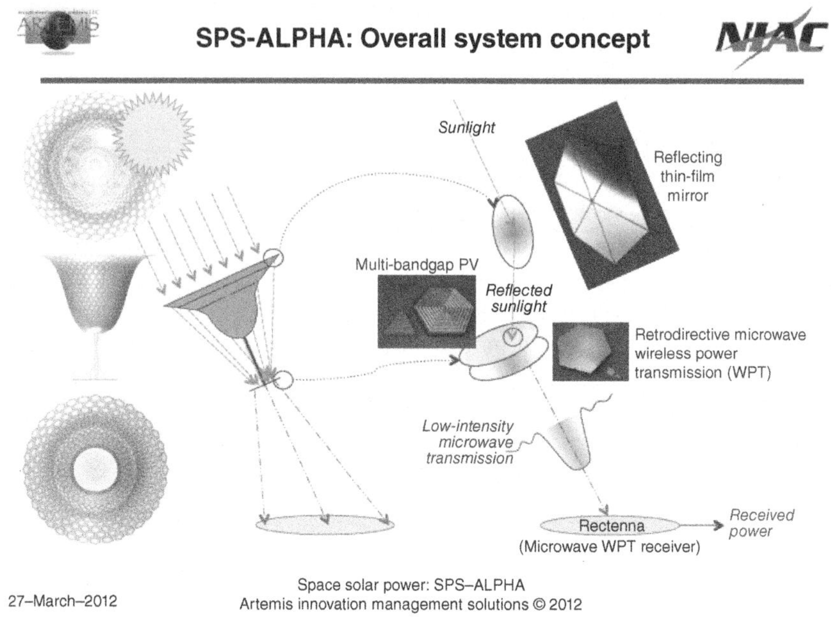

Figure 8.6 The SPS-ALPHA idea is funded by NASA Innovative Advanced Concepts (NIAC). Modular optical reflectors are used to focus sunlight on a smaller bank of PV cells. The optical modules are mounted in two geometric shapes: half-ellipsoid and sigmoid. The microwave transmission modules are on the opposite side facing the Earth. *Source:* Ref. [5] / NASA / Public Domain.

Figure 8.6 illustrates the working of an SPS-ALPHA system from the space segment to the ground segment. Innovative ideas have been on the satellite platform as the ground segment is relatively straightforward, and its size is primarily determined by the microwave frequency.

8.5 Laser as Wireless Power Transfer

Laser is an alternative for transmitting power from the satellite to the ground. Here we briefly describe the functional flow.

The main difference is that microwave generators and transmitting antennas are replaced by laser and optical subsystems. Continuous wave (CW) lasers are usually assumed.

Solar power cannot be directly converted to laser power. PV panels or optical reflectors to collect sunlight are still needed. Hence, the efficiency issue with PV cells remains the same. Modern solid-state lasers can now reach power conversion efficiency as high as 80%.

The transmitting portion of a laser system is much smaller than its microwave counterpart. The transmitter is only several meters in diameter. Atmospheric absorption of laser power from satellite to ground is higher and affected by weather patterns, with loss ranging from 60% to 90%. Unlike a microwave beam, a laser beam does not broaden that much. Therefore, the receiving end on the ground is also small in size. One concern of laser systems is that the power density within the narrow beam is high and could present a safety issue.

8.6 Buildup of Space Solar Power

How many SSP systems would we need as a primary power source for the world? This would be an important question in the second half of this century when the supply of fossil fuels runs low and if SSP proves to work.

Most proposals up to now have been for 1 GW at the grid. Therefore, the above question is the same as asking how many 1 GW power plants we would need on the ground to service, say, half the population in the world.

To answer that question, let us make some assumptions. The first is that the total population will reach 10 billion in the second half of this century, so we are looking to provide power to 5 billion people via SSP to power grids. We further assume that 5 billion people live in industrialized societies based on current trends, and their consumption habits are similar to that currently in the United States. We note earlier a 1-GW plant can power 800,000 American households. One can generalize this estimate to all industrialized nations.

Assume households have improved their power consumption habits by 25%. This means that a 1-GW plant could power about 1 million households. If an average household has 4 members, we are now looking at 1.25 billion households in the industrialized world of 5 billion people. Therefore, about 1,250 1-GW plants would be needed. The number may vary from 1,000 to 3,000, depending on the assumptions and projected consumption.

Today there are about 440 fission nuclear power plants (at different ages) in the world producing a total power of nearly 400 GW, so each average about 1 GW. If 1,000 1-GW fission power plants were to be constructed today at an average cost of US$5 billion each, it would cost US$5 trillion in total plus maintenance and eventual decommissioning. Nuclear power is a viable alternative to fossil fuel plants in the mid to near term. New technology, such as thorium reactors and mini nuclear power plants, is also coming into the picture. Space solar power is an important alternative to explore.

Using current technologies and reusable rockets, an SSP system from satellite to receiving antenna would cost about three to six times more than a nuclear power plant. Therefore, the size of investment would run in tens of trillions of US dollars to build over 1,000 SSP systems. Emphasis should be made that the launch cost ($ per kg) to the GEO orbit is a major contributing factor to the total cost.

8.7 Technological Challenges

The size enormity of an SSP satellite, its subsystems, and therefore the project cost have made the SSP idea a challenge to proceed with.

We will state a few of these challenges, some of which counter to one another.

1) Low total mass: The satellite mass directly impacts the launch cost. The ultimate goal is to reduce the total satellite mass if possible. Typical SSP proposals show the satellite mass in the order of 10 million kilograms to generate nominally 1 GW on the ground. The current launch cost easily takes up over 30% of the total budget. But reusable high-capacity upper stages with on-orbit refueling capability should drive down the cost per unit weight by as much as a factor of 10.
2) Conversion efficiency: The cnversion efficiency of each subsystem in the satellite translates to heat generation within the subsystem. Low PV efficiency means a larger area of PV panels would be needed to collect sufficient solar radiation. This, in turn, adds mass to the satellite. Hence, high PV efficiency has a cascade impact on the total cost.
3) Device efficiency at high temperatures: With large amounts of heat generated, devices from PV cells to microwave amplifiers (or laser modules) will run hot. Hence, device performance at high temperatures is a key requirement.
4) Thermal management: Each SSP system has its way to conduct large amounts of heat to one or multiple reservoirs to radiate the energy away. Thermodynamics requires that the heat reservoir to be massive and efficient for radiating heat away. These reservoirs add weight to the system.
5) Wireless power transfer: Delivering power wirelessly by microwaves and lasers is not new. What is new is the gigawatts of continuous power over a long distance of 36,000 km. Currently, several laboratories are testing wireless microwave power transfer with moderate power over distances of tens of kilometers.
6) Superstructure: Kilometer-sized structures are unavoidable for an SSP satellite. Recent proposals offer different innovations in constructing the superstructure to support PV panels and transmission equipment. Modern proposals use autonomous machines to assemble modules in orbit. It is clear that a team of astronauts working in high orbits, as originally proposed decades ago, is no longer a practical proposition. Instead, autonomous and semiautonomous machines will be the solution.

7) Services and warehouse in space: SSP satellites such as the ALPHA design will be composed of many mass-produced parts and modules. Autonomous machines are the best tools to perform services such as repair, replacement, and maintenance of different modules in the satellite. Warehouse types of depots in nearby orbit or similar solutions need to be considered to store spare parts. This means that service vehicles would need to go between the SPS satellite and the depot(s) as well as performing other services. Servicing a SSP satellite in space is a logistics exercise.
8) Mass production: The large number of PV panels, microwave amplifiers, and components must be mass produced, reliable, and cost-effective. The Cislunar alternative is to manufacture the components needed on the Moon using lunar resources and transports them back to Earth orbit for assembly. This would require advanced industrialization of the Moon.
9) Megaproject in space: An SSP satellite could be made from thousands of modules of suitable sizes that can be delivered by rockets and then assembled in high orbit. Missions of this size, especially in high orbits, have never been done. Such an undertaking spells the beginning of a new chapter in space construction.
10) Pointing accuracy: The pointing accuracy is between the transmitting antenna and the receiving antenna on the ground from a distance of 36,000 km. This issue applies to both microwave and laser, with the laser being more challenging due to its small beam size. Pointing errors result in loss of power and possibly safety concerns. For a laser beam, the pointing requirement is to be within $5 \times 10^{-6\circ}$. A nominal solution has been suggested by Japan's space agency JAXA. The ground sends a pilot beam to the satellite followed by a beacon beam from the satellite. The satellite and ground are now locked in.

8.8 Looking Ahead

Space solar power systems are a viable solution as a reliable energy source in the space age. All renewable energy sources must be considered owing to impending shortages of resources and climate effects. Solar power is an obvious energy resource for civilization.

The solar power density in space of 1.37 kW/m^2 is a hard number. It determines the collection area in space near Earth. As device efficiency improves from research efforts, the satellite's solar collection area can be reduced or the amount of power increase at the grid on the ground but still within safety limit on the ground.

An SSP satellite is a super-sized structure with dimensions measured in kilometers with weight millions of kilograms. The cost of each SSP project is tens of tens of billions of US dollars. The launch cost per kg is a big factor. It is incumbent upon the launch industry to provide reusable vehicles with high payload capacity to GEO. This in turn suggests on-orbit refueling to be realized as part of space transportation infrastructure. The space industry will also benefit from the increase in demand for launches.

It would make no sense to put just one SSP satellite in orbit. The question is more about how many, which is related to orbit allocation, system reliability, and the total cost of an SPS project. The total cost of an SSP project translates to the cost per kilowatt of electricity on consumers' bills. We cannot expect the initial cost per kilowatt to be the same as what we are paying today. It is a matter of how many times more consumers are willing to accept. Government's role and subsidy are instrumental.

SSP supporters point to special groups of early users who are willing to pay a premium price for electricity. These groups include remote communities, businesses that operate in remote areas or require uninterrupted energy supply, military operations, and disaster-related operations. These groups individually do not need 1 GW of power. Hence, the transmission beam from an SSP satellite above needs to be steerable. High-end groups may well be the initial users.

Building an SSP satellite in earth orbit lays the technological foundation for assembling mega structures in deep space in the future. As we envision the infrastructure that needs to be built for a Cislunar economy, mega structures in space will become routine. SSP satellites can be among the first.

In principle, an SSP satellite could also be placed in lunar orbits and Earth-Moon Lagrange point L1 as a source of energy because the power density at these locations is essentially the same as the Earth. However, solar power intensity on Mars is only about half of that around Earth, so the choice is not obvious for Mars.

Reading List

Energy Consortium (2023). *Energy of the Future: What Is Space Solar Power*. Royal Hawaiian Press [cited 2025 Feb 21]. [online]. Available from: https://www.nextbigfuture.com/2012/04/olar-power-satellite-via-arbitrarily.html.

Flournoy, D.M. (2011). *Solar Power Satellites (SpringerBriefs in Space Development) 2012th Editiion*. Springer.

Mankins, J.C. (2008). *Space Solar Power: The First International Assessment of Space Solar Power*. International Academy of Astronautics.

NASA Otps, Zimmerman (2024). *Space-Based Solar Power (Space Power)*. Nimble Books.

National Research Council (2001). *Division on Engineering and Physical Sciences, Aeronautics and Space Engineering Board. Laying the Foundation for Space Solar*

Power: An Assessment of NASA's Space Solar Power Investment Strategy. National Academies Press.

Naval Postgraduate School (2014). *Spaced-Based Solar Power System Architecture*. CreateSpace Independent Publishing Platform.

References

1 Invictus S. Multi-Rotary Joints SPS - 2015 SunSat Design Competition. YouTube [cited 2025 Feb 21] [online]. Available from: https://www.youtube.com/watch?app=desktop&v=XhgJwnpYRGc&t=4s.

2 Hou, X., Li, M., Niu, L. et al. (2021). SunSat design competition 2014-2015 first place winner – team CAST: Multi-rotary joints SPS. Online Journal of Space Communication *11* (18): 4.

3 Mankins JC. SPS-ALPHA: The first practical solar power satellite via arbitrarily large phased array (A 2011-2012 NASA NIAC Phase 1 Project). In NASA Innovative Advanced Concepts Program; 2012 [cited 2025 Feb 21] [online]. Available from: https://www.nasa.gov/wp-content/uploads/2017/07/niac_2011_phasei_mankins_spsalpha_tagged.pdf?emrc=28bc0b.

4 David L. Proposed satellite would beam solar power to earth. Space; 2012 [cited 2025 Feb 21] [online]. Available from: https://www.space.com/15189-solar-power-beaming-satellite.html.

5 Mankins J, Kaya N, Vasile M. SPS-ALPHA: The first practical solar power satellite via arbitrarily large phased array (a 2011-2012 NIAC project). In: 10th International Energy Conversion Engineering Conference; 2012. p. 3978.

9

Space Tourism

9.1 Introduction

The term "space tourism" entered the public's eyes with the growth of commercialization of space.

The Oxford Dictionary's definition of space tourism is: "The practice of traveling into space for recreational purposes."

Tourism is a commercial activity for civilians. The keyword is "recreation." This is our starting point for this chapter. The necessary tools and services are to be provided by or via for-profit enterprises. This chapter covers suborbital and orbital space flights powered by propulsion. High-altitude balloon rides are excluded.

A suborbital flight is one in which the vehicle reaches outer space but does not have enough tangential speed to orbit Earth and usually does not complete a full orbit.

Currently, civilians who have gone to space are wealthy individuals or sponsored by corporations. The per-person price varies depending on the flight profile and flight duration. For example, a ride to the edge of space on a Virgin Galactic rocket plane experiencing weightlessness for several minutes would cost at least US$450,000 as of this writing.

According to the market research firm Precedence Research, the global space tourism market size was estimated at approximately US$876.5 million in 2022 and is projected to reach US$21.2 billion by 2032 [1]. The global tourism market was US$8.9 trillion in 2023 and is projected to grow to US$12.3 trillion by 2032. Therefore, on a relative basis, space tourism is a novel venture.

9.2 Success Factors

Several factors are needed for space tourism to proliferate: 1) affordability, 2) passenger capacity, 3) destinations, and 4) preflight qualification. Factors 1), 2), and

New Space: From Low Earth Orbit to the Moon and Beyond, First Edition.
Edited by Stephen W. Cheung and Chih-Yung Wen.

3) are interconnected and influence one another. Preflight qualification sets a bar on a prospective passenger's health.

9.2.1 Affordability

Reusable vehicles, including boosters, upper stages, and crew capsules, are key to driving down cost. The commonly cited economic measure of cost per unit weight to low earth orbit (LEO), which runs between US$2,000 and US$10,000 per kg, only applies to cargo payloads but not to human passengers. The current price per person for an orbital flight for a tour to the International Space Station (ISS) via a SpaceX Crew Dragon is approximately US$50 million. Japanese billionaire Yusaku Maezawa reportedly paid $80 million for a Russian Soyuz spacecraft to the ISS for 12 days in 2021. One can surmise that crewed orbital flights will remain expensive using current rockets and crew capsules.

Market analysts use the term "Total Addressable Market (TAM)." It is the maximum size of the opportunity if every person who can afford a product decides to purchase it, i.e. 100% market share. TAM is a measure of potential revenue obtained by multiplying the maximum number of potential users by the price of the product at the time.

The "Wealth Management Report, World Report Series 2023" by Capgemini Research Institute [2] offers the number of individuals in the world as measured by their wealth bands as of 2023, as shown in Table 9.1. According to this categorization, one can discern that groups with "very high net worth" (US$5 to 30 million) and "ultrahigh net worth" ($30+ million) can afford suborbital flights but not necessarily the "high net worth group."

For space tourism to reach the same extent as regular tourism, transportation costs for passengers must be lowered to what an average household can afford, i.e. comparable to commercial airline tickets. This means that the launch cost must go down by several orders of magnitude. In 2024, a ticket priced at about US$50,000 as a guestimate, would be very enticing to passengers on suborbital flights.

The message is clear; the launch cost for crewed flights is a big barrier to entry.

Table 9.1 Number of people in high-income wealth bands as of 2023.

Wealth band	Investible assets (US$)	Number of individuals
Ultrahigh net worth	30+ million	210,000
Very high net worth	5–30 million	1,977,000
High net worth	1–5 million	19,524,000

9.2.2 Passenger Capacity

The passenger capacity is the number of passengers a vehicle can carry. Currently, orbital and suborbital crew capsules such as the Crew Dragon of SpaceX and the New Shepard can carry at most 6–7 passengers. For space tourism to be industrialized like commercial aviation, a cabin with a nominal capacity of 100 passengers would be economical. This would be an increase of one order of magnitude. A heavy-lift reusable rocket system with high lifting power and passenger capacity is needed.

At present, the SpaceX Starship system fits this bill. The Starship upper stage has a stated passenger capacity of 100 persons to LEO. On-orbit refueling adds distance beyond LEO. Current reusable crew capsules need to evolve to propulsive landing, like aircraft, instead of parachuting to sea and land. Hence, reusable technology should apply to all rocket stages and crew capsules.

9.2.3 Destinations

Tourism prospers on Earth due to affordable transportation and many destinations. There are cities and natural wonders to attract frequent travelers. Affordability and destinations complement each other.

There are only a few natural destinations in space for the public, from LEO to the Moon and Mars in the future. A round-the-Earth trip to LEO would be a good experience for first timers, and the total addressable market could be quite large. Going around the Moon (with and without landing) is another attraction. The Earth–Moon Lagrange point L1 is believed to be an excellent orbit that provides an excellent view of the Earth and Moon.

Multiple destinations drive up travel frequency for leisure and business. In turn, more flights drive down transportation costs and further stimulate desires to travel. Man-made destinations are crucial as part of space infrastructure. These man-made destinations are usually space stations of various sizes located in different orbits, as well as structures on the Moon and Mars. Matching infrastructure such as transfers stations (similar to airports) and refueling will also be needed. Commercial destinations include orbiting factories, research facilities, hotels, warehouses, spacecraft repair and maintenance services, and even universities and healthcare facilities. See Chapter 4 on space stations.

9.2.4 Preflight Qualification

Most people are medically qualified to fly on commercial airplanes. The g-load that most people experience is that of a commercial aircraft taking off or elevators going up. The g-load on the human body during a rocket launch is much higher when running between 2g and 4g for several minutes. An average person can take 2–5g for a short duration. The question of preflight qualification and training needs to be addressed.

Preflight training can be taken as part of the experience for adventurists, but it is an inconvenience for impulse travelers. These issues may become less important with data from more people going to space, advances in spacecraft design, and the emergence of related medical technologies.

9.3 Suborbital and Orbital Flights

This section examines three companies that already provide launch services to civilians. The first two are suborbital flights, and the last one is orbital.

9.3.1 Virgin Galactic

Virgin Galactic is unique in that no rocket booster (first stage) is used but is replaced by a modified jetliner as shown in Figure 9.1. The jetliner is known as a Virgin Mother Ship (VMS). A spaceplane, known as the Virgin Space Ship (VSS), is mounted to the bottom of the VMS fuselage for ascent into orbit. The VSS has a rocket engine for propulsion. It is a glider when returning to Earth.

9.3.1.1 Horizontal Takeoff

The Virgin Mother Ship and the spaceplane each has two pilots. The mother ship takes off from the runway like a jetliner. At an altitude of 15 km, the VSS is released.

Figure 9.1 Image of the Virgin Space Ship mounted on the bottom of the twin-fuselage jetliner Virgin Mother Ship. *Source:* Ref. [3] / Virgin Galactic.

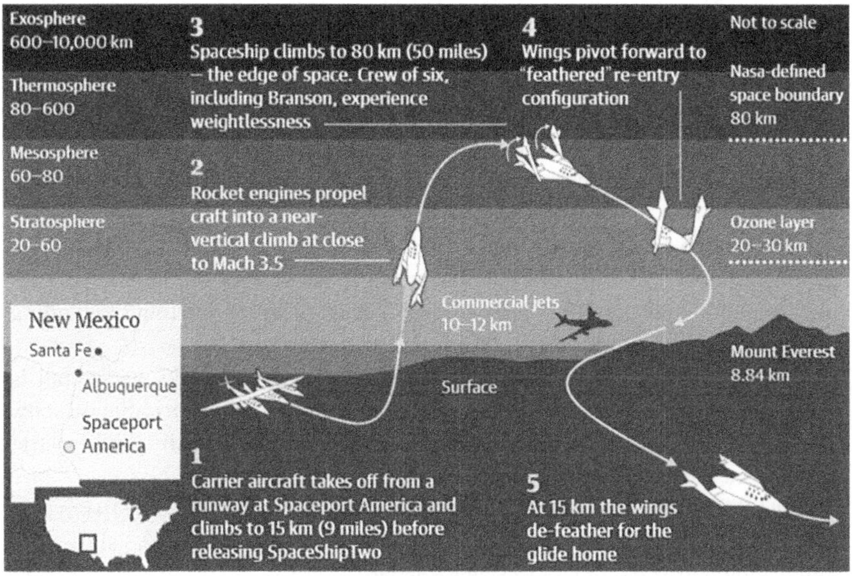

Figure 9.2 Flightpath of the Virgin suborbital flight [4]. *Source:* Guardian / Virgin Galactic / Reuters.

The spaceplane then fires its rocket engine and is propelled almost vertically to about 80 km altitude. The plane's forward acceleration reaches about 4g and speed nearing Mach 4 when the rocket engine shuts down at near 40 km altitude. The plane continues to go up by inertia. Passengers begin to feel weightless from this point on as the plane rises to its highest point at about 80 km altitude.

The VSS begins to drop after staying in the apogee for 3–4 minutes. The g-level is gradually picking up. The wings of the VSS are uniquely designed to move in a forward vertical configuration. This design is known as "feathering" and resembles a falling badminton shuttlecock. The purpose is to provide the freefalling rocket plane with stability and extra drag to slow down the fall sufficiently for reentry into the thicker part of the atmosphere so as to minimize heat generation.

At an altitude of about 15 km, the rocket plane "de-feathers" to the original position and glides to the home base at dissipating speed. The step-by-step process is illustrated in Figure 9.2.

9.3.2 Blue Origin

The New Shepard system by Blue Origin is reusable and designed for suborbital flights. The booster has a single rocket engine running on LH2 and LO2. The system is intended for space tourism for "up and down" flights. This means that the vehicle virtually gains no horizontal speed.

The crew capsule can house six passengers. Its interior is shown in Figure 5.15. Figure 5.16 shows the flight path. The main-engine-cutoff is followed by the crew capsule detached from the booster at about 60 km altitude. The crew capsule has limited propulsion capability for emergency detachment purposes only. After separation, the crew capsule and the booster each climb further by inertia. Both reach their respective apogees just above the Kármán line, with the crew capsule going a little higher.

As the system has no horizontal speed, the booster is essentially freefalling, reaching at most 0.75 km/s. No re-entry burn is needed. Fins on the booster provide navigation and stability. The booster is slowed down by starting the landing burn and lands approximately at 3 km from the launch pad.

The crew capsule comes down from a slightly higher altitude. Its freefall is slowed down by the thicker part of the atmosphere and reaches terminal velocity. Parachutes are deployed to land the capsule. The capsule typically lands a little further from the launch pad than the booster.

The total flight time is about 10 minutes reaching an apogee of 100–110 km altitude. The weightlessness experience lasts about 3 minutes. Figure 9.3 shows the launch and booster landing of the New Shepard.

9.3.3 SpaceX-Crew Dragon

Chapter 5 covers the Falcon 9 reusable booster. Here, we focus on the crew capsule due to its multiple usages, including that of space tourism. The Dragon is designed to shuttle crew and cargo from Earth to the ISS. The capsule rides on top of Falcon 9 booster and second stage. Figure 9.4 shows the full rocket with the Crew Dragon at the launchpad.

Standing over 8 m tall, the Crew Dragon capsule is composed of a cabin and trunk as shown in Figure 9.5. Up to seven passengers can be seated in the cabin although a trip to the ISS only seats up to four due to capacity limitations of the ISS. The pressurized cabin is for passengers and cargo with such requirements. There are 16 gas thrusters in the service section for various maneuvers, including the necessary orientation and adjustment as it approaches the ISS for docking. The trunk is unpressurized and for transporting cargo without pressurization requirements. It also has solar panels and fins to provide stability in the event of an emergency ejection during launch.

The flight path of the Crew Dragon launch to the ISS is shown in Figure 9.6. The Dragon docks with the ISS from its fore with the cover lifted, as shown in Figure 9.7.

When ready to return to Earth, the Dragon capsule detaches itself from the space station and then ditches its trunk. The trunk will slowly deorbit in low orbit and burn up in the atmosphere. The Dragon capsule fires its thrusters to slow itself down enough to start the deorbit process. The capsule orients itself with

Figure 9.3 The New Shepard booster returns near its launchpad [5]. *Source:* Ref. [6] / Blue Origin LLC.

the bottom facing the atmosphere during descent. The bottom of the capsule is designed to tolerate heat up to 1,600 °C during reentry. It then slows down further by parachutes and splashes into the ocean.

The shuttling service to the ISS is not intended for tourists. However, civilians from commercial firms such as Axiom Space have made trips to the ISS as part of the plan to launch commercial space stations from the ISS.

Figure 9.4 The SpaceX Crew Dragon sits atop the Falcon 9 booster for the Inspiration4 mission in 2021 [7]. *Source:* National Aeronautics and Space Administration / Wikimedia Commons / Public domain.

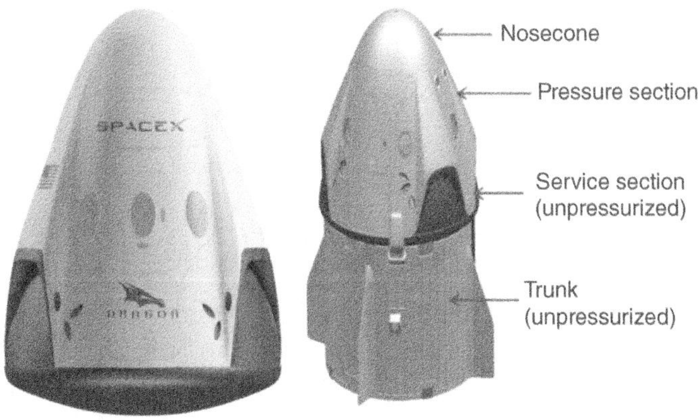

Figure 9.5 Crew Dragon capsule. The crew are seated in the pressurized cabin. The service section is where the reaction control systems, propellants, and other instruments are located. The thrusters are for maneuvering the capsule when it is near the ISS to dock. The trunk carries the unpressurized cargo. *Source:* Ref. [8] / SpaceX.

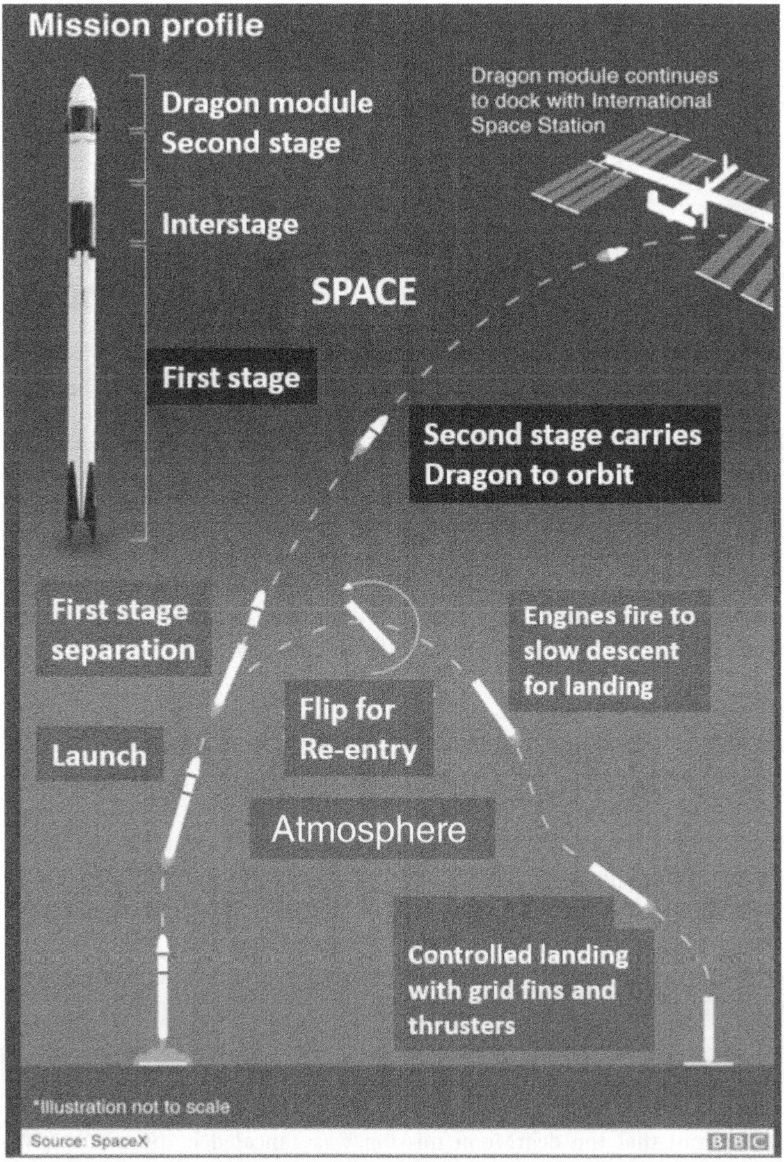

Figure 9.6 Flight profile of the SpaceX reusable booster and the Crew Dragon to the ISS. *Source:* Ref. [9] / SpaceX.

Figure 9.7 The Crew Dragon approaches and gets ready to dock with the ISS with the cover in its fore opened [10]. *Source:* Paopano / Adobe Stock Photo.

9.3.3.1 Inspiration4 Mission

Inspiration4 is an all-civilian mission on a Crew Dragon in September 2021. It was chartered by billionaire Jared Isaccman for reportedly US$220 million ($55 million per seat) while raising money for St Jude Children's Hospital. Donations topped $240 million. One of the four civilians was a former NASA pilot. The Inspiration4 mission is unique in that the capsule reached an orbit of 585 km, which was the highest altitude of crewed missions in the LEO region. The four-person crew orbited Earth for 3 days and splashed down on the Florida coast.

Inspiration4 can be interpreted as a step toward space tourism. It demonstrated the experience of orbital flight without a space station.

9.3.4 dearMoon Mission

We state upfront that the dearMoon mission was canceled in June 2024. We include it here to highlight the mission's objective. Originally scheduled to launch in 2023, various delays with the Starship program resulted in the cancelation of the mission.

The dearMoon project would have been the first all-civilian mission going around the Moon and was considered a lunar tourism project. The development and publicity of the SpaceX Starship system in the late 2010s led Japanese entrepreneur Yusaku Maezawa to sponsor this mission on board the Starship for a circumnavigation of the Moon. It was to be an art project as Maezawa selected eight civilian artists from the world to join the journey [11].

9.3.5 Starship Capacity

Current crew capsules such as the Crew Dragon and the New Shepard and the cabin of Virgin Galactic have passenger capacities for six to seven persons. The Starship would be a quantum leap in payload capacity (passengers and cargo). When proven flight worthy, Starship will be a vehicle for many new applications. Its impact on space tourism is expected to be disruptive. Starship's current crew capacity to the LEO is 100. When on-orbit refueling becomes routine, a ride on a Starship to the Moon, carrying tens of tourists and their belongings, will become reality.

While SpaceX has not officially signaled using Starship as a tourism vehicle, some space enthusiasts have put forth their own ideas. These ideas can be found online. Figure 9.8 illustrates a design based on the current dimensions of the Starship cargo bay. The idea is to divide the space into several decks for quarters and various activities. Offering amenities and room to move about should make deep space travel bearable.

9.4 Space Hotels

Man-made space destinations refer to space stations and developed areas on the Moon and Mars. In principle, space stations can be placed in any orbit if safety concerns such as radiation are addressed and the supply–demand justified. Currently, space hotels in LEO have been proposed. The Von Braun station and the Gateway station are covered in Chapter 4. They are gigantic structures serving as hotels and housing several hundred guests and crew.

9.5 Returning to Earth

What goes up must come down. Both suborbital and orbital flights must go through the re-entry process to return to Earth.

In suborbital flights, both the booster and crew capsule reach their respective apogees and free fall back to Earth through the thicker atmosphere. The New Shepard booster, for example, executes landing burns to slow down and land, while the crew capsule lands by parachutes.

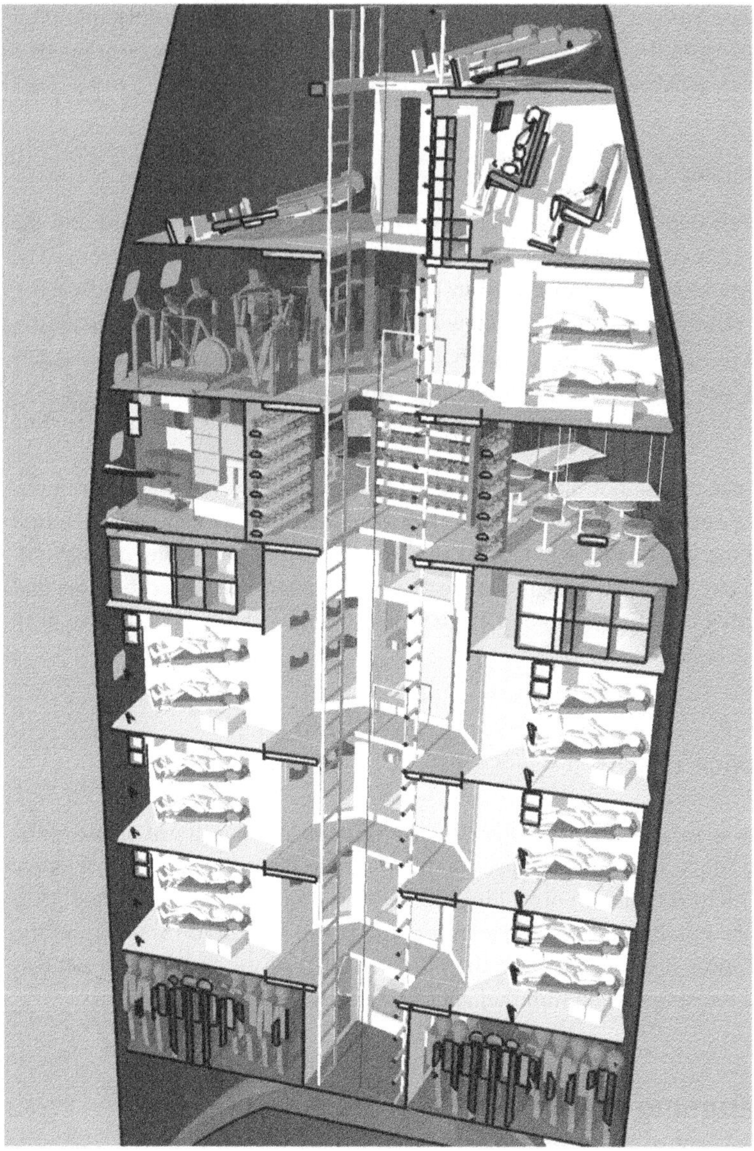

Figure 9.8 SpaceX enthusiast suggests the living arrangement for multiple crews and facilities. This arrangement would be useful for long-distance missions. *Source:* Adapted from Refs. [12, 13].

In orbital flights and suborbital flights with significant horizontal distance such as half way around the Earth, spacecraft such as the Dragon capsule and the Starship have orbital speeds of 8 km/s before applying some decelerating burns to slow down to prepare for re-entering the atmosphere. An orbiting spacecraft's re-entry comes at an angle to maximize the distance it travels in the atmosphere to dissipate its speed.

Plasma formation around the spacecraft in the upper atmosphere is covered in Chapter 5. Crewed spacecraft are equipped with heat shields to protect their structures. The flames are sometimes visible to the crew inside. Crewed flights on Starships provide valuable data for space travel on high-capacity vehicles.

9.6 Conclusion

Space travel encompasses both tourism and business-related travel. Space travel opens the door for the public to experience outer space. It is still a novel sector of the space industry. In dollar terms, space travel is only a small percentage of the global tourism industry and will likely grow at substantial rates in the next two decades. Point-to-point suborbital travel can become a big business, depending on flight time.

More destinations are needed for space tourism and space travel to prosper. Space business travel is driven by activities such as factories, logistics, and high-end laboratories in space. Hence, commercial space stations in different orbits around the Earth and the Moon are man-made destinations. The Moon is an obvious destination. Given its harsh environment and long day-night cycle, it will take some time to develop the Moon. The same situation is true with Mars.

Affordability is the barrier to entry to expand the "total addressable market" to include more potential consumers. Affordability is also closely linked to vehicle capacity.

Reusable rocket stages and on-orbit refueling are critical for space travel beyond LEO in the future.

Reading List

Byte N. *Space Tourism: Word Searches with Easy to Read Print about Planets, Spacecrafts, Space Stations and More*. Independently Published; 2025.

Guven U. *Space Tourism: Suborbital Flights – Commercial Flights – Spaceports*. Independently Published; 2020.

Irmici C. *The Future of Space Tourism: Exploring the Next Frontier*. Independently Published; 2024.

Sandeepa BB. *Space Tourism: Legal and Policy Aspects* (1st Edition). Edition). Routledge India; 2024.

Toivonen A. *Sustainable Space Tourism: An Introduction* (Volume 3). Channel View Publications; 2020.

Van Pelt M. *Space Tourism: Adventures in Earth Orbit and Beyond Hardcover* (2005th Edition). Copernicus; 2005.

Webber D. *Space Tourism Business: The Foundations*. Curtis Press; 2021.

References

1 Space tourism market size to worth around USD 21.17 Bn by 2030. Precedence Research [cited 2025 Feb 21] [online]. Available from: https://www.precedenceresearch.com/press-release/space-tourism-market.

2 Wealth Management Report. World Report Series 2023 [cited 2025 Feb 21] [online]. Available from: https://www.capgemini.com/wp-content/uploads/2024/05/WWR-2023_web.pdf.

3 Virgin galactic selects boeing subsidiary aurora to build new motherships. Virgin Galactic [cited 2025 Feb 21] [online]. Available from: https://www.virgingalactic.com/news/virgin-galactic-selects-boeing-subsidiary-aurora-to-build-new-motherships.

4 Devis D. Virgin numbers: Branson makes it to space (maybe). COSMOS [cited 2025 Feb 21] [online]. Available from: https://cosmosmagazine.com/space/branson-in-space-by-the-numbers/.

5 Wood C. Blue Origin brought the first official tourists to space. Popular Science [cited 2025 Feb 21] [online]. Available from: https://www.popsci.com/science/blue-origin-space-tourism/.

6 Foust J. Blue Origin flies next-generation New Shepard vehicle. SpaceNews [cited 2025 Feb 21] [online]. Available from: https://spacenews.com/blue-origin-flies-next-generation-new-shepard-vehicle/.

7 Kremer K. Up close with Inspiration4 Falcon 9 Rocket and Crew Dragon prelaunch on pad 39A: Photos. Space UpClose [cited 2025 Feb 21] [online]. Available from: https://www.spaceupclose.com/2021/09/up-close-with-inspiration4-falcon-9-rocket-and-crew-dragon-prelaunch-on-pad-39a-photos/.

8 Bouchard A. SpaceX's Crew Dragon capsule suffers anomaly during testing. Labroots [cited 2025 Feb 21] [online]. Available from: https://www.labroots.com/trending/space/14584/spacex-s-crew-dragon-capsule-suffers-anomaly-testing.

9 Rincon P. What is the SpaceX Crew Dragon? BBC News [cited 2025 Feb 21] [online]. Available from: https://www.bbc.com/news/science-environment-52840482.

10 Crew Dragon. Gunter's Space Page [cited 2025 Feb 21] [online]. Available from: https://space.skyrocket.de/doc_sdat/dragon-v2.htm.

11 Home page. dearMoon [cited 2025 Feb 21] [online]. Available from: https://dearmoon.earth

12 King P. VR tour of SpaceX Starship interior concept. HumanMars [cited 2025 Feb 21] [online]. Available from: https://www.humanmars.net/search/label/Starship%20interior.

13 SpaceX Starship interior concept for 100 passengers. HumanMars [cited 2025 Feb 21] [online]. Available from: https://www.humanmars.net/2019/11/spacex-starship-interior-concept-for.html.

10

Mining Asteroids

10.1 Introduction

Asteroids contain unimaginable amounts of raw materials that can provide for our civilization for a very long time with no concern about resource depletion. Successful extraction of minerals from asteroids can stimulate the Earth's economy and fuel the booming space age toward the Cislunar economy.

10.1.1 Astronomy Unit

The astronomy unit (AU) is a convenient unit for measuring interplanetary distances. One AU is the distance between the Earth and the Sun and is ~150 million km. Light would take 500 seconds to travel 1 AU. As the Earth orbits the Sun, our distance to other planets changes. It is, therefore, common to set the distance reference from the Sun. Mars is 1.5 AU, and Jupiter is 5.2 AU from the Sun.

This chapter focuses on some basics about asteroids. Most large asteroids reside in the Asteroid Belt (the Belt). Near-Earth asteroids' orbits come close to our neighborhood. Comets are also included here because they come around periodically.

10.2 Cosmic Background

Both comets and asteroids are remnants bearing witness to the formation of the early solar system. Contemporary theories suggest that soon after the Sun was formed about 4.6 billion years ago, large amounts of primordial dust particles and volatiles were attracted around the Sun. These materials coalesced via accretion, forming the planets that we know today. Accretion means growth by the accumulation of more matter. The inner four planets (Mercury, Venus, Earth, and Mars) are rocky. Those lumps of dust that did not accrete enough mass became thousands of remnants, or "planetesimals."

New Space: From Low Earth Orbit to the Moon and Beyond, First Edition.
Edited by Stephen W. Cheung and Chih-Yung Wen.

Jupiter is the biggest planet in the solar system, with a mass of about 0.1% of the Sun. After the Sun, Jupiter is a gravity powerhouse and has a significant influence on nearby planets and planetesimals. One theory suggests that there should have been a planet between Jupiter and Mars. The reason the hypothetical planet did not come into being is due to gravitational interactions primarily by Jupiter and Saturn.

Instead, the Asteroid Belt was formed where a major portion of the large and small remnant planetesimals gathered in the region between Jupiter and Mars. The Asteroid Belt is also known as the Main Belt. Subsequent and frequent collisions among the planetesimals produced rocky fragments of all sizes. Such gravity shoveling sent rocks toward the Sun as well as the outer edge of the Solar System. Jupiter could also influence the asteroids by hurling some of them in every direction, including toward the inner planets.

The asteroids in the Belt orbit between 2.2 and 3.2 AU around the Sun. The Belt is the transition from the four inner rocky planets to the outer icy planets known as the gas giants. The rocks in the inner belt move faster than those in the outer belt. Asteroids comprise rocks, metals, and volatiles. The Asteroid Belt contains millions of asteroids ranging in size from as small as pebbles to a diameter as large as 1,000 km (over 7% that of Earth). Ceres is the largest asteroid in the Belt with a diameter of 940 km.

The fashion the Belt was formed suggests that the asteroids' orbits have a variety of eccentricities and inclinations that cause the rocks to collide with one another at relative speeds of multiple km/s. Such collisions produce thousands of fragments.

10.2.1 Near-Earth Asteroids

Near-Earth asteroids (NEAs) are found in the inner solar system. These objects likely originated from the Belt and were disturbed by the gravitational attraction of nearby planets to change course that brought them to Earth's neighborhood. Asteroids are the source of the many meteorites that rain down our sky today and pelt Earth, the Moon, and Mars during the early days of the solar system.

Asteroids are over 1 m in size, while meteoroids are under 1 m in size. Meteorites are the remaining rocks that make it to the ground. When encountering a planet like Earth, meteoroids come through the atmosphere at speeds of tens of km/s that result in streaks of vaporized materials. Over 90% of the meteorites that fall to Earth have compositions similar to rocks on Earth. Their compositions contain iron and magnesium silicates, thereby supporting the fact that they are the most abundant minerals on the inner planets.

Meteorites found on the ground bear witness to the materials in the early solar system. They were small boulders or pieces broken from larger planetesimals millions of years ago. The atmosphere has protected the Earth from frequent bombardments. Mars and the Moon have a thin to no atmosphere and are subject to more direct impacts by asteroids and meteoroids, forming craters large and small.

10.2.2 Comets

Asteroids together with comets that occasionally cross our Cislunar skies are collectively called near-Earth objects (NEOs). Comets differ from asteroids in their compositions. Both were formed billions of years ago. Comets are made up of ice, volatiles, dust, and rocks and originated in the outer solar system where the Kuiper Belt and the Oort Cloud are located. The Kuiper Belt is just beyond the eighth planet Neptune and is 30–35 AU from the Sun, while the inner edge of the Oort Cloud is 2,000–5,000 AU and its outer edge 10,000–100,000 AU from the Sun. Pluto is in the Kuiper Belt.

The Kuiper Belt shapes like a doughnut and contains hundreds of thousands of icy bodies bigger than 100 km and over a trillion smaller ones, as illustrated in Figure 10.1. The Oort Cloud, on the other hand, is believed to be the true edge of the Solar System and is a spherical cloud of icy objects encompassing the

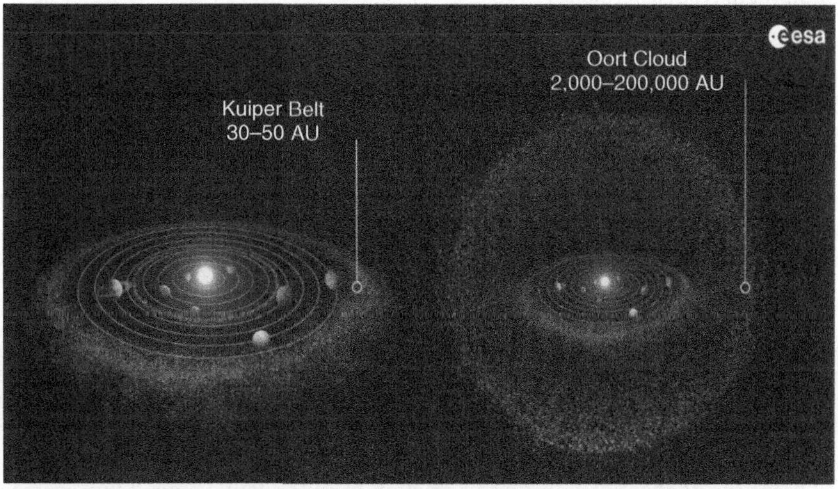

Figure 10.1 The diagram on the left shows the solar system out to the Kuiper Belt beyond Pluto. The Kuiper Belt has comets that orbit the Sun with periods less than 200 years. These Kuiper comets are close to the ecliptic plane. Further out toward the outer boundary of the solar system is a spherical distribution of icy objects called the Oort Cloud on the right of this figure. These icy comets orbit the Sun with periods over 200 years and follow inclined orbits. *Source:* Ref. [1] / European Space Agency (ESA).

whole solar system. The number of comets in the Oort Cloud is believed to be in the billions. Objects from the Kuiper Belt and the Oort Cloud sometimes get perturbed by the gravity of the outer planets and turn into comets coming through the inner solar system.

The comet size ranges from a few hundred meters to tens of kilometers (or much bigger). As of 2021, approximately 4,600 comets are known to Earth. When coming toward the Sun, the solar radiation causes the volatiles on a comet to warm and "outgas." A long tail of gas and dust called "coma" is formed.

Comets from the Kuiper Belt are referred to as "short-period" comets with orbital periods less than 200 years. Comets from the Oort Cloud are called "long-period" comets with orbital periods of more than 200 years. Orbits of long-period comets are also found to be highly inclined relative to the ecliptic plane due to the Oort Cloud's spherical shape. Haley's comet is a short-period comet coming around every 70–80 years with an inclination 18°. Hale-Bopp is the most famous long-period comet with an orbiting period of about 2,530 years and inclination of 89.43°.

While comets do come near Earth, they are small in number by far. We focus our interest on the rocky NEAs since they bear plenty of materials that can be used for Cislunar industrialization, albeit some of the materials are of little value to Earth to obtain them from asteroids.

10.3 The Asteroid Belt

All the big asteroids are found in the Belt. The total mass of the asteroids in the Belt adds up to $\sim 2.3 \times 10^{21}$ kg or about 3% of the Moon's mass. The Belt contains four large asteroids—Ceres, Vesta, Pallas, and Hygiea—and millions of smaller rocks. The four large asteroids together account for 50% of the total mass of the Belt.

Jupiter's gravity has also created regions known as the Kirkwood gaps within the Belt, as shown in Figure 10.2. These are zones where asteroids are few. Asteroids in these gaps are perturbed by Jupiter's gravity regularly. Their orbital periods are simple fractions (such as 1/2, 1/3) of Jupiter's orbital period, known as orbital resonance. Such resonance increases the eccentricity of the affected asteroids and heightens the chances of collisions or ejection from the Belt. Some of the ejected rocks end up being "planet-crossing" and becoming NEAs.

Asteroids orbit the Sun like all planets do. Ceres, for example, have an eccentricity of 0.079 and an inclination of 10.6°. The eccentricity and orbital inclination of the asteroids in the Belt are shown in the histograms in Figure 10.3. The gravitational influence of Jupiter caused asteroids in the Belt to follow elliptical orbits with eccentricity between 0 and 0.4, clustering around 0.15. Unlike the planets,

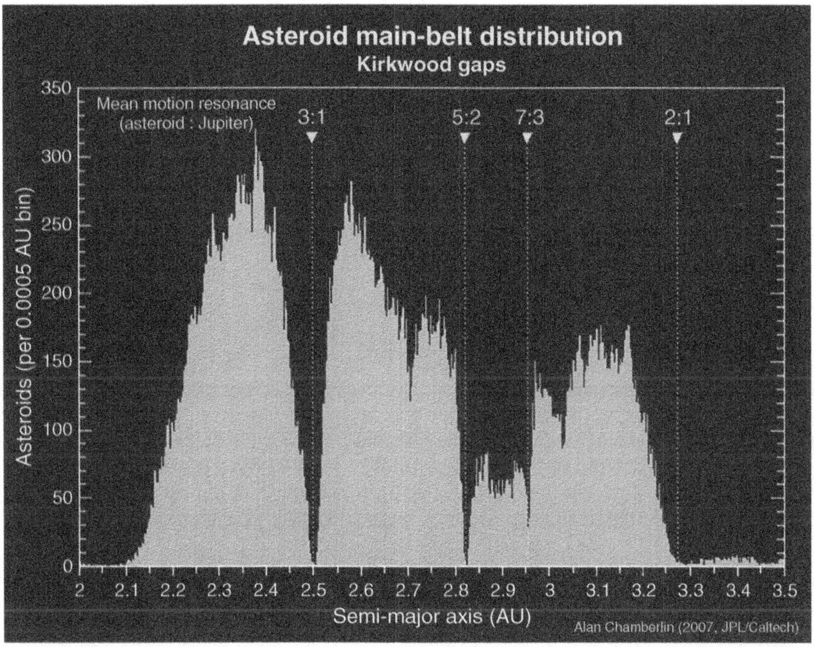

Figure 10.2 Asteroids are not distributed evenly from the inner edge to the outer edge of the Asteroid Belt. Gaps are known as Kirkwood gaps in which asteroids are either absent or unstable. In the latter case, the asteroids in these gaps are influenced by Jupiter and ejected into different orbits. The *X*-axis represents the distance from the Sun. The asteroid nominally begins from 2.2 to 3.2 AU. *Source:* Ref. [2] / NASA / Public Domain.

frequent collisions might have sent these rocks to orbit with inclinations from 0° to 20° and some even higher. The third largest asteroid Pallas, for example, has an eccentricity of 0.23 and a highly inclined orbit of 34.8°, relative to the ecliptic plane.

Some asteroids become moons of planets or bigger asteroids. The two moons around Mars are believed to be asteroids. Only large asteroids are near-spherical in shape such as the larger planets and moons. Most asteroids are odd shaped and many look like potatoes.

About 2 AUs away from the Asteroid Belt, there are two large collections of asteroids at the Sun–Jupiter Lagrange points L4 and L5. Known as the Trojans, they share Jupiter's orbit around the Sun. There are about 1 million Trojan asteroids larger than 1 km in diameter. The total mass of the Trojans is about 1/5 of the mass of the Belt. The Trojans are in stable orbits and are likely to have been there for billions of years. Hence, NEAs do not come from the Trojans.

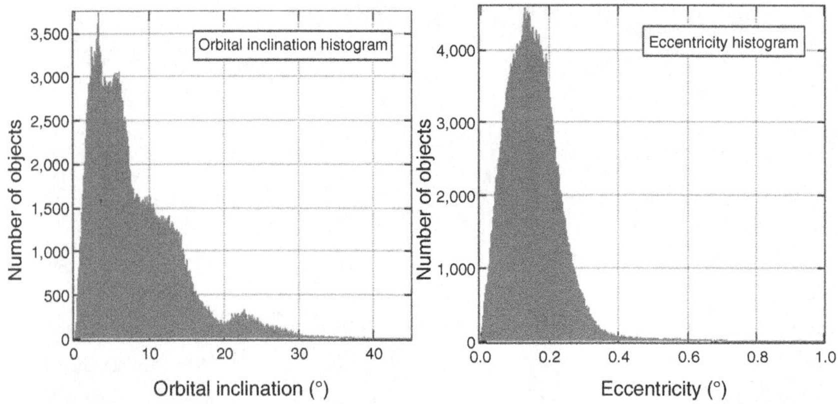

Figure 10.3 Histograms of asteroids in the Asteroid Belt by their eccentricity on the left and orbital inclination on the right. Many asteroids in the Belt have eccentricities between 0.0 and 0.3. Most asteroids' inclinations range from 0° to almost 20°. Most asteroids in the Belt orbit close to the ecliptic plane, but their eccentricity is not circular but is likely influenced by the large planet Jupiter. *Source:* Ref. [3] / NASA / Public Domain.

10.3.1 Differentiation Process

The accretion process in planets and big planetesimals builds up internal heat so that their cores often become molten. This heat comes from radioactive materials, tectonic frictions, and collisions. A process known as "Differentiation" occurs in these molten cores. In geology, materials inside a molten body are separated from one another by their relative densities. Heavy materials such as iron sink into the molten core, while low-density magmas such as aluminum and silicates rise to form the crust.

Some heavy planetesimals have gone through differentiation. The Earth is still going through differentiation. Most planets have gone through differentiation in their history. Differentiated asteroids come from differentiated planetesimals, and their compositions contain evidence of once molten metals. Similarly, undifferentiated asteroids are aggregates of materials that bear evidence of the early solar system. Some planetesimals might have started the differentiation process but did not accrete enough matter to become planets. Asteroids over 1 km in diameter are examples of partially molten metals.

10.3.2 Asteroids by Size and Number

The asteroid belt is estimated to have 1–2 million asteroids larger than 1 km in diameter and many millions of smaller ones with sizes range from 1 m and up. Approximately 30 asteroids are known to be larger than 200 km in diameter. Ceres is the largest asteroid in the Belt with a diameter of 940 km. The average

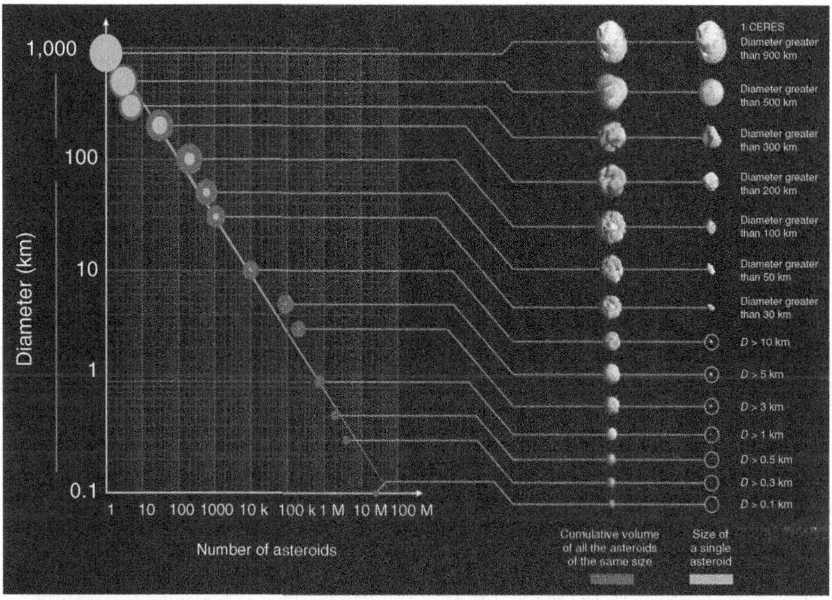

Figure 10.4 This chart is generated by Marco Colombo of the DensityDesign Research Lab as shown on Wikimedia. The general indication is that the number of large asteroids is inversely related to their size. In essence, large asteroids are small in number and small asteroids are high in number. *Source:* Ref. [4] / Clmb.marco / Wikimedia commons / public_domain.

distance between asteroids in the Belt is almost 1 million km. Smaller asteroids, especially those difficult to detect from Earth, could be closer to one another.

Asteroids in the Belt normally pose no hazards to Earth and the Cislunar space. When perturbed gravitationally by the heavy gas giants, their orbits would change and enter the inner solar system. This is when these asteroids could cross paths with the inner planets.

Figure 10.4 shows a very interesting plot of asteroid size versus their estimated number by the DensityDesign Research Lab. For example, there are only a few hundred asteroids over 100 km, but about 1 million asteroids are 1 km in size, according to Figure 10.4. The figure provides hints for asteroid mining in the future, that is, we do not need to mine the biggest asteroids in the Belt. Rather, we can go after the larger number of rocks that are nominally 1–10 km in size.

10.3.3 Asteroid Types

Asteroids are classified by type. NEAs follow the same classification although it is not clear whether they share the same distribution statistics. Asteroids range from

Table 10.1 Asteroids in the belt fall into three main types: C-, S-, and M-together with other minor types.

Type	Approximate percentage	Albedo	Differentiation	Density (kg/m^3)	Color
C-Type	70–75	0.03–0.09	Undifferentiated	1,380	Coal black
S-Type	17	0.1–0.22	Undifferentiated	2,710	Moderately bright, slightly reddish
M-Type	8	0.1–0.18	Differentiated	5,320	Grayish

The C-type dominates in number.

metallic to rocky. Metals are often found in asteroids, and most are nickel and iron. Composition and percentages vary from one asteroid to another and indicates the unique past each asteroid has gone through.

Asteroid detection and identification are achieved through optical means based on their reflected sunlight. Asteroids are classified in accordance with their spectral properties, that is, colors and albedo (light reflectivity). A regular object has an albedo between 0 and 1, with 0 being totally unreflective. Earth's albedo, for example, is 0.3.

There are three basic types of asteroids in the Asteroid Belt: Carbonaceous (C-type), silicaceous or stony (S-type), and metal-rich (M-type). Their quantitative distributions and characteristics are shown in Table 10.1.

The C-types are "coal black" and almost nonreflective. The albedo of both the M-type and S-type is about the same but different in colors. M-types are typically differentiated and should have molten metal cores. C-types and S-types are undifferentiated. Besides composition, there are other important differences in the internal structure of the asteroids. Many asteroids are solid, suggesting that they could have been molten at some point in their existence. Others are loose "rubble piles" held together by the force of their gravity and are often formed in collisions.

Among the three types, the C-types have the lowest density and the M-types have the highest. There are other types of asteroids. They are either stand-alone or subgroups of the C, S, and M types. Examples are A-, D-, E-, F-, P-, Q-, and R-types. They are uncommon and few within the Belt.

Figure 10.5 shows the distribution of asteroids by type versus distance from the Sun in AU. The origin in Figure 10.5 starts with 1.5 AU from the Sun where Mars is located. The Y-axis of Figure 10.5 represents the types of asteroids found at that distance from the Sun by percentage but not necessarily the actual asteroid count.

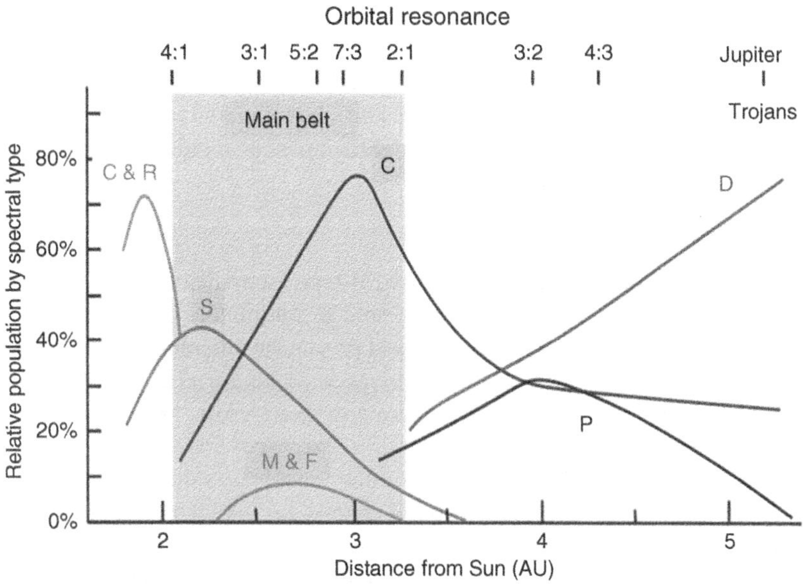

Figure 10.5 Statistical distribution of different types of asteroids in the asteroid belt. The main types are C, M, and S. The asteroid belt is the shaded area. Asteroids are found on the interior and exterior edges of the asteroid belt. *Source:* Ref. [5] / NASA / Public Domain.

For example, at 3 AU, C-types make up for almost 80% of the asteroids found there, S-types 10–15%, and M-types make up for about 5%.

C-, S-, and M-types are all found in the Belt. C-type peaks in percentage toward the outside edge of the Belt. At the outer edge of the Belt, nearly 80% of the asteroids are C-type. D-types are also found at the outer edge of the Belt and beyond when the share of the C-type drops steadily. D-types are the most abundant asteroids (by percentage) beyond the Belt. The Jupiter Trojans are mostly D-types. D-types are dark and are both carbon and volatile rich with water ice in their interiors. These D-types could be the intermediate between the primitive asteroids in the Belt and the comets in the Kuiper Belt.

About 70% of the asteroids found between Mars and the inner edge of the Belt are the rare E- and R-type, while the S-type takes up around 20%. S-types dominate on the inner edge of the Belt.

C-types contain large amounts of carbon molecules and other rocks and metals. Lying mostly on the outer edge of the Belt, they are rich in volatiles. C-types are among the most ancient objects from their primordial days and are undifferentiated. These asteroids are predominantly clay and silicate rocks, mixed with rocks, metals, and small quantities of volatiles such as nitrogen, oxygen, and hydrogen.

Some C-types have water. For example, Ceres and Hygiea, which are the biggest and the fourth largest asteroid in the Belt, are C-type.

S-types are found between the inner edge and the middle of the Belt. The composition includes silicates, which are oxygen and silicon, as well as metals such as iron, nickel, and magnesium. S-types are brighter and appear reddish in color.

10.3.4 M-Type Asteroids

Relatively few in number, there are about 90 M-type asteroids in the Belt larger than 30 km in diameter. They are mostly found in the middle part of the Belt. M-types are believed to come from the cores of previously differentiated asteroids because of collisions with other asteroids. M-type asteroids are reflective. The iron core is sometimes exposed after they have lost their crusts from collisions. Their volatile contents are low.

M-type asteroids have created dreams of unfathomable wealth in the minds of many entrepreneurs and investors. M-types contain up to 80% iron with nickel being the second most abundant by weight. The rest is a mixture of cobalt, gold, and elements of the platinum group, including platinum, iridium, osmium, rhodium, and ruthenium—all of which are considered critical materials for modern technology industries. Their compositions depend on their distance from the Sun when they are formed. Some have partially melted cores in which molten iron had sunk to the core.

10.3.4.1 16-Psyche

The largest and the most prominent M-type asteroid is 16-Psyche, located near the outer part of the Asteroid Belt. 16-Pschye's orbital distance is between 2.5 (perihelion) and 3.3 AU (aphelion) from the Sun, with eccentricity 0.134 and inclination 3.1°. It is the only M-type with a metal core body found so far. With a mean diameter of 220 km and mass 2.3×10^{19} kg (~1% of the total weight of the Belt), it is the most massive M-type known to date and is among the top 10 biggest asteroids. 16-Psyche's composite picture is shown in Figure 10.6.

16-Psyche's core is differentiated and composed of 95% nickel and iron. The asteroid is believed to be a fragment of a metallic core that originated from a differentiated planetesimal. It is estimated to contain metals worth US$10–100 quintillion (1 quintillion is 10^{18}). The economic impact and management of such amounts of metals have been actively debated as the global economy is only about US$110 trillion in 2024.

In October 2023, NASA launched a mission to visit 16-Psyche. The spacecraft will take 6 years to reach 16-Psyche, and the mission is planned for the Psyche spacecraft to orbit and study the asteroid for about 21 months, collecting data to help scientists understand its composition and history.

Figure 10.6 A composite picture of the 16-Psyche asteroid. *Source:* Ref. [6] / NASA / Public domain.

10.3.5 Asteroid Values

M-type asteroids have gathered the most attention because of the high-value metals that can benefit the Earth's industries. That is not to say that other materials, such as silicates and carbon compounds, have no economic value. All these materials are useful off-world resources in the construction of shelters and spacecraft away from Earth.

The estimated values of asteroids in dollar terms have been "wildly" speculated. Estimates mostly use the current market price of the materials and multiply by the tonnage that can be extracted from a target asteroid. It is highly unlikely that the market value would hold if that much materials were introduced to circulation on Earth. However, there seems to be no other fair way to place realistic values at present.

Given the number and size of the asteroids in the Belt, mining asteroids in the Belt will create large economic values. We use quintillion (qt) as the unit of measurement. In 2016, the website Visualcapitalist put forth an infographic of the top 50 most valuable asteroids [7]. The values stated in the infographic are for the readers' reference only.

Each of the top 50 asteroids on the Visualcapitalist list is valued over US$1 qt. We list the top 10 and the bottom 3 of the 50 asteroids in Table 10.2. Note that the value of these asteroids is not necessarily about the rare metals in their compositions, but the importance of their contents to space settlers and infrastructure builders. Oddly, 16-Psyche is not on the list with no reason given. Interested readers are referred to the website to determine the full list.

Table 10.2 Listing of the top 10 and last 5 valuable asteroids in the top 50 in the asteroid belt.

Rank	Asteroid name	Size (km)	Ore value ($ qt)
1	Davida	326	27
2	Diotima	209	7
3	Alauda	195	5.7
4	Palma	189	5.2
5	Lachesis	174	4.1
6	Winchester	172	3.9
7	Stereoskopia	168	3.7
8	Chiron	166	3.6
9	Siegena	165	3.5
10	Gyptis	163	3.4
48	Charybdis	114	1.2
49	Hersilia	113	1.1
50	Messalina	117	1.1

Their size varies from 100 to 300 km. Ore values are listed in quintillions (qt) of US dollars.
Source: Adapted from Ref. [7].

10.4 Statistics of Near-Earth Asteroids

Accessing asteroids in the Asteroid Belt from Earth presents huge technological challenges. On average, the inner asteroid belt is over 1 AU away from Earth and Mars is 0.5 AU away. Going to Mars takes 7–8 months (Chapter 13). The NASA Psyche mission will take 6 years, as stated before. Additionally, mining an asteroid requires spacecraft with a high payload capacity to bring equipment there and then tons of minerals back. Hence, the more practical route to access the Belt is from Mars. That will be possible after Mars is industrialized.

Until humans are ready to travel to the Belt for commercial explorations, NEAs are within reach of Earth now. NEAs either come close to or intersect with Earth's orbit. Some NEAs carry precious metals worth trillions of US dollars.

Near-Earth objects and near-Earth asteroids are often used interchangeably. NEAs come from the Asteroid Belt. NEOs include NEAs and a small number of comets. As a matter of definition, NEOs are objects with an orbital perihelion less than a predetermined distance. This predetermined distance is set at 1.3 AU from the Sun. Therefore, NEOs are found within 0.3 AU (45 million km) on either side of the Earth's orbit. Objects more than 0.3 AU from Earth's orbit are not NEOs.

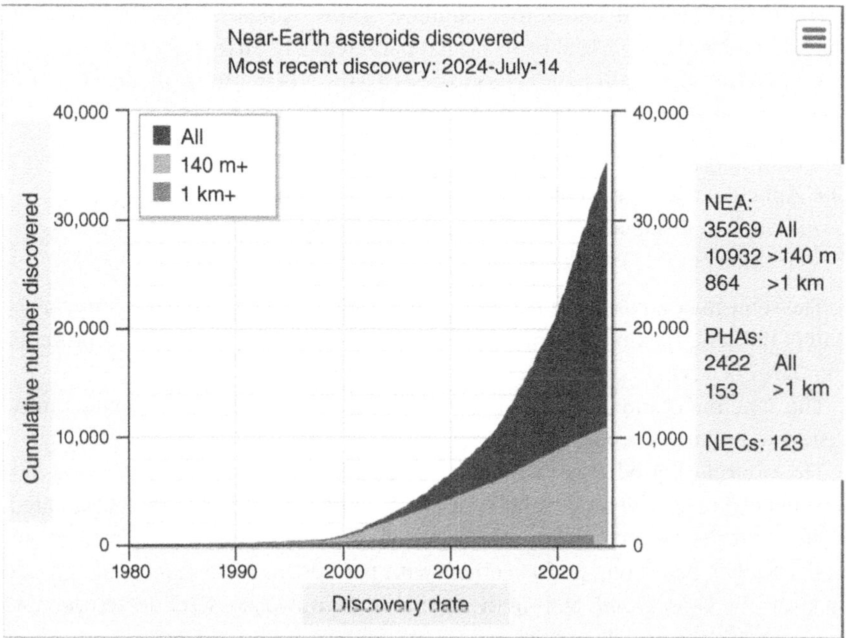

Figure 10.7 The number of near-Earth asteroids being discovered has been increasing due to improved equipment. The number of NEAs larger than 1 km remains small, although more potential PHAs have been discovered. *Source:* Ref. [9] / NASA / Public Domain.

Another boundary is the closest distance between a NEO and Earth and is set at 0.017 AU (2.6 million km) in asteroid family classification below, also on either side of the Earth's orbit. A NEO coming within 0.05 AU (7.5 million km) from Earth is considered too close for comfort. This object is further classified as a potentially hazardous asteroid (PHA), depending on its size (see below).

The Jet Propulsion Laboratory's Center for Near-Earth Object Studies (CNEOS) [8] maintains an active database of NEAs. Figure 10.7 shows that the total number of NEAs exceeded 35,000 as of July 2024. Over 10,900 NEAs are bigger than 140 m and 864 more than 1 km in diameter. Over 120 are short-period near-Earth comets (NEC). There are 2,422 PHAs, of which 153 are over 1 km in diameter as of July 2024.

10.4.1 Near-Earth Objects Families

For a long time, much attention was paid to the possibility of asteroids crashing into Earth, leading to global extinctions similar to what happened to the dinosaurs. Systematic studies ensued.

NEOs are classified under four families: Amos, Atiras, Atens, and Apollos. Each family is characterized by its orbital parameters relative to that of Earth.

We first introduce four parameters to describe the NEOs' orbital shapes. Three of these quantities are distances, and the fourth is time:

a = Semi-major axis of an ellipse
Q = Aphelion distance (farthest distance from the Sun)
q = Perihelion distance (closest distance to the Sun)
P = Orbiting period around the Sun for comets

The semi-major axis is the maximum length of the ellipse. The quantity "a" is half of this length akin to the radius of a circle. Hence, the quantity "a" represents the "size" of a NEO's orbit.

The aphelion Q and perihelion q of an orbiting asteroid represent the proximity distances to Earth's orbit.

The orbiting period P is chosen specifically for comets. P indicates how far a comet can reach the outer solar system—or simply where the comet originated from. Short-period comets have orbital periods under 200 years and at most go out to the Kuiper Belt. Long-period comets with periods over 200 years could go out as far as the Oort Cloud. Setting the period limit at 200 years delineates the size of a comet's orbit and its origin.

Referencing an article on NEOs by CNEOS, the four families are listed in Table 10.3. NECs and PHAs have been added to the list [10].

Table 10.3 Near-Earth objects are classified by their orbital parameters relative to Earth.

Group	Definition	Description
Amors	$a > 1.0$ AU, $1.017 < q < 1.3$ AU	Orbits exterior to Earth's orbit but interior to Mars' orbit
Atiras	$a < 1.0$ AU, $Q < 0.983$ AU	Orbits fully contained inside Earth's orbit
Atens	$a < 1.0$ AU, $Q > 0.983$ AU	Semi-major axis smaller than Earth's orbit; aphelion may come close to Earth
Apollos	$a > 1.0$ AU, $q < 1.017$ AU	Semi-major axes larger than Earth's orbit, but perihelion may come close to Earth
NECs	$q < 1.3$ AU, $P < 200$ years	Near-Earth comets with a period under 200 years
PHAs	MOID ≤ 0.05 AU Diameter ≥ 140 m	Coming closer than 0.05 AU from Earth. See Section 10.4.2

Near-Earth comets come from the outer solar system. A special class of NEA is potentially hazardous asteroids whose size and orbital proximity could present a danger to Earth. MOID stands for minimum orbit intersection distance.

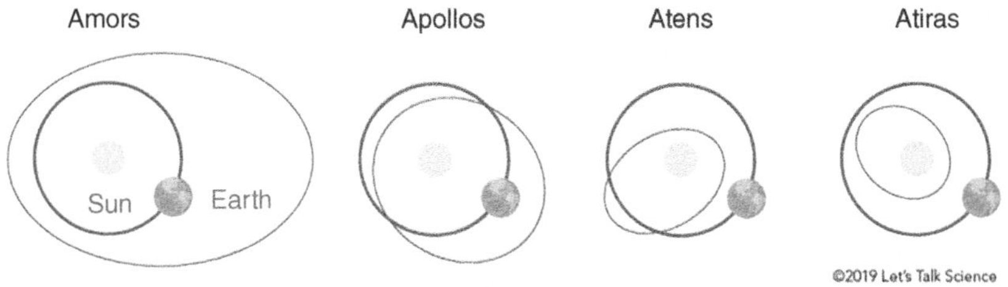

Figure 10.8 Graphical illustration of the four NEO families. The AMORs' orbits are larger than and fully contain that of Earth; and their proximity distances are sufficiently far away but within 0.3 AU to qualify as a NEA. Atiras' orbits are interior to, and do not intersect with, that of Earth; their closest distances are within 0.3 AU. Orbits of both Apollos and Atens cross-paths with that of Earth in some way. *Source:* Ref. [11] / Let's Talk Science.

Figure 10.8 helps us visualize the orbital paths of the four families: Amors, Apollos, Atens, and Atiras.

Amors have orbits larger than 1 AU ($a > 1.0$ AU), and their paths normally do not cross that of Earth. To be safe, the perihelion of an Amor should also be more than 0.017 AU (2.6 million km) but less than 0.3 AU from Earth to qualify as an NEO.

Apollos have orbits larger than that of Earth ($a > 1.0$ AU), but their perihelion comes close to or actually intersects with Earth's orbit within 0.017 AU. Since Apollos may come from the Asteroid Belt, there is no upper limit placed on their aphelions.

An Aten's orbit is smaller than that of Earth ($a < 1.0$ AU), but its orbital shape is such that its aphelion Q comes close to or intersects with Earth's orbit within 0.017 AU.

An Atira's orbit is fully contained inside Earth's orbit ($a < 1.0$ AU). Since an Atira may follow an elliptical orbit, its aphelion Q is set to be less than 0.983 AU from the Sun. This means that Atira's closest distance to Earth is more than 0.017 AU.

In summary, the Atens and the Apollos intersect with or at least come close to Earth's orbit within the limit of 0.017 AU. The Amors and Atiras are sufficiently far away and do not cross paths with Earth unless their orbits are disturbed gravitationally. Asteroids in all 4 families may qualify as PHA depending on their size.

10.4.2 Potentially Hazardous Asteroids (PHAs)

There are two criteria to qualify as a PHA. The minimum orbit intersection distance (MOID) draws a boundary at 0.05 AU (7.5 million km), which is almost 20 times the distance between Earth and the Moon. Asteroids coming closer than 0.05 AU pose a potential threat to Earth. Hence, an asteroid is a PHA as soon as it gets closer than 0.05 AU. The second criterion sets the asteroid size at 140 m and larger. An asteroid of this size crossing this MOID boundary would present a potential hazard to the Earth. For reference, asteroids between 40 and 140 m in size are considered "city killers," while those 1.5 km or larger are "planet killers." More than 99% of known potentially hazardous objects (PHAs plus comets) pose no impact threat in the next 100 years, and only 17 could not be ruled out as threats in the same period.

Referencing the CNEOS website [9], we list the numbers of NEOs on January 1, 2014; January 1, 2024; and July 25, 2024. The breakdowns are summarized in Table 10.4. The second and third columns record the counts of each category in 2014 and 2025. and the net change is shown in the fifth column. The fourth columns represents the total counts in July 2024, and the net change over 7 months in 2024 is shown in the sixth column.

Table 10.4 shows that the number of NEOs is increasing all the time due to new discoveries but not necessarily because more asteroids are coming our way. The change over 10 years from 2014 to 2024 is substantial in all categories. The increase in the number of km-size NEAs in general and PHAs in particular is noticeable.

PHA candidates may come from all four families (Amors, Atiras, Atens, and Apollos) sometime in the future. Most of the known PHAs are currently from the Apollo family.

10.5 Missions to Near-Earth Asteroids

Asteroids are important mineral resources in the grand scheme of the Cislunar economy. We need to know what asteroids have to offer us economically. NEAs that come within reasonable distances from Earth are opportunities due to their proximity, thereby making it possible to intercept them by spacecraft from Earth. NEAs are reachable from Earth using current rocket and spacecraft technologies. NEAs could provide the precious resources we need in building up the Cislunar infrastructure, while asteroids in the Belt are more for the future.

Compared to certain large asteroids in the Belt that are worth quintillions of dollars, NEAs have materials worth trillions of dollars and are more meaningful to our industrial world at this time. The media often cites wealth valued in trillions when a NEA passes by missing Earth by a safe distance. In July 2015, an S-type NEA 2011-UW158 passed by Earth about 2.4 million kilometers away. The 300-m rock reportedly packed US$5.4 trillion worth of platinum [12].

NEAs are also the most realistic and the best fertile ground to prove the technologies and feasibility of mining asteroids. NEAs have resources such as volatiles, hydrogen, oxygen, and industrial metals. Figure 10.7 and Table 10.4 show that about 10,000 NEAs are over 140 m in size and over 800 more than 1 km. These asteroids are sufficiently large to contain economically attractive amounts of resources.

Table 10.4 The number of NEAs being discovered has been rising.

	Number (2014/01/14)	Number (2024/01/01)	Number (2024/07/25)	Net change (2014/01/01–2024/01/01)	Net change (2024/01/01–2024/07/25)
Total NEO	10,625	34,136	35,393	23,511	1,257
NEC	104	122	123	18	1
NEA	10,521	34,014	35,270	23,493	1,256
NEA-140 m*	5,849	10,738	10,933	4,889	195
NEA-km**	807	864	864	57	0
NEA	10,521	34,014	35,270	23,493	1,256
Amors	4,015	12,086	12,462	8,071	376
Atiras	12	32	32	20	0
Atens	801	2,685	2,800	1,884	115
Apollos	5,693	19,211	19,976	13,518	765
PHA subtotal	1,419	2,383	2,422	964	39
PHA-km**	147	153	153	6	0

The numbers in each of the four families between 2014 and 2024 are shown [9].
* "NEA-140m" are NEAs with diameters roughly 140 m and larger.
** "PHA-km" and "NEA-km" are PHAs and NEAs with diameters one kilometer and larger.
Source: Ref. [9] / NASA / Public Domain.

10.5.1 Sample Collection Missions

Turning asteroid materials into industrial commodities in volume is a long process. The first step is to reach and land on a target NEA to collect samples to identify or verify its contents. Sample collection is a must to corroborate with the optical data of a target asteroid.

Asteroids describe elliptical orbits. Its speed varies along its orbit. A spacecraft rendezvousing with a target asteroid must steadily form an orbit that matches that of the target and intercepts it at the right speed.

To date, there have been four successful missions in which spacecraft made direct contact with their target NEAs for specific objectives. The missions are Hayabusa-1, Hayabusa-2, OSIRIS-REx, and DART. The first three missions successfully rendezvoused with their targets, mapped them, and then collected and returned samples to Earth.

Hayabusa means falcon in Japanese, while OSIRIS-REx stands for *O*rigins, *S*pectral *I*nterpretation, *R*esource *I*dentification, *S*ecurity, *R*egolith *Ex*plorer. Table 10.5 summarizes the characteristics of the three asteroids and the key information of the respective missions. All three NEAs are PHAs of the Apollo family.

Table 10.5 Three sample collection missions to NEAs.

Item	Hayabusa-1	Hayabusa-2	OSIRIS-REx
Agency	JAXA	JAXA	NASA
Asteroid name	Itokawa (S-type), PHA	Ryugu (C-type), PHA	Bennu (C-type), PHA
Family	Apollo	Apollo	Apollo
Size (estimate)	535 × 270 × 210 m	900 m	500 m
Mass (estimate)	30 million tons	450 million tons	73 million tons
Average orbital speed (km/s)	25	17–25	28
Orbit inclination and Eccentricity	1.62°, 0.28	5.87°, 0.19	6.03°, 0.20
Key dates			
Launch	May 2003	Dec 2014	Sept 2016
Rendezvous	Sept 2005	June 2018	Dec 2018
Collect samples	Nov 2005 (<1 gm)	Feb 2019 (5 gm)	Oct 2020 (60 gm)
Return samples	June 2010	Dec 2020	Sept 2023

These missions take an average of 6 years from the launch to the samples returned.

DART stands for *D*ouble *A*steroid *R*edirection *T*est. The mission aimed to demonstrate the feasibility of colliding with a target asteroid to change its trajectory as a tactic of planetary defense. The target asteroid Dimorphos is the "moonlet" of a bigger asteroid called Didymos, which is a PHA of the Apollo family.

Referencing Table 10.5, these spacecraft took nominally 2 years to rendezvous with their targets using ion thrusters after launch to low earth orbit (LEO). Their initial orbits steadily converged with the targets' orbits.

OSIRIS-REx will be discussed in detail next. Readers are referred to online information about DART and the Hayabusa-1 and 2 missions.

10.5.2 OSIRIS-REx Mission

A C-type PHA of the Apollo family, Bennu was believed to have been formed 4.5 billion years ago. Figure 10.9 is a photograph of Bennu. With an orbital eccentricity of 0.20, its perihelion and aphelion are 0.897 and 1.36 AU from the Sun, respectively. It comes close to Earth every 6 years at an orbital inclination of 6.0° relative to the ecliptic. Compared to the histograms in Figure 10.3 for asteroids in the Belt, Bennu's eccentricity and inclination are between normal and "subnormal." Bennu orbits the Sun with a period of 435 days (1.2 years). As of 2022, it will be as close to

Figure 10.9 Image of asteroid Bennu captured in 2018 by NASA's OSIRIS-REx spacecraft. *Source:* Ref. [13] / NASA / Public domain.

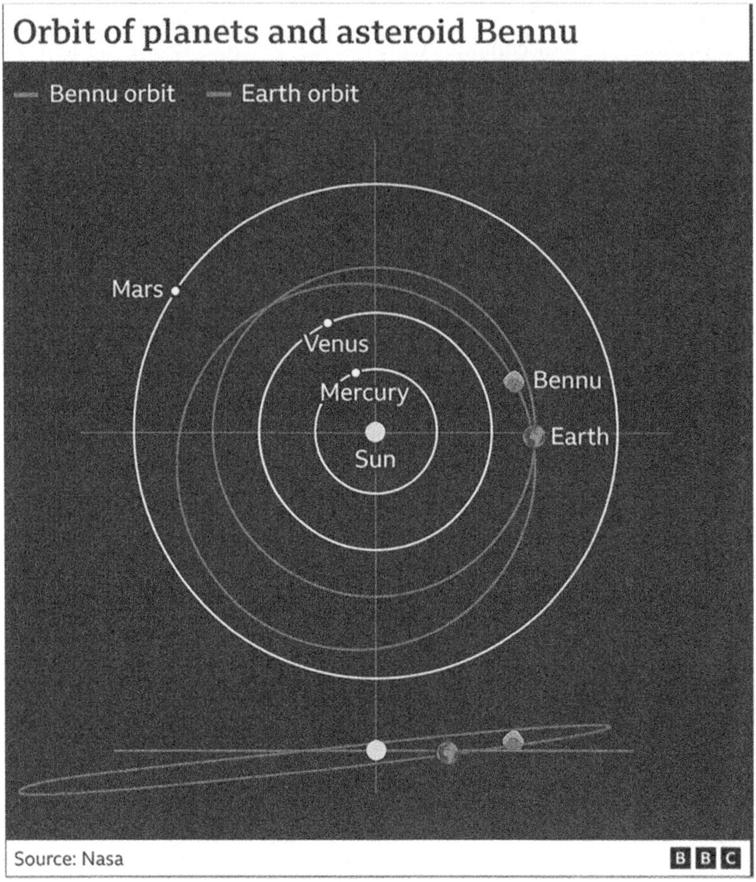

Figure 10.10 Bennu is a C-type asteroid of the Apollo family. This orbit's inclination is 6.03°, which is relatively low. *Source:* Ref. [14] / BBC.

Earth as 480,000 km (0.0032 AU). Bennu's orbit relative to the Sun and the inner planets is shown in Figure 10.10.

Composed of carbonaceous material, Bennu originally came from the breakup of a much larger parent asteroid. The parent asteroid is believed to have been formed 4.5 billion years ago. Bennu's material makeup is of scientific interest.

Bennu was selected because it met several mission requirements:

- Proximity to Earth: needing small ΔV from LEO
- Low eccentricity and low inclination so no large extra ΔVs would be needed to change the orbital plane
- Ease of sample collection: Bennu's surface matter should be loose gravel

Figure 10.11 Photo of the OSIRIS-REx satellite. *Source:* Ref. [15] / NASA / Public domain.

- Low spin rate: Small asteroids tend to spin/tumble, making physical contact more challenging. A low spin rate makes sample collection less challenging.

The OSIRIS-REx spacecraft weighs 880 kg with dimensions $2.44 \times 2.44 \times 3.15$ m, as shown in Figure 10.11. To see how the spacecraft's orbit matched with Bennu, interested readers are referred to an animation by NASA available in reference [16].

Figure 10.12 is a snapshot of the animation taken in May 2018 showing the relative locations of Earth, the spacecraft, and Bennu. The spacecraft's orbit changes steadily. Figure 10.12 shows that the spacecraft left Earth around the 6-o'clock position going in a counterclockwise direction in September 2016. It then takes about 1.75 orbits to reach its position in Figure 10.12. Since Bennu's orbital period is about 1.2 years, the duration is $1.2 \times 1.75 = 2.1$ years, which is consistent with the rendezvous timeline in Table 10.5. The inclination change is not shown in the animation.

In December 2018, the spacecraft's speed matched Bennu's from about 19 km away. Then, it moved to about 6.5 km and performed spectroscopic surveys of the rock's surface. The spacecraft then spent 18 months studying the asteroid up close. It was lowered to 1.75 km away to survey the asteroid to select a sample

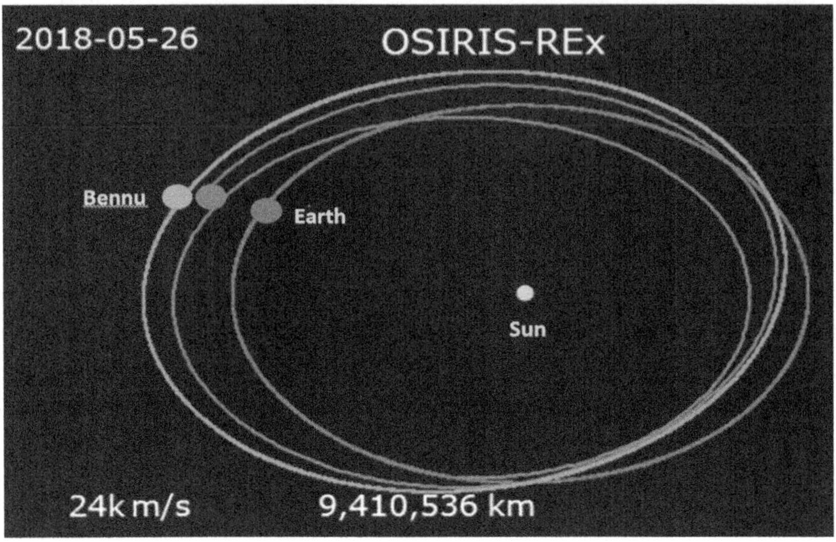

Figure 10.12 Snapshot of an animation describing the flight path of the OSIRIS-REx to rendezvous with the asteroid Bennu. The animation is by Phoenix7777 in his contributing article based on NASA data. Labels have been added to this image to denote the Sun, Earth, and Bennu. *Source:* Adapted from Ref. [16].

collection site. Upon lowering to an orbit around the asteroid about 1 km away by small thrust, the spacecraft descended on the selected site slowly while extending a boom 3 m long. When contact was made with the asteroid, a burst of nitrogen was fired and caused a crater about 8 m in diameter on the loose regolith. A mechanism collected the ejected regolith particles. The spacecraft then backed away. After verifying that enough samples had been collected, the spacecraft made moves to start the return journey. It took 3 years for the spacecraft to drop off the samples during a flyby in 2023, almost 5 years after the rendezvous.

In June 2024, NASA released early results of the samples [17]. The NASA team found the asteroid's dust to be rich in carbon, nitrogen, and organic compounds, so Bennu contains "the original ingredients that formed our solar system." Readers are referred to a short video released by NASA recording the boom touching down on Bennu [18].

10.5.3 Accessing Near-Earth Asteroids

Access to NEAs requires more quantitative evaluations for mission planning purposes. Two quantities are most important: Flight duration and ΔV. Flight duration is the time it takes for a spacecraft to rendezvous with the target NEA. The term ΔV represents the energy consumption for the spacecraft to go from a starting point (usually the LEO) to rendezvous with the target. Flight duration and energy consumption do not go together, and it is usually a tradeoff.

CNEOS provides a table listing the important characteristics of NEAs, including flight duration and ΔV for each asteroid [19]. These NEA candidates are deemed "suitable for future human space flight round-trip rendezvous missions." As will be elaborated again in Chapter 13, human missions to NEAs and Mars must be concerned about flight time versus energy consumption.

Among other data, two pairs of estimates are provided for each NEA candidate in the CNEOS table. The first pair is for minimal ΔV and the corresponding flight duration. The second pair is for the minimal flight duration and the corresponding ΔV with a different total mission time.

The return journey usually takes longer because the distance of travel is not optimal. Therefore, commercial mining ventures need to develop the acceptable time allowed to collect samples unless technology becomes available to conduct the analysis in situ.

10.5.3.1 Easily Recoverable Objects

A subclass of NEAs has been under consideration as good candidates for trial mining. Known as easily recoverable objects (EROs), they are 12 small-sized asteroids identified in 2013 as candidates for a different orbital transfer approach. These asteroids are Apollos, and some are PHAs as well [20].

The details of these 12 asteroids can be found online. The size of the 12 EROs ranges from 2 to 20 m. As calibration, a perfectly round asteroid 10 m in diameter would weigh over 500 tons (C-type), 900 tons (S-type), and 1,600 tons (M-type). Hence, the amounts of materials in these EROs are noticeable.

The ERO candidates are close to Earth's orbit. Their small size and proximity make them accessible and even transportable to other orbits using current propulsion technology. The orbital change amounts to applying a ΔV on the object. One idea is to move these asteroids to the Sun–Earth Lagrange points L1 and L2 and hold them there to be mined later by dedicated missions.

In summary, the thousands of NEAs provide perfect starting points to test out asteroid mining technologies and could even generate respectable revenues.

10.6 Infrastructures for Mining Asteroids

The authors do not take any position in debates about the possible economic disruption caused by an oversupply of mineral commodities. We need the materials to build up the Cislunar economy and will find ways to use the resources. Industrializing and building economic values require infrastructure to mine asteroids for large volumes of minerals. This section considers some of the infrastructures needed to facilitate eventual asteroid mining in volume.

For mining asteroids, we are still at the exploration stage, having done missions to collect samples. Unlike lunar missions with clearly stated objectives, there are no similar objectives set for asteroids as of this writing. Advocates of mining asteroids come from the private sector.

10.6.1 Asteroid's Local Gravity

Asteroids differ from celestial bodies such as the Moon in one aspect: nonexistent local gravity. The Moon's gravity is 1/6 of Earth's, or 0.17g. How much is the local gravity on an asteroid?

Say we have a spherical asteroid of radius R_1 with density d_1. We would like to estimate the local gravity g_1 on this asteroid compared to that of Earth. Using Newton's gravitational equation,

$$g_1 = g \times \left(\frac{R_1}{R_E}\right) \times \left(\frac{d_1}{d_E}\right) \tag{10.1}$$

where g, R_E, and d_E are the gravity, radius, and density of Earth.

We estimate the local gravity of Mars as a check. From online information, the radius and density ratios of Mars to Earth are 0.53 and 0.71, respectively. We can work out g (Mars) to be 0.37g.

Next, we estimate the local gravity of Ceres, the largest asteroid in the Belt. Its radius is under 500 km, and density (obtained online) is ~2,000 kg/m^3, whereas Earth's radius and density are 6,400 km and 5,500 kg/m^3, respectively. The local gravity of Ceres can now be estimated to be ~0.03g, or 3% of Earth's local gravity.

We can safely conclude that asteroids essentially have no local gravity to speak of. The implication is that a person or a mining machine would have no firm footing on an asteroid. This lack of footing impacts mobility and the choice of mining methods on the surface of asteroids. Any machine operation that works by pushing on the ground would receive a reaction force that could push the equipment away unless the surface consists of just loose regolith and pebbles. Hence, anchoring mechanisms must be implemented to prevent a person or equipment from drifting away.

Volume mining means that large amounts of materials will be excavated and collected, in the order of tens to hundreds of tons. Methods of mining and material collection need to account for this weightlessness.

10.6.2 ΔV to Asteroids

Unless otherwise noted, the missions described here have ΔVs relative to LEO as a common practice. Bear in mind that asteroids in the Belt and NEAs typically describe inclined elliptical orbits, as shown in Figure 10.3. Going from Earth to rendezvous with an asteroid requires the spacecraft to match the asteroid's orbit in terms of speed, position, and inclination. Inclination change consumes ΔV. Hence, ΔV values vary with inclinations, making each asteroid mission unique.

A quick survey of the Assessable NEA Data Table provided by CNEOS [19] via the NASA-funded near-Earth object Human Space Flight Accessible Targets Study (NHATS) gives a range of total ΔV assuming a certain stay time on the target asteroid. The total ΔV range is from 4 to 6 km/s from the LEO. This data table has two ΔV columns next to each other. One column provides the minimum total ΔV with an assigned flight duration shown. The other column is the total ΔV with the shortest flight duration. These two conditions are each accompanied by the best time to depart Earth. Hence, suggested departure dates for energy-efficient and duration-efficient missions are different.

Often, the timing of a true minimum ΔV does not happen often, while windows with acceptable ΔV open approximately every two years. For example, we can look at the first NEA 2000 SG344 on the data table. The details are available on the website [21]. The ΔV (minimum energy) is 3.56 km/s, requiring 153 days to and 185 days back to the LEO. Tagging on the 8-day stay time to survey the asteroid, the total mission time adds up to 346 days.

The ΔV (minimum duration) is 5.95 km/s requiring 49 days to and 57 days back; the total mission duration including the 8 days stay adds up to 114 days only. The departure dates for the two cases are also different. We can see the substantial increase in ΔV but corresponding decrease in the mission duration from this example.

What is the impact on payload capacity due to higher ΔV consumption? According to the Rocket equation, an increase in ΔV to reduce flight duration necessitates a smaller final mass in an exponential way. Since the payload is part of the final mass, the payload mass also decreases in an exponential way to make room for more fuel. This point is intuitive. If we want to get there faster, we need to burn more fuel at the expense of the payload mass. Therefore, time efficiency directly impacts the energy efficiency and payload capacity.

In addition, the mining spacecraft arrives at the target with equipment and may return with equipment and tons of excavated materials. In that case, the return journey would be a new calculation of ΔV and flight duration.

In summary, ΔV and mission duration play big roles in mission planning to a target NEA.

10.6.3 Transportation Infrastructure

In mining NEAs, the immediate infrastructure we need is transportation and energy. We pay attention to vehicles and support structures to facilitate the high-volume mining of asteroids.

The transportation infrastructure for mining asteroids covers several key items: 1) transport vehicles, 2) refueling stations, and 3) off-Earth processing plants and asteroid warehouses (if applicable).

10.6.3.1 Vehicles

We may accept the current mode of sending a spacecraft to investigate and collect samples from a target NEA, taking 5 or so years from start to finish. However, the goal is mining minerals in volume and delivering them to select destinations. We need high-capacity vehicles to transport mining equipment and return large quantities of raw or processed materials. We need the space equivalents of trains, trucks, and container ships, towing small asteroids or carrying tons of mined materials.

Taking our imagination further, we may wonder if we should consume large fuels (ΔV) to maneuver a large spacecraft (like a Starship) to land on an asteroid. Or would it be more practical to send shuttles back and forth from the mothership to the mining site on an asteroid? In addition, surface conditions may not be conducive for such an operation. These logistics issues and specialization of space vehicles will be resolved as more experience is gained from more missions and higher volumes of minerals extracted.

High-capacity vehicles need more than ion drives to provide propulsion due to the vehicles' mass and the cargo they hold. Some versions of the SpaceX Starship is a good option. Hybrid propulsion is the recommended option for human missions to Mars in Chapter 13. The kind of propulsion suitable for NEA mining may be the compromise between going to the Moon and going to Mars.

10.6.3.2 Refueling

Currently planned for refueling in LEO, high-capacity vehicles rely on refueling to deliver cargo from Earth to the Moon. On-orbit refueling, though not yet demonstrated as of this writing, is an important component of lunar missions to be covered in Chapter 11. Current sample collection missions need no refueling and rely on ion drives to provide the thrust without sacrificing the mission duration. On-orbit refueling will change this paradigm, and high-power thrusters can be used, allowing high-capacity vehicles to do the mining.

10.6.4 Energy Infrastructure

Industrial operations consume plenty of power to run their machinery. Here we look at the energy sources for vehicles, mining operations on a target asteroid, and the processing of raw materials either in situ on the asteroid or in off-Earth facilities.

There are three reliable sources of energy in space: solar, nuclear, and hydrogen fuel cells. Fission nuclear power and hydrogen fuel cells will be covered in

Chapter 12 as viable power sources on the Moon. They are closed systems that can operate anytime and anywhere. The amount of power needed to power a high-capacity vehicle and operate the mining equipment will need to be determined. A nuclear reactor can provide more power than a fuel cell.

Without local gravity, moving soil and drilling on an asteroid's regolith and then collecting the output depend partly on the soil's compactness and partly on the composition of the asteroid's surface. In some cases, thermal energy is needed to mine the frozen volatiles. Hence, the power consumption requirements vary with the target asteroids.

Solar power is freely available in space. Referencing Table 10.3, NEAs are defined by their closest and farthest distances from Earth. We are mostly concerned about the solar power intensity available to space vehicles and mining operations in the worst case—the farthest away according to the definition of a NEA.

Recall that the solar power intensity outside the Earth's atmosphere is $1.37\,kW/m^2$. The solar power intensity at 1.3 AU from the Sun relative to Earth's 1 AU can be estimated, giving a factor of $1/1.3^2 = \sim0.6$. Therefore, the solar intensity at 1.3 AU is $1.37 \times 0.6 = 0.81\,kW/m^2$. In other words, the same-sized solar panel on Earth's orbit will receive 40% less solar power at 1.3 AU. Extra-large solar panels would be needed to generate enough power.

10.7 Conclusion

Asteroids constitute a key part of the mineral resources available in our Solar System to support the Cislunar economy. Civilization can tap into this large pool of natural resources as a solution to the Earth's mineral resources shortage. Some resources can serve advanced industries on Earth immediately. Other resources will supplement those found on the Moon and Mars.

We can summarize the state of readiness of the mining asteroids as follows:

1) We on Earth are technologically not ready for the Asteroid Belt. It is more practical and economical to target near-Earth asteroids. The economic returns of mining NEA range from acceptable to attractive. Even the smallest of NEAs weighing several hundred to thousands of tons with the right compositions can bring materials worth in the trillions.
   Since lunar missions can utilize local resources, the initial demand for asteroid materials is likely to be precious metals to be used on Earth. This is the economic perspective we should keep in mind.
2) Our current rocket and spacecraft technologies allow us to rendezvous with and collect samples from target NEAs. Such missions still take 5–7 years from start to completion. Is this prospecting period too long for business ventures?

The Asteroid Belt has plenty more asteroids than the NEAs by number and size. NEAs are reachable, and appropriate space vehicles will be needed to shorten the mission time and excavate materials in volume. In parallel, the investigatory missions of the Asteroid Belt can start.

We are in the early phase of going to the Moon to be followed by building up the lunar infrastructure as well as the transportation capabilities between Earth and the Moon. Going to LEO is now routine. On-orbit refueling as a capability, once proven, will also become routine in spaceflights. The industry will invest in excavating and transporting large volumes of materials. High-capacity space vehicles that can carry many tons of cargo are vital to both lunar development and material extraction from the NEAs.

In addition, facilities to process tons of raw materials are part of the space infrastructure supporting asteroid mining. Some cases may process the excavated materials in situ for the target contents. Other cases may transport the raw materials to designated facilities. These facilities could be on the Moon and Mars or in select orbits.

It is conceivable that one or two ventures with attractive financial outcomes from NEAs will accelerate the commercial development of the ecosystem and relevant technologies via innovations.

Reading List

Brooks A. *Asteroid Mining: The Future of Energy*. Independently Published; 2020.

Diego SW. *Why Asteroid Mining Could Save the Earth?* Independently Published; 2024.

Lewis JS. *Mining the Sky: Untold Riches From the Asteroids, Comets, and Planets*. Basic Books; 1997.

Sivolella D. *Space Mining and Manufacturing: Off-World Resources and Revolutionary Engineering Techniques* 2019 Springer Praxis Books; 2019.

References

1 Kuiper Belt and Oort Cloud in context. European Space Agency [cited 2025 Feb 21] [online]. Available from: https://www.esa.int/ESA_Multimedia/Images/2014/12/Kuiper_Belt_and_Oort_Cloud_in_context.

2 Asteroid main-belt distribution. NASA Jet Propulsion Lab [cited 2025 Feb 21] [online]. Available from: https://ssd.jpl.nasa.gov/diagrams/mb_hist.html.

3 Asteroids. Journal Through the Galaxy [cited 2025 Feb 21] [online]. Available from: https://jtg.sjrdesign.net/advanced/asteroid.html.

4 Colombo M. Asteroids by size and number. DensityDesign Integrated Course Final Synthesis Studio [cited 2025 Feb 21] [online]. Available from: https://commons.wikimedia.org/wiki/File:Asteroids_by_size_and_number.svg.

5 Gradie J, Tedesco E. Compositional structure of the asteroid belt. Science. 1982;*216*(4553):1405–1407.

6 Exploring the metal-rich asteroid psyche. NASA [cited 2025 Feb 21] [online]. Available from: https://www.nasa.gov/image-article/exploring-metal-rich-asteroid-psyche/.

7 Desjardins J. There's big money to be made in asteroid mining. Visual Capitalist [cited 2025 Feb 21] [online]. Available from: https://elements.visualcapitalist.com/theres-big-money-made-asteroid-mining/.

8 NEO earth close approaches. Center for Near Earth Object Studies [cited 2025 Feb 21] [online]. Available from: https://cneos.jpl.nasa.gov/ca/.

9 Discovery statistics. Center for Near Earth Object Studies [cited 2025 Feb 21] [online]. Available from: https://cneos.jpl.nasa.gov/stats/totals.html.

10 NEO basics. Center for Near Earth Object Studies [cited 2025 Feb 21] [online]. Available from: https://cneos.jpl.nasa.gov/about/neo_groups.html.

11 Taylor K. The orbits of the four groups of near-Earth asteroids. Let's Talk Science [cited 2025 Feb 21] [online]. Available from: https://letstalkscience.ca/educational-resources/backgrounders/where-can-we-find-asteroids-and-comets.

12 Howell E. 'Trillion-dollar asteroid' zooms by Earth as scientists watch (video). Space [cited 2025 Feb 21] [online]. Available from: https://www.space.com/30074-trillion-dollar-asteroid-2011-uw158-earth-flyby.html.

13 Asteroid Bennu. The Planetary Society [cited 2025 Feb 21] [online]. Available from: https://www.planetary.org/space-images/asteroid-bennu.

14 Amos J. Osiris-Rex: Asteroid Bennu 'is a journey back to our origins'. BBC [cited 2025 Feb 21] [online]. Available from: https://www.bbc.com/news/science-environment-66844738.

15 Nowakowski T. We will find organic materials on Asteroid Bennu, says OSIRIS-REx principal investigator. Phys.org [cited 2025 May 1] [online]. Available from: https://phys.org/news/2015-07-materials-asteroid-bennu-osiris-rex-principal.html.

16 OSIRIS-REx. Wikipedia [cited 2025 Feb 21] [online]. Available from: https://en.wikipedia.org/wiki/OSIRIS-REx.

17 Surprising phosphate finding in NASA's OSIRIS-REx asteroid sample. NASA [cited 2025 Feb 21] [online]. Available from: https://www.nasa.gov/missions/osiris-rex/surprising-phosphate-finding-in-nasas-osiris-rex-asteroid-sample/.

18 Recording the boom touching down on Bennu. NASA [cited 2025 Feb 21] [online]. Available from: https://svs.gsfc.nasa.gov/vis/a010000/a013700/a013745/13745_TAG.mp4.

19 Accessible NEAs – Data Table. Center for Near Earth Object Studies [cited 2025 Feb 21] [online]. Available from: https://cneos.jpl.nasa.gov/nhats/.
20 These 12 asteroids are the easiest ones to bring back to Earth. Universe Today [cited 2025 Feb 21] [online]. Available from: https://www.universetoday.com/104084/these-12-asteroids-are-the-easiest-ones-to-bring-back-to-earth/.
21 Accessible NEAs – object/trajectory details. Center for Near Earth Object Studies [cited 2025 Feb 21] [online]. Available from: https://cneos.jpl.nasa.gov/nhats/details.html#?des=2000%20SG344&dv=6&dur=360&stay=8&launch=2020-2045.

11

Going to the Moon

11.1 Introduction

> When we send astronauts to the surface of the Moon in the next decade, it will be in a sustainable fashion…This time, when we go to the Moon, we will stay.
>
> *Jim Bridenstine, Head of NASA (2019)*

NASA's Artemis lunar program aims to land astronauts on the Moon sometime in the mid to the late 2020s and through the 2030s. Russia has announced plans to land cosmonauts on the Moon and build a base by the early 2030s. China has active lunar programs and intends to send humans to the Moon by 2030. Their objectives are similar.

The Moon is about 384,000 km from Earth and takes 3 days of space travel. By the solar system standards, the Moon is close by, yet it is a very different environment for humans to live in. It is the obvious celestial body to go next if humans want to venture beyond Earth. A different perspective can be taken. There is simply no other celestial body in the solar system with living conditions like Earth. Humans must overcome the challenges of living in different environments when we step out of the Earth.

This writer takes the optimistic position that Cislunar space is civilization's future, and we need to know what these new destinations are like, what they have to offer, and what it takes to travel and settle there. Settling on the Moon will lead to many opportunities, such as industrialization, special materials, scientific advancement, and serving as a waystation to other planets.

New Space: From Low Earth Orbit to the Moon and Beyond, First Edition.
Edited by Stephen W. Cheung and Chih-Yung Wen.

The science magazine *Nature* in a 2018 article "How to build a Moon base" [1] provides a simple indication of how things would go on the Moon. The choice locations for lunar settlements can be anywhere from the equator to the poles on the Moon, but severe constraints exist. The location of the first outpost must be chosen carefully.

The Moon is more than an outpost to learn how to live in an off-world environment for a few explorers. The next phase is infrastructure building so that large-scale industrial activities can take root to support one or more communities, as well as extract large quantities of materials to have meaningful economic impacts. Such infrastructures include transportation, energy, and communication. Indeed, the US has begun such a process via the Luna-10 study (Chapter 12), in which technology and economics have an equal bearing on the development of the Moon.

Lunar missions in the 21st century are different from those 6 decades ago. Advanced technologies such as computers, microelectronics, robotics, and new materials are making orbital missions routine and lunar missions more ambitious. Reusable vehicles are driving down launch costs. Upcoming lunar missions will see autonomous and semiautonomous machines do much of the heavy lifting and routine work on the Moon, such as data recording and producing materials for shelters.

This chapter covers the general lunar environment and the lunar missions of this decade. The next chapter will address the challenges and opportunities to settle on the Moon.

11.1.1 Lunar Environments

The current theory is that the Moon was formed when another massive rock collided with the still molten Earth about 4.5 billion years ago and resulted in the Earth and the Moon we know now. Therefore, the expectation is that the Moon should have a similar material composition to Earth, making the Moon a great source of resources that could benefit Earth and lunar settlers.

The lunar environments can be summarized as follows: 1) gravity about 1/6 of Earth, 2) weak to nonexistent magnetic field, 3) Rotation axis tilt of 1.5° relative to the ecliptic plane, 4) temperature from 127 °C (sunlit areas) to −130 °C (dark) at the equator, −250 °C in deep polar craters, 5) day and night length about 14 days each, 6) super thin to negligible layer of atmosphere, 7) frozen water in shadowed areas and in other forms, 8) radiation about 200 times higher than on Earth, 9) most minerals similar to that found on Earth, and 10) soil different from that on Earth.

By all accounts, the lunar environment is harsh, unfriendly, and deadly without protective measures. There is no air to breathe, and it is either very hot or very cold.

11.2 About the Moon

11.2.1 Landscape

High temperatures when sunlit and the lack of atmosphere explain the lack of liquid water on the lunar surface. When viewed from Earth, the shaded areas in Figure 11.1 are termed maria (plural for mare). They are the low areas and are likely made flat and smooth from ancient lava flows that cover craters from asteroid impacts earlier. The near side of the Moon is covered with multiple large maria—about 31% of the surface.

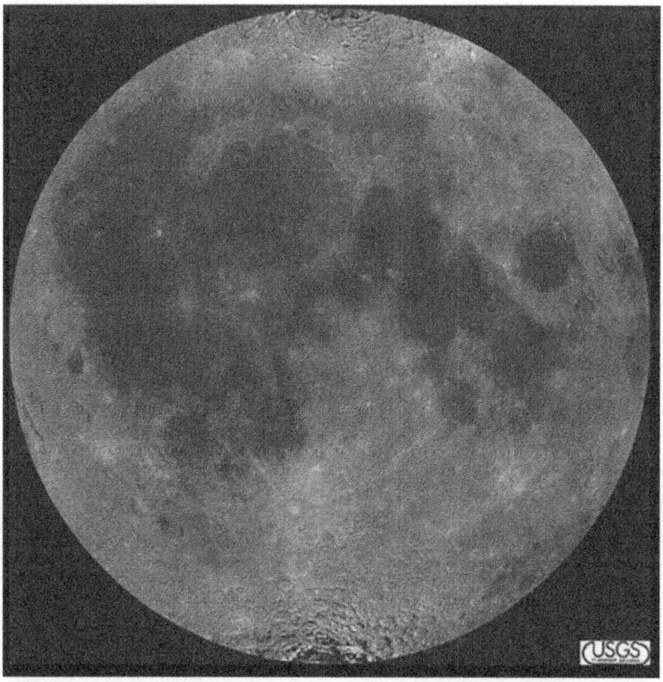

Figure 11.1 Maria (lowlands) appears as dark areas while highlands are bright. *Source:* Ref. [2] / United States Geological Survey / Public domain.

The bright areas in Figure 11.1 are called the highlands. They were formed when the ground was lifted due to multiple asteroid impacts billions of years ago. The highlands are heavily cratered and prominent on the lunar far side.

Impacts of asteroids and meteorites over the millennia have also produced a layer of debris consisting of rocks and dust. This layer covers the entire lunar surface and is known as the lunar regolith. The top layer of the fine matter is referred to as lunar soil, but it contains no organic materials (as the term "soil" usually implies).

Moon dust is formed from the pulverization of rocks and soils when meteoroids bombard the lunar surface, as well as from steady exposure to charged particles from the Sun. Moon dust is ultrafine and is hazardous to human health and machine operation.

11.2.2 Apollo Landing Sites

The landing sites of the Apollo program and the unmanned spacecraft prior were all on the near side of the Moon. The north and south poles have not been visited by humans or rovers. Continuous communications with the Earth were important and could not be guaranteed from the poles.

11.2.3 Upcoming Landing Site: The Lunar South Pole

Lunar missions this century aim ambitiously to establish outposts to be followed by exploration and large-scale activities. Pioneers will stay for days in the early phase and for weeks and months as experience builds.

The best place to land should be suitable for a longer stay than the Apollo astronauts and have the right resources to steadily fulfill the larger objectives of the Artemis series of missions.

The locations of choice must have 1) access to sunlight for solar power, 2) access to water and oxygen, or materials containing them, and 3) suitable shelters for pioneers. These requirements led to the lunar south pole being the choice location. The lunar equator can provide solar power but has no water due to its direct exposure to sunlight that can reach over 100 °C. The 14-day-long lunar night also causes a lack of sunlight in non-polar areas for a long time. Water is believed to exist in the permanently shadowed areas of craters where the temperature is extremely low. The poles meet this criterion because sunlight only grazes the polar surface and is not affected by the diurnal cycles, which is made possible by the small axial tilt of 1.5°. The radius of the Moon's polars is less than 50 km.

Solar power collection stations can be erected such that at least one is always exposed to sunlight. Erecting solar power panels at the peak of the highlands in the south pole is one option. This makes solar power the primary source for a lunar outpost soon after landing.

11.3 Artemis Program

Currently, the United States, China, Russia, and India have each advanced their plans to land humans or machines on the Moon. Other countries are sending satellites to collect various data from the above.

The Artemis program is a series of lunar missions by NASA. The program is an international collaboration through the Artemis Accords signed in October 2023. The United States will be the champion and member countries will contribute their resources and know-how. The Artemis program has three phases. Its main goals include 1) setting up a lunar space station termed Gateway, 2) landing astronauts on the Moon, and 3) setting up an outpost to collect scientific data about lunar resources.

The Artemis teams will stay on the Moon for a significantly longer period than the Apollo astronauts. Missions will progressively deliver humans, equipment, and supplies to the destination. The Gateway will serve as a waystation for vehicles from Earth and for lunar landing vehicles that will shuttle between the space station and the lunar surface.

11.3.1 Lunar Orbits

The lunar space station needs to describe a stable polar orbit in order to minimize fuel consumption. Whereas space stations orbiting Earth see residual atmospheric drag, the Moon has no atmosphere. Another requirement is a reasonable distance to the lunar south pole for physical access and observation. Hence, a special orbit for the lunar space station should bear these two requirements in mind. Many possible orbits exist. We will examine some options.

A low lunar orbit (LLO) is circular or elliptical and is typically 100 km or less from the lunar surface, with a period of about 2 hours. A polar LLO would be great for access to the lunar poles since there is no atmospheric drag. However, the Moon's density is not homogeneous, so there are variations in local gravity. Such influences at this low altitude are quite significant and will require frequent fuel consumption to maintain a stable orbit, making the LLO not suitable for Gateway's long-term presence.

Alternate orbits come from leveraging the Earth–Moon Lagrange points. A distant retrograde orbit (DRO) is a stable circular orbit that circles the Moon every two weeks. The Artemis 1 observation spacecraft (below) followed a DRO that went between Lagrange points EML1 and EML2. But, it is quite far from the lunar surface. EM-L1 is 63,000 km from the Moon and is not suitable for regular commuting between the lunar station and the surface.

A near rectilinear halo orbit (NRHO) is a special halo orbit that can be selected to pass close to the lunar region of choice, such as one of the poles. NRHO looks like an ellipse but not exactly. "Near rectilinear" means that some segments of the orbit have a larger curvature than an ellipse of the same major axis, and some segments have a smaller curvature. A halo orbit is a three-dimensional orbit set around one of the EM-L1 and EM-L2.

Figure 11.2 shows a snapshot of an animation of NRHO. In this case, a spacecraft in this orbit will be closest to the lunar north pole and farthest from the south pole. This NRHO is stable with an orbiting period of 6.5 days, with its low point (also known as perilune) at about 1,500 km and high point (also known as apolune) at 70,000 km from the lunar surface. An important advantage is that a spacecraft in this orbit would have an unobstructed view of Earth, making it possible for the observation of the Moon's far side and continuous communications with Earth.

An NRHO is a good choice for the Gateway station, which is designed to stay in this orbit for 15 years, so minimal fuel consumption is vital. Figure 11.3 shows the two families of NRHO-L1 and NRHO-L2 referring to the two Lagrange points. Each family has two orbit options that offer the closest distance to the north or the south pole.

NASA places the Gateway in a NRHO South L2, i.e. the perilune occurs at the north pole at an altitude of 1,500 km and the apolune at the south pole at an altitude of 70,000 km. The orbital period is 6.5 days. Recall that the speed at the apogee of an orbit is the slowest. In this orbit, the time over the south pole would be the longest. This would allow the most time for astronauts to study the south pole from the Gateway. Second, a lunar lander undocking from the Gateway over the north pole would have the fastest speed and could apply thrust to change the orbit to an LLO and steadily descend on the south pole.

11.3.2 Transportation Vehicles

Artemis missions are propelled to Earth orbits by a heavy-lift rocket system known as the Space Launch System (SLS). The current plan is for the lunar payload to be inserted into an NRHO-South L2 orbit around the Moon to rendezvous with the Gateway. From there, a separate lander will shuttle to and from the lunar surface.

11.3.2.1 Space Launch System

The Space Launch System is NASA's most potent heavy-lift launcher. The SLS is a two-stage rocket with a powerful first stage and an upper stage. The payload is

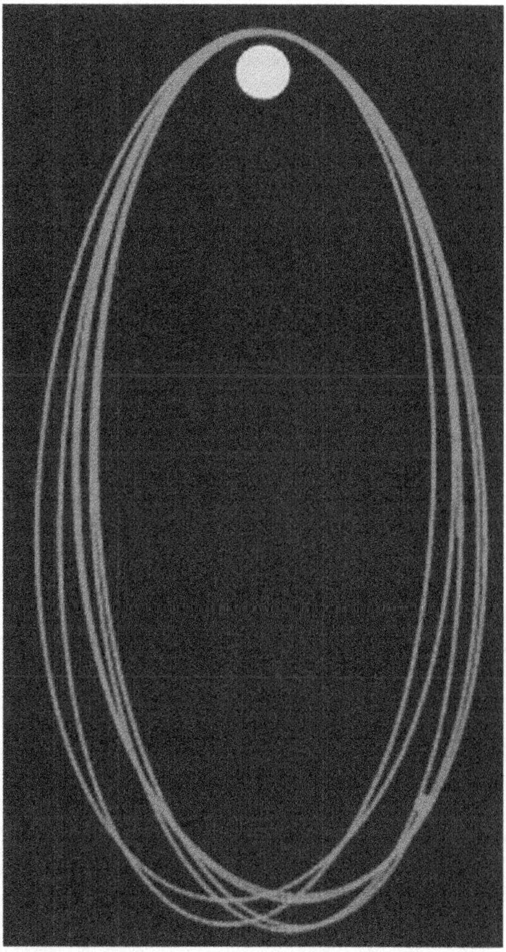

Figure 11.2 The link shows an animation of the Gateway space station around the Moon following a near-rectilinear halo orbit (NRHO). The article contributor Phoenix7777 produced the animation using the ESA data. *Source:* Ref. [3] / Phoenix7777 / Wikimedia Commons / Public Domain.

a crewed or cargo spacecraft. The SLS first stage is common for crew and cargo missions, but the upper stage is different. The crewed vehicle is known as the Orion.

Figure 11.4 shows the crew configuration of the SLS system. The first stage comprises a core stage and two side boosters. The core stage has four powerful RS-25 engines. The two side boosters provide extra thrust to push the payload to the LEO. Together, the first stage's lifting capacity is 130 metric tons. The engines on the SLS and the Orion crew vehicle run on liquid hydrogen and liquid oxygen, while the side boosters use solid fuels. Readers are referred to the NASA SLS fact sheet for a more detailed description of the SLS [6].

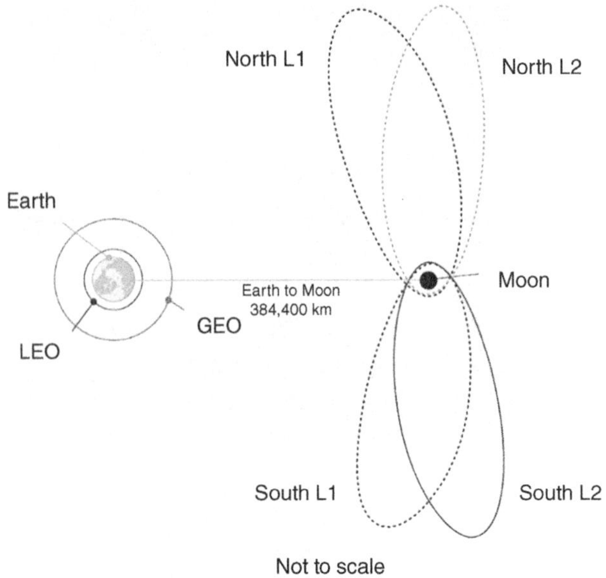

Figure 11.3 The diagram shows two families of the near-rectilinear-halo orbit NRHO-L1 and NRHO-L2 using the two Earth–Moon Lagrange points. The orbit of each family can be chosen with the closest point near the lunar north pole or south pole. The closest distance is 1,500 km, and the farthest is 70,000 km. *Source:* Ref. [4] / Maxar Technologies.

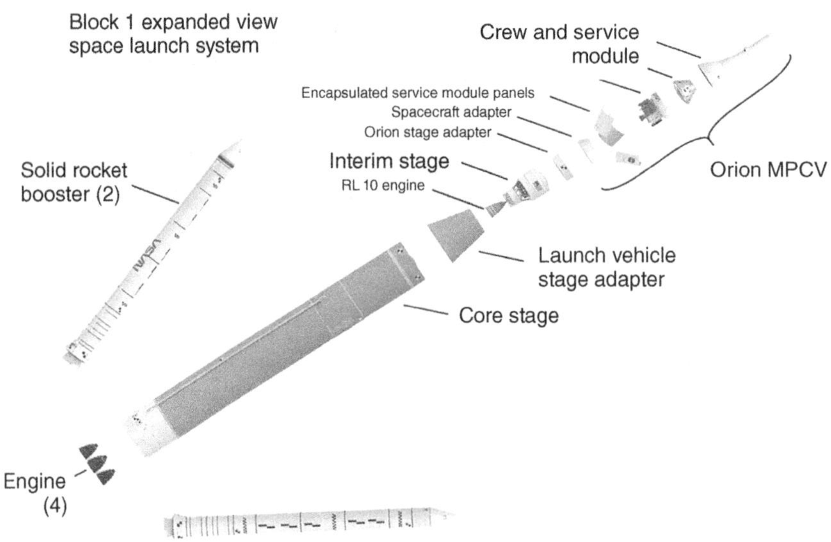

Figure 11.4 Construction of NASA's Space Launch System (SLS) for crew. The Orion crew vehicle, which comprises a service module and a crew module, is the payload. *Source:* Ref. [5] / NASA / Public domain.

The second stage of the crew configuration is called the "interim cryogenic propulsion stage" (ICPS) with one engine and sits atop the core stage, as shown in Figure 11.4.

The cargo configuration needs more lifting thrust. The extra thrust comes from a different upper stage with four rocket engines. The cargo configuration is therefore taller. The core stage and the two side boosters are single use, i.e. not reusable, and fall into the ocean.

11.3.2.2 Starship Human Landing System (HLS)

In 2021, NASA awarded a $3 billion contract to SpaceX to build the lunar variant of the Starship spacecraft known as Starship HLS to take astronauts to the Moon from the Gateway. The first two HLS missions are to be launched in the second half of the 2020s.

The current Starship spacecraft can carry 100 passengers or over 100 tons of cargo to LEO. The caveat is that the Starship needs to be refueled after it reaches LEO in order to go further to high orbits or the Moon.

There is no plan in the current Artemis missions for the Starship HLS to return to Earth. Therefore, the HLS has no flaps or heat shields due to the lack of a lunar atmosphere. The raptor engines will likely not be used to land due to concerns about kicking up too much Moon dust, which may stay suspended in space for a long time. Landing and lateral maneuvers are to be propelled by hot-gas reaction control system (RCS) thrusters placed around the midsection of the Ship, as illustrated in Figure 11.5.

Figure 11.5 Artist's picture of the Starship HLS landing on the Moon. Hot gas reaction control systems (RCSs) in the ship's midsection are used to provide vertical propulsion for slow down and landing maneuvers. *Source:* Ref. [7] / SpaceX.

A point of interest is that the Starship is designed to run on methanox and not LH2/LO2. The common view has been that LH2/LO2 can be obtained by dissociating water on the Moon. As will be seen in Chapter 12, the Moon lacks carbon and carbon dioxide or in very low concentrations. This makes refueling for the Starship on the Moon a separate technological issue for SpaceX.

11.3.3 Orion Multi-purpose Crew Vehicle

The full name of the Orion is Orion Multi-purpose Crew Vehicle or Orion MPCV. Its construction is shown in Figure 11.6. Orion is the NASA spacecraft that will transport four astronauts to the Gateway and later return to Earth. Orion is designed for deep space exploration to the Moon and Mars from low Earth orbit. It is equipped with various capabilities to support the crew. The crew module will return to Earth by splashdown.

Orion comprises two modules: the crew module and the service module. The service module, sometimes referred to as the European Service Module, provides propulsion. Upon launch, the solid boosters will drop out after about 2 minutes. The core stage continues on to LEO. At about 160 km altitude and 8 minutes after

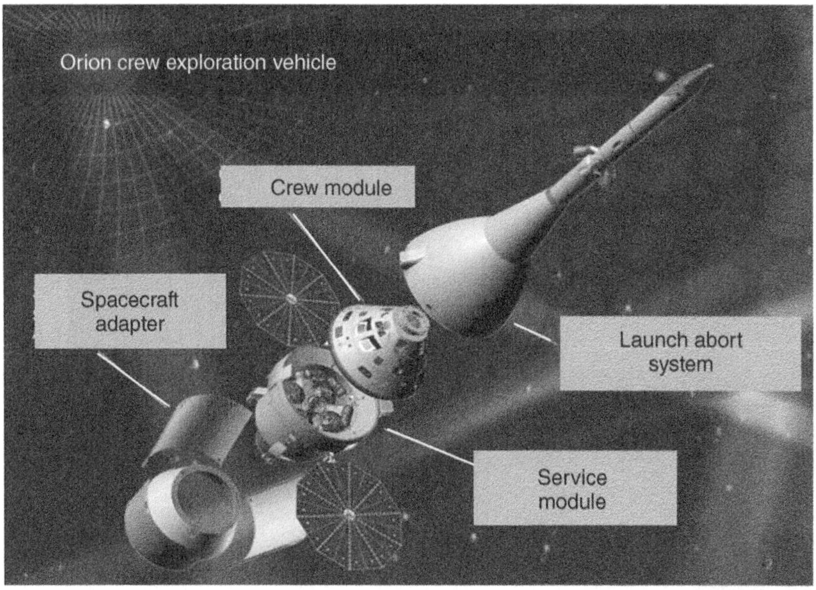

Figure 11.6 Construction of NASA's Orion crew exploration vehicle. The vehicle is designed for human deep-space missions to the Moon and Mars. The crew and cargo are housed in a module atop the service module, which has propulsion capabilities. *Source:* Ref. [8] / NASA / Public domain.

liftoff, the core stage reaches 7.6 km/s when stage separation occurs. The upper stage continues toward a higher orbit.

Orion detaches from the SLS upper stage at approximately 3,700 km altitude; the service module propels Orion toward the Gateway orbit. The service module is equipped with a main engine, 8 auxiliary thrusters, and 24 reaction control system thrusters. The main engine and the auxiliary thrusters are at the spacecraft's aft, as shown in Figure 11.7. The RCS thrusters are on the side for various precision maneuvers such as docking.

The Artemis program calls for Orion to only go to the Gateway, where it will stay docked until it is ready to return to Earth. The crew module can last 21 days if undocked and 6 months if docked with the Gateway. It also has a life support system and heat shields for reentry to Earth. The pressurized volume is 20 m^3, of which 9 m^3 is habitable.

Since the Gateway is not ready in the early phase of Artemis, the plan is for Orion to dock with a Starship HLS.

The service module will propel Orion on its return journey to Earth. It will be detached from the crew module before reentering Earth and burning up in the atmosphere. The RCS thrusters on board the crew module will reorient the vehicle to point the heat shield forward during re-entry. The crew module will parachute down to the ocean after slowing down in the atmosphere.

Figure 11.7 The European Service Module is the service module that provides the propulsion for Orion toward the lunar space station. The main engine is in the center. Eight auxiliary thrusters are next to the main engine near the perimeter. Twenty-four RCS thrusters are used for lateral maneuvers for Orion. The service module also propels Orion from the lunar space station back to Earth. *Source:* Ref. [9] / NASA / Public domain.

11.3.4 Lunar Gateway Space Station

The Lunar Gateway, or simply Gateway, is a key piece of infrastructure connecting the Earth and the Moon. The Gateway is intended to be an orbiting outpost for astronauts and spacecraft commuting to the lunar surface as well as a laboratory for scientific experiments on board. It is a mini version of the ISS with a designed life expectancy of 15 years or more. The Gateway's projected weight is 63 tons with a pressurized volume of over 125 m^3 (versus the ISS's 420 tons and 900 m^3). The Gateway will only be occasionally occupied by astronauts, although experiments can run autonomously. Current plans call for an annual 42-day visit by four-member crews.

The Gateway consists of modules called "elements," which ESA and JAXA will collaborate with NASA to build. The elements' functions are aligned with the Artemis objectives, as shown in Figure 11.8. The Orion MPCV is shown being docked to the assembly on the lower right. Assembled in orbit, the elements represent key functions such as habitation, power and propulsion, laboratory, supplies, and docking with other vehicles.

Some of the Gateway modules will be built and operated by contractors. They are identified in Figure 11.8.

The elements and their functions are listed in Table 11.1.

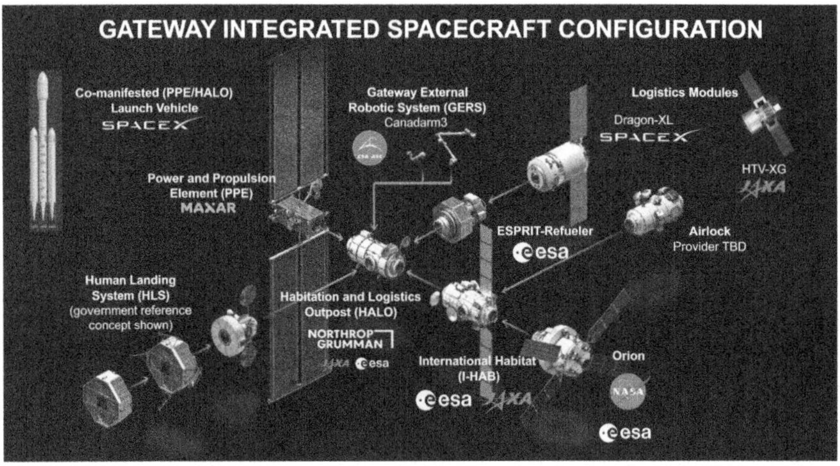

Figure 11.8 The lunar space station Gateway is made up of several segments known as Elements. The PPE, HALO, I-HAB, ESPRIT, Canadarm-3, and Airlock are connected permanently together. Vehicles such as the Orion dock to the I-HAB while the HLS and Logistics Module dock to the HALO from the side. The organizations responsible for the segments and vehicles are also shown in [10]. *Source:* NASA/Wikimedia Commons/Public domain.

Table 11.1 The Gateway elements and their respective functions.

Element	Function
Power and propulsion element (PPE)	60 kW total power for the propulsion systems, station keeping, and operating onboard equipment. Propulsion comes from solar electricity.
Habitation and logistics outpost (HALO)	A habitation module; used in research, command center, power distribution, and docking ports
International habitation (I-Hab)	Second habitation module with multiple docking ports
The European System Providing Refueling Infrastructure and Telecommunications (ESPRIT)	Provides fuel to the PPE, storage facilities, and habitation
Airlock module	Payload transfer from the habitat modules to the extravehicular space
Canadarm-3	– Advanced robot, capture visiting spacecraft – Assist astronauts during spacewalks – Tool for scientific research – Inspect, maintain, and repair the station
Logistics module	Commercial vehicle for logistics, carrying cargo and materials from Earth. Not intended to return to Earth in early missions.

The elements will be transported to the lunar orbit in Artemis 4, 5, and 6 missions. The current timelines for the Gateway elements to be delivered are outlined in Table 11.2 based on the information available.

Readers are referred to various websites [10] for pictures and details of each of the elements. We will briefly describe each.

Referencing Table 11.1, HALO is one of the two habitation elements. It is a pressurized habitat for astronauts and preparation for trips to the lunar surface, as well as conducting research on board the station.

Like HALO, the I-Hab serves as living quarters with ports for docking by visiting spacecraft.

ESPRIT is a fuel depot as well as a storage warehouse. The fuel will serve Gateway's own use and the lunar lander.

The Canadarm-3 robotic arm comprises a large arm 8.5 m long and a smaller one, as shown in Figure 11.9. The smaller arm is designed for dexterity. The system works both autonomously and by control from either Gateway or the ground control in Canada. The robotic arm serves multiple functions. According to the

Table 11.2 Schedule of different Gateway elements being delivered to the lunar orbit for constructing the Gateway.

Year	Mission objective	Mission name	Launch vehicle	Human/robotic elements
2027	Launch PPE and HALO	Gateway	Falcon Heavy	Robotic (unmanned)
2026	Orion MPCV with crew for the first lunar landing on Starship	Artemis 3	SLS	Crewed
2028	Orion MPCV and I-HAB module; 2nd Moon landing	Artemis 4	SLS	Crewed
2030	Orion MPCV and ESPRIT Refueling Module (ERM)	Artemis 5	SLS	Crewed
2031	Orion MPCV and the Crew and Science Airlock module	Artemis 6	SLS	Crewed
2031	(Proposed) Orion MPCV and Logistics Module	Artemis 7	SLS	Crewed

Source: Ref. [11] / NASA / Public Domain.

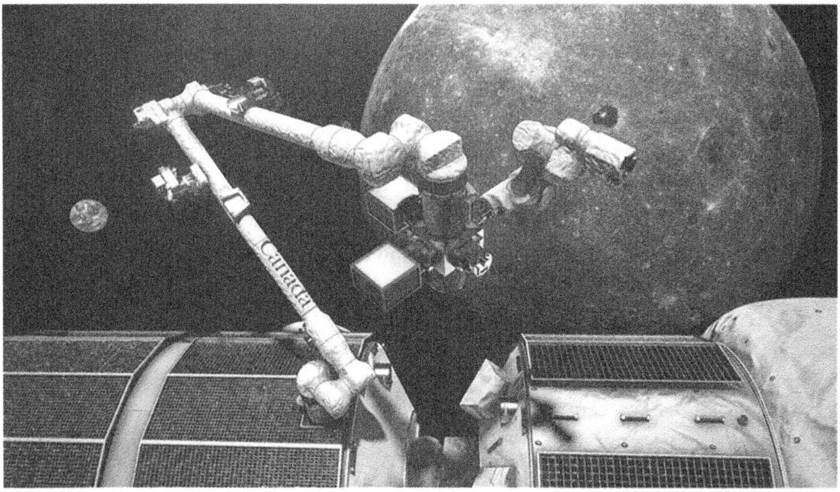

Figure 11.9 Canadarm-3 is a robotic arm. It is not attached to a specific element on the Gateway but to the exterior of the lunar space station. It can move freely between different elements and perform different tasks. *Source:* Ref. [12] / Canadian Space Agency.

Canadian Space Agency [13], the robotic system is designed to 1) maintain, repair, and inspect the Gateway, 2) relocate Gateway modules, 3) help astronauts during spacewalks, 4) enable science in lunar orbit, and 5) catch spacecraft visiting the Gateway.

The Logistics Module is not a permanent part of the Gateway. Its role is to deliver pressurized and unpressurized cargo from Earth to the Gateway. Supply missions will dock to and undock from ESPRIT. The Logistics Module is to stay docked for 6–12 months. At the end of this period, the module is not designed to return to Earth in the early phase of the Artemis Program. It will remove the spent materials, undock, and send itself to a disposal orbit.

NASA states that the logistics of sending supplies to the Gateway will follow the precedent set for the ISS by subcontracting to commercial vendors. The vendors will design and demonstrate these vehicles. Each supply mission will deliver at least 3,400 kg of pressurized cargo and 1,000 kg of unpressurized cargo to the Gateway.

SpaceX has been awarded contracts for two cargo missions. A Dragon XL spacecraft will be launched on a Falcon Heavy rocket. Its total cargo capacity is 5,000 kg. Public information suggests that the Dragon XL is based on the Crew Dragon capsule. Interested readers are referred to an online video for a more detailed description [14].

11.4 Artemis Missions

The Artemis program is the 21st-century version of the Apollo program (Artemis is the twin sister of Apollo). We will cover Artemis 1 to Artemis 5. The Artemis program is planned to go into the 2030s.

11.4.1 Artemis 1 Mission

Artemis 1 was an unmanned mission to test the SLS launching system and the Orion vehicle. Formerly called Exploration Mission-1, Artemis 1 went around the Moon without landing. Artemis 1 was launched in November 2022 and splashed-down back on Earth in December 2022. The mission duration was 25.5 days from start to completion. The launch part of the mission is illustrated in Figure 11.10 as several steps.

Steps 1–6 describe the flight up to the MECO and the booster falling into the ocean. Of interest is that the core stage's MECO (Step 5) happens at a much higher altitude of 162 km. Steps 7–10 show the upper stage taking Orion to high orbits to a high altitude of over 3,700 km before separation. The Orion vehicle disengaged from the upper stage and continued to go around the Moon. As there was no plan for a lunar landing, only a distant-retrograde orbit (DRO) was chosen.

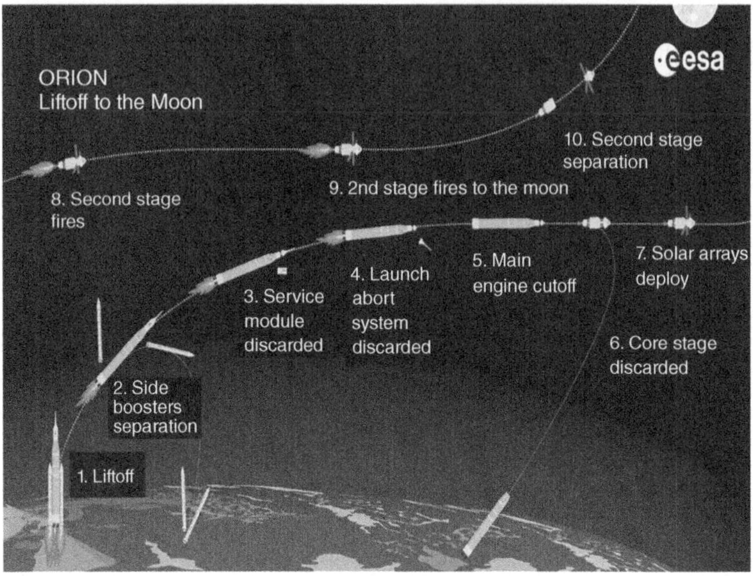

Figure 11.10 Mission description of the unmanned Artemis 1 mission. The Orion vehicle sits atop the SLS launching system. Steps 1–10 describe from the launch to the second stage propelling Orion to a translunar journey. *Source:* Ref. [15] / European Space Agency.

The complete flight path of Artemis 1 is shown in Figure 11.11. The first path (Bubbles 1–6) describes from the launch on Earth to the lunar orbit. The second path (Bubbles 6–9) describes the stay in lunar orbit. The third path (Bubbles 9–13) is the return journey to Earth. The lunar orbit came as close as 130 km from the

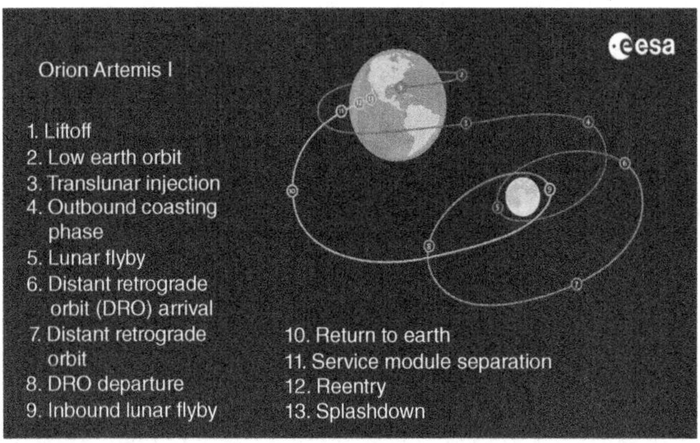

Figure 11.11 The flight path of the Artemis 1 mission has three segments: from Earth to the lunar orbit (red path), around the Moon (blue path), and from the Moon back to Earth (orange path). *Source:* Ref. [15] / European Space Agency.

lunar surface. A variety of experiments and data collection were conducted in deep space.

Artemis 1 successfully demonstrated the capabilities of the SLS and the Orion.

11.4.2 Artemis 2

Scheduled to launch in 2026, Artemis 2 will be a crewed flight without a lunar landing. The Orion vehicle will take a crew of 4 to 8,900 km beyond the Moon. After a lunar flyby, Orion then returns to Earth. Flight data and other support systems will be tested in this 8-to-10-day mission. The flight path is illustrated in Figure 11.12.

Bubble 6 is when the ICPS and the Orion separate in a high Earth orbit. Notation A in Figure 11.12 is "Proximity Operations Demonstration." This is for the crew on board the Orion to evaluate the manual handling of the Orion vehicle. Notation B after Bubble 7 means that the ICPS itself will follow a disposal trajectory back to Earth (dotted line in Figure 11.12). After separation, the Orion continues its translunar outbound trajectory around the Moon. The return journey is straightforward.

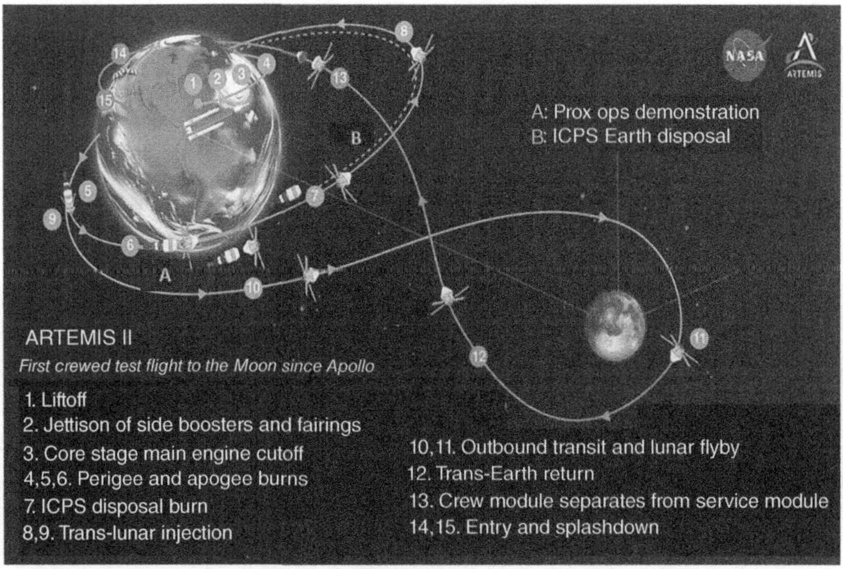

Figure 11.12 Flight path of the Artemis 2 crewed mission. There is no lunar landing or space station in this mission. After launching from Earth, the Orion goes on a lunar flyby around the Moon and then returns to Earth *Source:* Ref. [16] / NASA / Public Domain.

11.4.3 Artemis 3

Projected to launch in 2027, Artemis 3 is a four-person crewed mission. Two astronauts will land on the Moon. The lunar stay is 6.5 days. Artemis 3 is unique and complex in that two spacecraft, Orion and the Starship HLS, dock with each other in lunar orbit. The first woman and a person of color will land on the Moon—a historic milestone for the space age. The mission is estimated to take 30 days from start to finish.

A noncrewed Starship HLS loaded with supplies will be launched from Earth ahead of the SLS launch. After refueling by fuel tankers already waiting in LEO, the HLS will proceed to the Moon. It will enter the NRHO designed for the Gateway. The HLS will wait there for Orion. The concept of the operation is shown in Figure 11.13.

Note that the HLS will be refueled in LEO, a key capability to be demonstrated before the mission.

We emphasize that the Gateway will not be in place for this mission. Therefore, the Starship HLS could be considered as the "acting Gateway." This Artemis 3 mission will begin with four astronauts on board the Orion and will be launched by the SLS. The Orion spacecraft will enter the NRHO, meet up, and dock with the Starship HLS. Two astronauts will then be transferred to the HLS. The other two astronauts will remain on board the Orion to perform experiments.

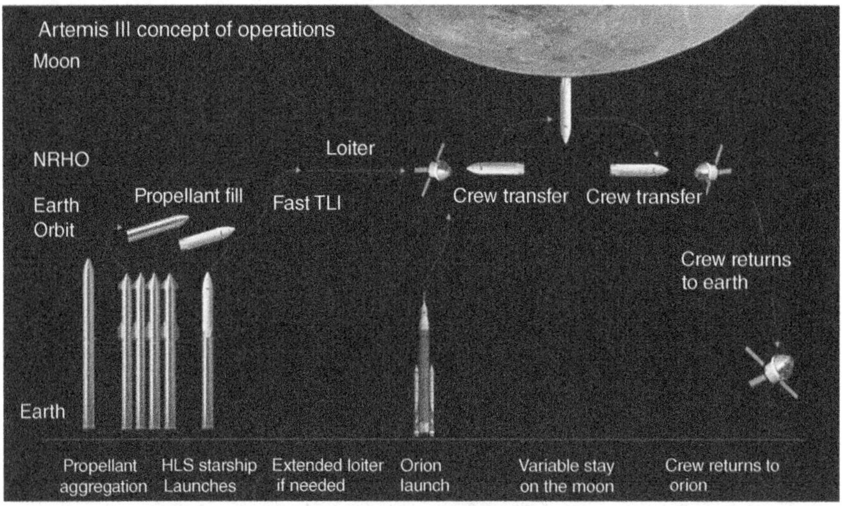

Figure 11.13 Artemis 3 is a human mission to land 2 astronauts on the Moon. Four astronauts will ride on the Orion atop the SLS launch system to a near-rectilinear-halo orbit. An unoccupied Starship HLS will be launched separately and will be waiting to rendezvous with the Orion. Two astronauts will transfer to the HLS and proceed to land on the Moon. The Starship will be refueled in LEO [17].

Upon undocking from the Orion above the north pole, the Starship HLS will change orbit and steadily descend to the lunar south pole while the Orion continues its orbit for one NRHO revolution, which is 6.5 days. As of this writing, there is no definitive information about HLS's landing site at the south pole.

The flightpath of Artemis 3 is illustrated in Figure 11.14 with the steps as numbered. The steps are as follows: 1) Launch, 2) LEO, 3) translunar injection, 4) outbound coasting phase, 5) outbound powered flyby, 6) Gateway orbit insertion burn (and dock with the HLS), 7) human lander landing on the Moon, 8) undocking and astronauts return to Earth in Orion, 9) inbound powered lunar flyby, 10) inbound coasting phase, 11) service module separation, 12) reentry, and 13) splashdown.

The Starship HLS's function in the Artemis program is as a commuter vehicle for lunar landing with the ability to dock with Orion. The Starship HLS in this particular mission will not be reused after Orion departs for Earth.

Upon landing, the two astronauts will go through a series of tests. Their temporary residence is the HLS as there is no plan to build any shelters in this mission.

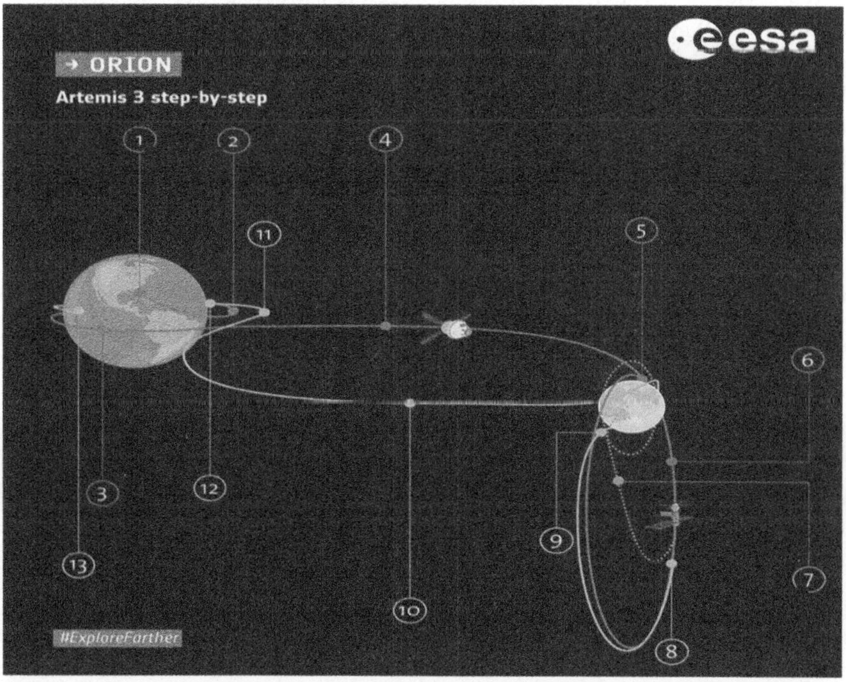

Figure 11.14 Flightpath of Artemis 3. Bubbles 1 to 5 show the outbound path from Earth to the Moon. Between Bubbles 6 and 7 indicate that the Gateway will be there, but it will not be in this mission. Instead, the Starship HLS will be waiting there to dock with the Orion vehicle where the crew transfer will occur in orbit. *Source:* Ref. [18] / European Space Agency.

When ready, the two astronauts will proceed to explore the lunar surface near the landing site. Their exploration includes recording the surroundings, performing geological surveys, and collecting samples. Water ice samples will be collected and analyzed. They will maintain constant communications, including data transmission, with mission control on Earth and the Orion.

The two astronauts will have 6.5 days to complete their tasks on the Moon. Their stay on the Moon is timed so that the HLS will lift off to the NRHO to reunite with the Orion. The astronauts will spend the next 5 days transferring samples to the Orion.

The Orion will undock from the HLS and depart for Earth (see Bubbles 8 and 9 in Figure 11.14). The Starship HLS will be disposed of by moving on to a heliocentric orbit.

11.4.4 Artemis 4

Currently scheduled for 2028, Artemis 4 has two objectives: 1) deliver the International Habitation (I-HAB) element to integrate with the power and propulsion element (PPE) and the habitation and logistics outpost (HALO) already in lunar orbit to form the Gateway and 2) deliver two astronauts to land on the Moon on board another Starship HLS. Two astronauts will remain with the newly assembled Gateway, and two astronauts will spend 6.5 days on the Moon and then return to the Gateway. Like Artemis 3, the mission is estimated to take 30 days from start to finish. The flight path is shown in Figure 11.15 in which the numbered steps are shown. The steps are straightforward.

Approximately a year prior to the Artemis 4 launch, two Gateway elements, PPE and HALO, will have been built and launched together on a Falcon Heavy rocket. The payload will take approximately 12 months to reach the designated NRHO around the Moon by solar-electric propulsion (ion drive), collecting data along the way. Once in orbit, the PPE and HALO will become the Gateway albeit still missing other elements. Its onboard computer will run a series of tests to prepare for the arrival of the Artemis 4 spacecraft.

A new Starship HLS and a Dragon XL Logistics module will be launched separately to rendezvous with the Gateway. The new version of the Starship HLS will be designed to dock with the newly assembled Gateway.

The Dragon XL vehicle will become the logistics module for the Gateway (see Table 11.1) in this mission. It will carry both pressurized and unpressurized supplies and scientific experiments. It will be launched on a Falcon Heavy rocket and dock with Gateway's HALO.

The cargo configuration of the SLS rocket will be used for Artemis 4 due to the increased payload weight, which will be the Orion and the I-HAB element. The cargo configuration has a more powerful upper stage and more cargo capacity.

The Orion and the I-HAB are not connected at launch. The upper stage will push the payload to a translunar orbit after separating from the SLS. When the boost is

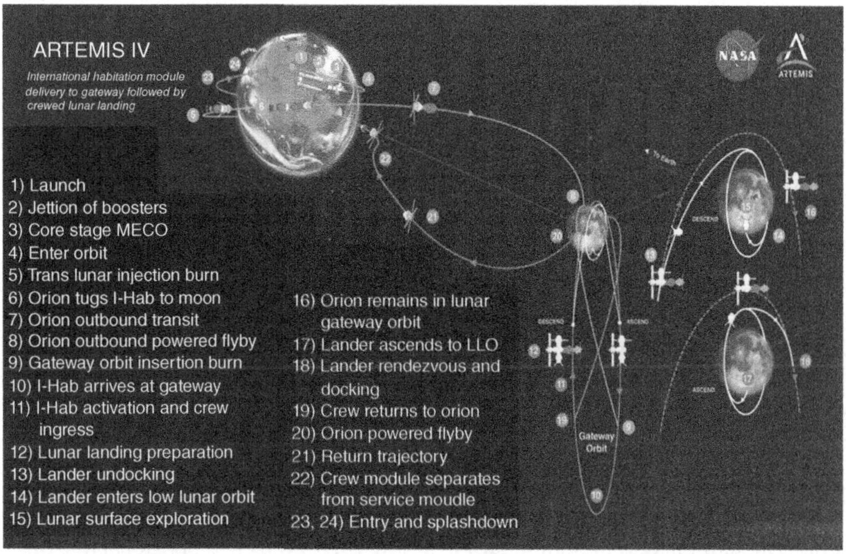

Figure 11.15 Artemis 4 is a crewed mission to the lunar Gateway. Gateway elements PPE and HALO were sent and assembled in the lunar orbit a year earlier. A separate Starship HLS will also be sent and waiting for the Orion. The Orion and the I-HAB module will be launched together and integrated in high earth orbit first. The Orion then pushes the I-HAB toward the Gateway. Two astronauts will transfer to the HLS and proceed to land on the Moon. *Source:* Ref. [19] / NASA / Public Domain.

complete, the Orion will detach from the upper stage utilizing the RCS thrusters on its service module to flip itself around and connect with the I-HAB module. Orion will then extract I-HAB from the upper stage and proceed to the Gateway orbit. The pair will be powered by Orion's service module. The journey will take about 4 days. I-HAB will dock with the HALO element with Orion serving as a "tugboat."

Humans will enter the lunar space station for the first time. The crew will spend several days activating and integrating the Gateway elements, followed by transferring equipment and conducting experiments. Two astronauts will then depart for the lunar south pole on the Starship HLS, which will have already docked with the Gateway. On the lunar surface, the two astronauts will explore, perform science experiments, collect samples to bring back to Earth, and deploy instruments for geology and other sciences. The HLS will be their temporary housing.

11.4.5 Artemis 5

Planned for 2030, Artemis 5 will have similar objectives as that of Artemis 4: 1) deliver the ESPRIT refueling element to the Gateway and integrate with the station and 2) deliver two astronauts to the lunar south pole. A European astronaut will be on board this mission. Their flight path and total mission contents are essentially the same as Artemis 4 and will not be shown. The mission will last 30 days from start to finish.

The ESPRIT element will be delivered in the same way as the I-HAB in Artemis 4 to integrate with the Gateway. Two astronauts on the Orion will board a lander and descend to the lunar south pole, where they will carry out exploration assignments.

The difference between Artemis 5 and Artemis 4 is that the lunar lander will be from a different vendor. A private consortium led by Blue Origin will build its own Human Landing System called the Blue Moon under a NASA contract awarded in 2023. The consortium aims to provide landers for recurring crewed expeditions between the Gateway and the lunar surface. Blue Origin plans to utilize its New Glenn rocket to carry the payload to LEO.

The lander to be developed is the Blue Moon Mark 2 (MK2), as shown in Figure 11.16. MK2 can transport 4 astronauts to the Moon's surface for up to 30 days. A cargo variant is also planned with a capacity of 30 tons (one way) and 20 tons (round trip) and is reusable.

A different launch system will deliver the Blue Moon lander to the Gateway orbit. The launch system is Blue Origin's New Glenn reusable booster. The "tug" that propels the Blue Moon to Gateway is called the Cislunar Transporter, which

Figure 11.16 Blue Moon MK2 by Blue Origin is a human lander system (HLS) as an alternative to the Starship HLS. It will shuttle between the Gateway and the lunar surface, carrying astronauts and cargo. *Source:* [20] / NASA / Public domain.

will be built by Lockheed Martin. Both the Blue Moon and the transporter will be propelled by liquid hydrogen and liquid oxygen, like the Orion.

The transporter has two modules: a propulsion module and a fuel module.

Figure 11.17 illustrates the sequence of events in which the MK2 lander and its transporter are transported to the Gateway. The left side of Figure 11.17 illustrates the sequence of events in which the MK2 lander and the transporter connect. In the middle of the figure, the Orion is launched to the Gateway in the same way as prior missions.

The sequence of events goes as follows.

Step 1a: Cislunar transporter and fuel module launched to the LEO on a New Glenn system
Step 1b: Fuel tanks launched to the LEO to fuel the transporter fuel module
Step 2: The Blue Moon MK2 launched on a New Glenn to LEO
Step 3: MK2 to rendezvous with the transporter and connect with the fuel module. Fuel to be transferred to MK2 and the transporter to deliver MK2 to the Gateway
Step 4: Orion and ESPRIT launched on SLS, same as Artemis 4, and arrived at the Gateway for integration
Step 5: After the transfer of crew and equipment, MK2 will land on the Moon
Step 6: After the astronauts complete their work, MK2 will ascend and return to the Gateway
Step 7: Orion to return the crew to Earth, like in previous missions

According to Table 11.2, Artemis 6 mission is scheduled in the early 2030s. Readers can search for information about the future Artemis missions when it becomes available.

11.4.6 Blue Ghost by Firefly Aerospace

In March 2025, the US private company Firefly Aerospace lands its first noncrewed spacecraft Blue Ghost lunar lander on the Moon carrying scientific instruments for NASA. The mission is part of the Commercial Lunar Payload Services (CLPS) initiative for the Artemis program, similar to the cargo missions to the ISS by commercial vendors.

The Blue Ghost I weighs 1,517 kg at launch with a dry mass of 469 kg, as shown in Figure 11.18. It is launched on a Falcon 9 rocket and follows the flight sequence shown in Figure 11.19. The Blue Ghost landed at Mare Crisium, a basin near a volcanic feature on the northeast of the Moon's near side.

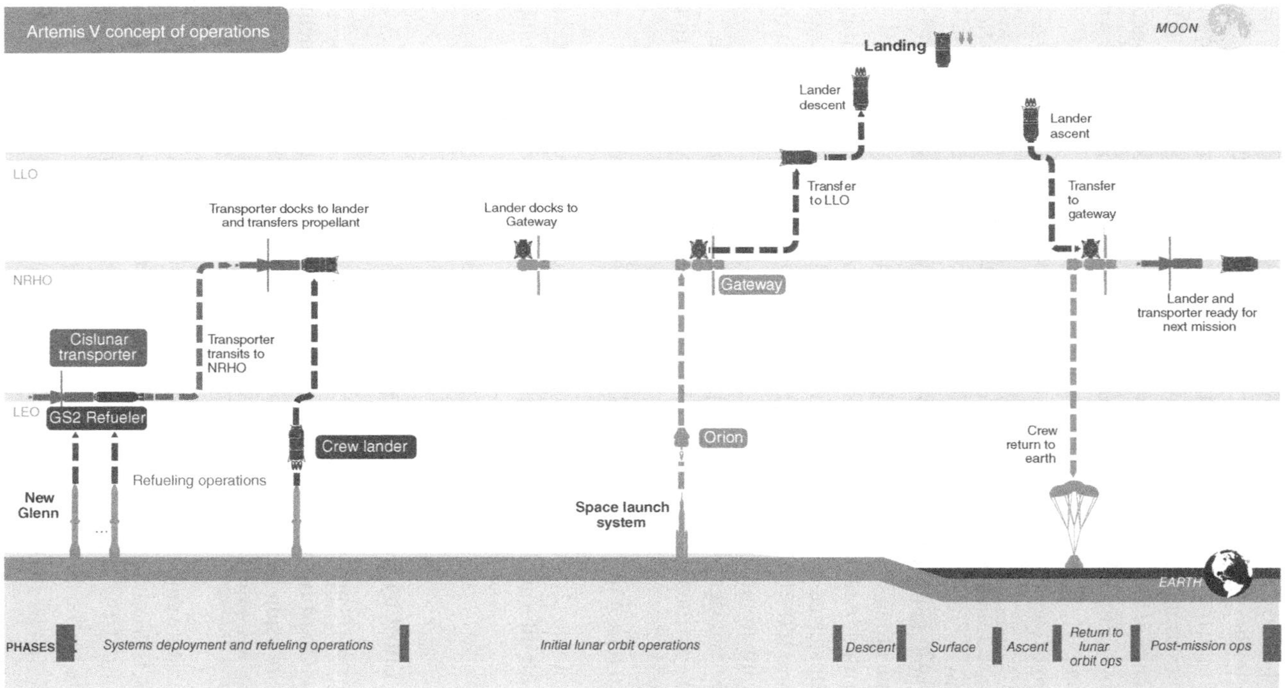

Figure 11.17 Concept of Artemis 5. It consists of several missions coming together. For convenience, we pasted enlarged labels to the key items of the figure. The Orion goes to the Gateway as usual. The element ESPRIT will have been delivered to the Gateway like Artemis 4. The Blue Moon lander will be sent to orbit by Blue Origin's New Glenn launch system. It will rendezvous with the Cislunar transporter in Earth's orbit by Lockheed Martin, which will be launched separately. The transporter will make the translunar journey rendezvous with the Gateway. *Source:* Ref. [21] / NASA / Public Domain.

Figure 11.18 Artist's image of the Firefly Aerospace's Blue Ghost lunar lander. *Source:* Ref. [22] / NASA / Public domain.

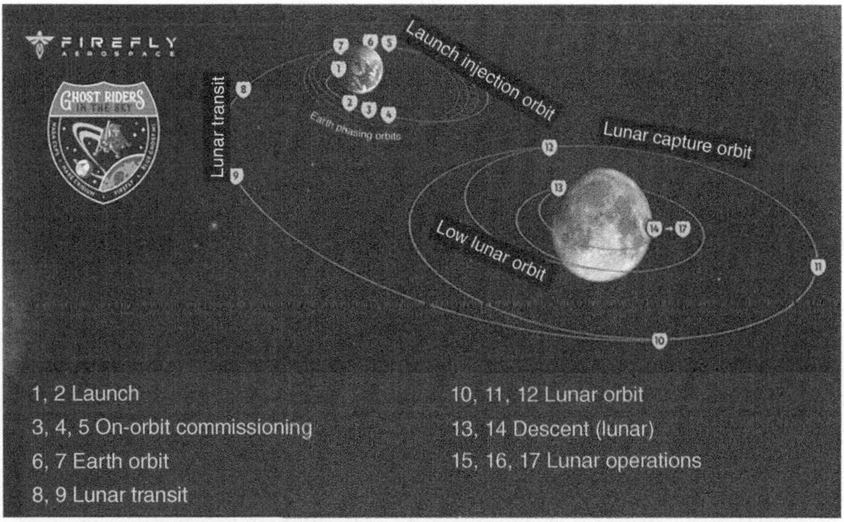

Figure 11.19 The flight path of the first Blue Ghost lunar mission. *Source:* Ref. [23] / fireflyspace.com.

11.5 China Lunar Program

The Chang'e program, officially called the Chinese Lunar Exploration Program, is a lunar exploration program under the China National Space Administration (CNSA). Chang'e is the name of the goddess of the Moon in Chinese mythology.

The Chang'e program consists of four phases of development, as shown in Table 11.3. Each phase represents a progressive advancement starting from orbiting the Moon and ending in conducting research on the Moon.

We use Chang'e 5 mission to briefly illustrate the mission design of the Chinese Lunar exploration program. The Chang'e 5 probe was launched in November 2020 on a Long March 5 rocket from Hainan, China.

11.5.1 Chang'e 5 Mission

Chang'e 5 is a noncrewed mission with the objective to collect scientific samples from the nearside of the Moon and return the samples to Earth. The mission was launched and completed in 2020. The payload is known as the Chang'e 5 probe. The sequence of events of the Chang'e 5's mission profile from launch from Earth to return is explained in Figure 11.20.

The Chang'e program does not have a lunar space station; the spacecraft lands on the Moon after separating from the assembly called the returner/orbiter for the translunar journey.

The Chang'e 5 probe comprises four modules: the ascender, the lander, the returner (reentry capsule), and the orbiter, as shown in Figure 11.21. The lander is designed to collect surface and subsurface samples using a scoop and a drill.

Table 11.3 The Chinese Lunar Exploration Program (commonly known as Chang'e) is composed of four phases.

Phase	Objective	Mission
Phase 1 "rao"	Orbiting the Moon	Chang'e 1 (2007), Chang'e 2 (2010)
Phase 2 "luo"	Soft landing and deployment of the rover on the Moon	Chang'e 3 (2013), Chang'e 4 (2018–2019)
Phase 3 "hui"	Returning lunar samples	Chang'e 5 (2020)
Phase 4 "kan"	Scientific research on the Moon	Chang'e 6 (2024), Chang'e 7 (2026, planned), Chang'e 8 (2028, planned)

From the initial phase of orbiting the Moon to conducting scientific research on the lunar surface in the final phase.

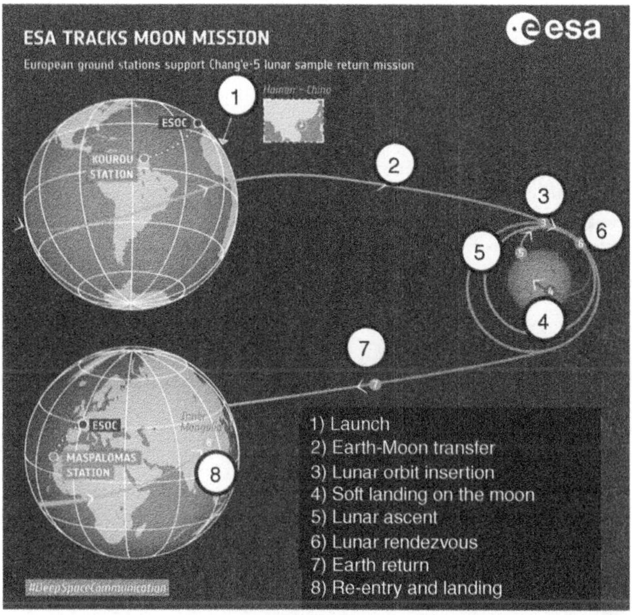

Figure 11.20 The planned flight profile of China's Chang'e 5 lunar sample mission. *Source:* Ref. [24] / European Space Agency.

After launch, the Chang'e 5 probe entered the transfer phase from Earth to the Moon. After executing the trajectory correction maneuvers, the probe entered a small near-circular lunar orbit about 200 km above the Moon's surface. The lander/ascender assembly was separated from the returner/orbiter. The orbiter assembly continued to orbit around the Moon. The lander then descended to land propulsively. The landing destination was at the Oceanus Procellarum, which is the largest lunar mare located on the northwestern edge of the nearside. Figure 11.22 shows the image of the lander/ascender assembly.

The lunar sample collection operation was carried out by the lander, which lasted approximately two days. These samples were stored in the ascender atop the lander.

In the return phase, the ascender took off from the lander, which was left behind. The ascender later docked with the returner/orbiter assembly to which the lunar samples were transferred. Then, the ascender separated from the returner/orbiter assembly. The ascender would eventually crash to the lunar surface. The returner/orbiter continued to orbit the Moon for several days and then executed a series of trans-Earth insertion maneuvers. Subsequently, three trajectory correction maneuvers were executed that brought the returner/orbiter

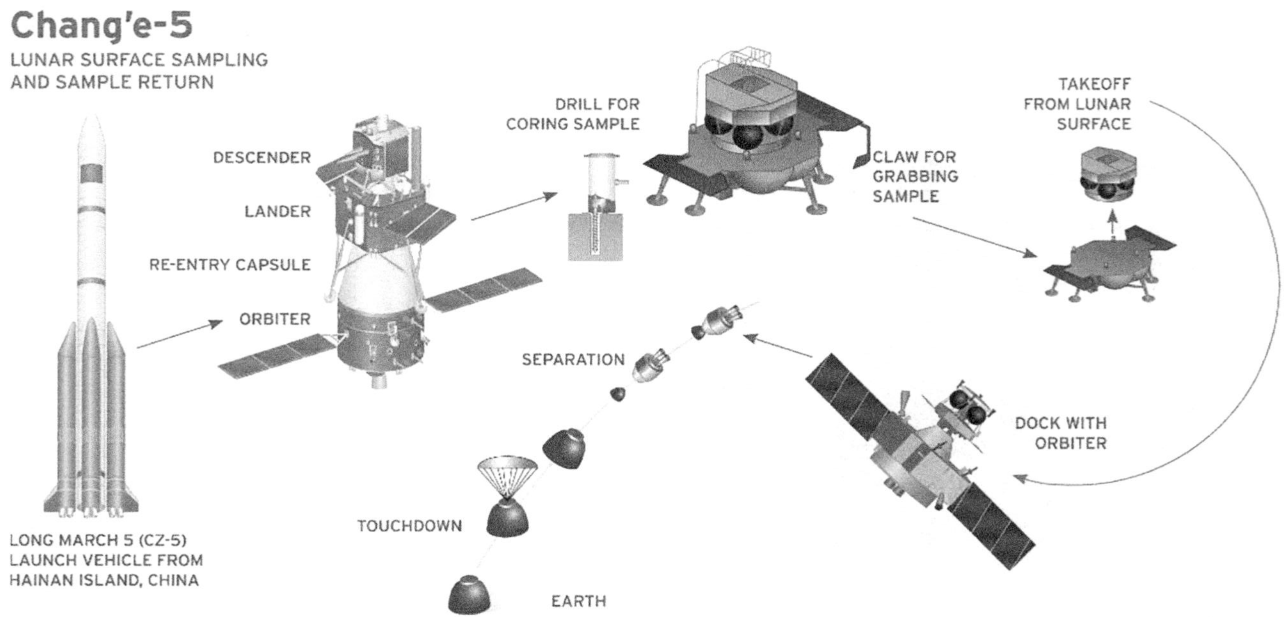

Figure 11.21 Overview of the various mission phases to return about 2 kg of lunar surface material to Earth. *Source:* Ref. [25] / eoPortal.

Figure 11.22 Image of the Chang'e 5 lander with the ascender on top. *Source:* Ref. [26] / The Planetary Society.

to Earth's atmosphere. The orbiter was dropped, and the returner re-entered the Earth's atmosphere in December 2020 and landed in Inner Mongolia.

11.6 Conclusion

Governments are going to the Moon while the private sector seeks to commercialize the low Earth orbit. Both the US and China have ongoing lunar missions with programmatic details.

The Moon is more than a good training ground for multiple reasons. This chapter focuses on getting to the Moon from Earth. By looking at the missions, we gain an understanding of the thinking of space agencies and put the early chapters of this book to good use. There are many ways to reach the Moon. The objective is to identify reliable approaches and progressively test the hypotheses behind the approaches.

The Gateway is a key piece of the Artemis program. Initial missions and space vehicles are to go through the Gateway. Vehicles used for the Earth-Gateway-Earth trips are different from the Gateway-Moon-Gateway ones. The Gateway is therefore a waystation in addition to being a research facility to study the Moon and its vicinity. The Gateway will be evaluated for the station's life expectancy and the unique (NRHO) orbit it is in. The Gateway space station concept will reportedly be replicated for Mars in due course.

High-capacity vehicles facilitate the transport of equipment to set up outpost(s). We can see from the SpaceX HLS and the Blue Moon MK2 that the on-orbit refueling capability in LEO has become a criterion for high-capacity vehicles to fulfill the objectives of the Artemis program. High-capacity HLS vehicles indicate that plenty of equipment and supplies will be sent to build up the lunar outpost(s). Also, NASA has included commercial contractors early in the design and full delivery of new vehicles servicing the "supply chain" from Earth to the Moon.

Reading List

Milione RP. *Return to the Moon: Orion Reference Guide (Artemis Project)*. Independently Published; 2021.

Rhatigan J, Adelman B. *Artemis Mission: Return to the Moon*. Lerner Publications; 2024.

Schmitt H. *Return to the Moon: Exploration, Enterprise, and Energy in the Human Settlement of Space* 1st. Copernicus; 2006.

Stakem P. The Artemis Missions: Return to the Moon and on to Mars. Independently Published; 2021.

von Ehrenfried M. *The Artemis Lunar Program: Returning People to the Moon* 2020 Springer Praxis Books; 2020.

References

1. Gibney E. How to build a Moon base. Nature [cited 2025 Feb 21] [online]. Available from: https://www.nature.com/articles/d41586-018-07107-4.
2. Moon HS. I: highlands & lowlands [cited 2025 Feb 21] [online]. Available from: https://www.simonhanmer52.ca/moon-i--highlands--lowlands.html.
3. Animation of lunar gateway around Earth - frame rotating with Moon - front. Wikimedia [cited 2025 Feb 21] [online]. Available from: https://commons.wikimedia.org/wiki/File:Animation_of_Lunar_Gateway_around_Earth_-_Frame_rotating_with_Moon_-_Front.gif.
4. Fritz MK, Tilley S, Lee T. What are cislunar space and near rectilinear halo orbits? MAXAR [cited 2025 Feb 21] [online]. Available from: https://blog.maxar.com/space-infrastructure/2019/what-is-cislunar-space-and-a-near-rectilinear-halo-orbit.
5. Space Launch System Exploration Upper Stage (EUS). NASA [cited 2025 Feb 21] [online]. Available from: https://www.nasa.gov/reference/space-launch-system-exploration-upper-stage-eus/.
6. SLS fact sheets. NASA [cited 2025 Feb 21] [online]. Available from: https://www.nasa.gov/humans-in-space/space-launch-system/sls-fact-sheets/.

7 Clark S. Companies release new details on human-rated lunar lander concepts. Spaceflight Now [cited 2025 Feb 21] [online]. Available from: https://spaceflightnow.com/2020/04/30/companies-release-new-details-on-human-rated-lunar-lander-concepts/.

8 NASA. Current 2009 Cev design. Wikipedia [cited 2025 Feb 21] [online]. Available from: https://en.wikipedia.org/wiki/Crew_Exploration_Vehicle#/media/File:File-_Current_2009_Cev_design.jpg.

9 Dinner J. Europe hands Artemis 2 moon ship's service module over to NASA (photo). Space [cited 2025 Feb 21] [online]. Available from: https://www.space.com/artemis-2-orion-service-module-handed-over-nasa.

10 Patrascu D. NASA shows all elements of lunar space station Gateway, and it's impressive. Autoevolution [cited 2025 Feb 21] [online]. Available from: https://www.autoevolution.com/news/nasa-shows-all-elements-of-lunar-space-station-gateway-and-it-s-impressive-1html.

11 Sloss P. Initial Gateway elements beginning systems integration while NASA works on a launch date. NasaSpaceFlight [cited 2025 May 1] [online]. Available from: https://www.nasaspaceflight.com/2024/05/gateway-launch-date/.

12 NASA. Canadarm3, Canada's robotic system for Gateway. Canadian Space Agency [cited 2025 Feb 21] [online]. Available from: https://www.asc-csa.gc.ca/eng/multimedia/search/image/12642.

13 About Canadarm3. Canadian Space Agency [cited 2025 Feb 21] [online]. Available from: https://www.asc-csa.gc.ca/eng/canadarm3/about.asp.

14 Dragon-XL. Gunter's Space Pagem [cited 2023 Jan 14] [online]. https://space.skyrocket.de/doc_sdat/dragon-xl.htm

15 Artemis I short mission overview. European Space Agency Studies [cited 2025 Feb 21] [online]. Available from: https://blogs.esa.int/orion/2022/11/15/artemis-i-short-mission-overview/.

16 Artemis II Map. NASA [cited 2025 Feb 21] [online]. Available from: https://www.nasa.gov/image-article/artemis-ii-map-2/.

17 Chojnacki K. Human landing system. NASA [cited 2025 Arp 28] [online]. Available from: https://ntrs.nasa.gov/api/citations/20220003725/downloads/22%203%207%20Kent%20IEEE%20paper.pdf.

18 Artemis 3 step-by-step. European Space Agency [cited 2025 Feb 21] [online]. Available from: https://www.esa.int/ESA_Multimedia/Images/2020/05/Artemis_3_step-by-step.

19 Hambleton K, Williams CE. NASA's artemis IV: building first lunar space station. NASA [cited 2025 Feb 21] [online]. Available from: https://www.nasa.gov/general/nasas-artemis-iv-building-first-lunar-space-station/.

20 O'Shea CA. NASA selects blue origin as second artemis lunar lander provider. NASA [updated 2023 May 19; cited 2025 Feb 28] [online]. Available from:

https://www.nasa.gov/news-release/nasa-selects-blue-origin-as-second-artemis-lunar-lander-provider/.

21 Watson-Morgan L. Human landing system. NASA [cited 2025 May 1] [online]. Available from: https://ntrs.nasa.gov/api/citations/20230010065/downloads/10%207%20HLS%20101.pdf.

22 Tralie J. NASA to study the Moon's interior. NASA [cited 2025 May 1] [online]. Available from: https://svs.gsfc.nasa.gov/14750.

23 Blue Ghost. Fireflyspace [cited 2025 Feb 21] [online]. Available from: https://fireflyspace.com/missions/blue-ghost-mission-1/.

24 ESA tracks Chang'e-5 Moon mission. European Space Agency [cited 2025 Feb 21] [online]. Available from: https://www.esa.int/ESA_Multimedia/Images/2020/11/ESA_tracks_Chang_e-5_Moon_mission.

25 Spacecraft. eoPortal [cited 2025 Feb 21] [online]. Available from: https://www.eoportal.org/satellite-missions/chang-e-5#spacecraft.

26 Chang'e-5: China's Moon sample return mission. The Planetary Society, CNSA [cited 2025 Feb 21] [online]. Available from: https://www.planetary.org/space-missions/change-5.

12

Settling on the Moon

12.1 Introduction

In August 2023, the term LunA-10 first appeared in the news when the US Defense Advanced Research Project Agency (DARPA) announced the kickoff of the study. LunA-10 stands for 10-Year Lunar Architecture. The study would take 7 months, and the goal was to develop a framework for developing an integrated lunar infrastructure over the next 10 years. Fourteen companies were selected to collaborate on the study. The names of the 14 companies can be found in online news [1]. They include SpaceX, Blue Origin, and Northrop Grumman, to name a few.

LunA-10 investigates 6 areas: 1) construction and robotics, 2) mining, 3) power, 4) in situ resource utilization (ISRU), 5) logistics, and 6) mobility. These six areas all contribute to infrastructure building. Market analysis is also included in the study from the supply–demand perspective. Mobility (local transportation), for example, is the needed infrastructure to move astronauts around the outposts as well as to transport large volumes of locally obtained materials. These materials may have been mined from somewhere nearby and are needed for constructing lunar facilities and shelters.

The keyword is infrastructure, which represents the government's agenda to go well beyond exploration. Infrastructure building serves economic purposes. The LunA-10 study allows the commercial sector to participate in the development of lunar infrastructure early. Such infrastructure may be commercially owned and operated. Such an initiative is a strong indication of the US Government's interest in building some form of lunar economy over time. As such, lunar missions in the 21st century go beyond exploration and scientific research.

We know for certain that the current lunar programs plan to set up camps at the lunar south pole. That is our starting point. This chapter examines what is on the Moon and how a community can settle there in a sustainable manner.

While the Earth rotates one revolution in 24 hours, the Moon completes one rotation every 27.32 days, which is also the same as its orbital period around Earth

New Space: From Low Earth Orbit to the Moon and Beyond, First Edition.
Edited by Stephen W. Cheung and Chih-Yung Wen.

(a lunar month). This phenomenon is known as synchronous rotation and is the reason why the same side of the Moon is always facing Earth. Therefore, the lunar day and night each lasts for ~14 days. Nonpolar areas must deal with large swings in temperatures as well as loss of sunlight for 14 days.

The lunar poles have an advantage. The Moon's rotational axis only tilts 1.5° versus Earth's 23.5°. Sunlight comes tangentially, and the polar surfaces are sunlit at least 90% of the time. "Peaks of eternal light" are those mountain peaks located in the polar regions that are exposed to near-constant sunlight. Research modeling suggests that the temperature at the sunlit areas around the poles is around −50 °C.

The rims of the polar craters are exposed to sunlight, while the shadowed crater floors are always dark and cold at temperatures between −170 °C and −230 °C. This unique advantage gives lunar pioneers access to sunlight and water at the same time—both are primary resources that enable humans to settle on the Moon.

12.2 Lunar South Pole and Water

Access to water is one reason why the south pole is favored as a landing destination. Water is a primary resource for human survival, plant growth, and many industrial processes. A common estimate is that there are about 600 million tons of water on the Moon. Much of the lunar water resides in the shadowed areas of craters in the form of ice. The current belief is that water was brought in by water-bearing asteroids in a fashion similar to Earth. But water in sunlit areas evaporated a long time ago, except for those bonded to chemicals and the soil.

Water and volatiles can also be dissociated to obtain elements such as hydrogen and oxygen on an industrial scale. Hydrogen and oxygen can also be used as rocket fuel or in fuel cells. Hence, water gives us the perfect start, and we can utilize solar energy to obtain vital resources as byproducts.

There are many craters in the poles with more on the south pole. Satellite data so far supports the existence of water in the shadowed areas of these craters. Figure 12.1 shows the imaging results indicating water ice in the north and south poles. The south pole has more water ice trapped on the cold floors of the unlit craters. A neutron spectrometer on the Lunar Prospector satellite has mapped the global abundance of neutrons on the Moon. High neutron counts indicate the presence of water molecules. The results also support the theory that water is also present in a different form in the regolith near the poles. This is shown in Figure 12.2.

12.2.1 Extracting Water from Ice

Most lunar missions plan to focus on finding water in its natural state, such as ice. There are two options to extract water from the ice on the crater floors and from

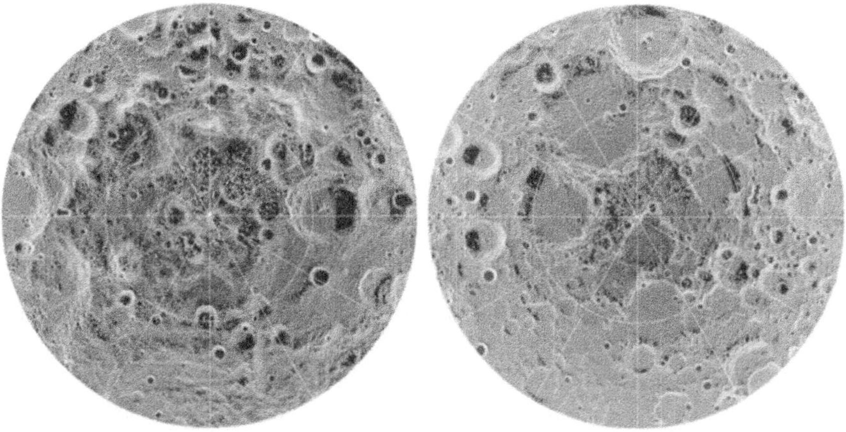

Figure 12.1 Water ice is identified by satellites on the Moon's south pole (left) and north pole (right). The ice is in the shadowed areas. *Source*: Ref. [2] / NASA / Public domain.

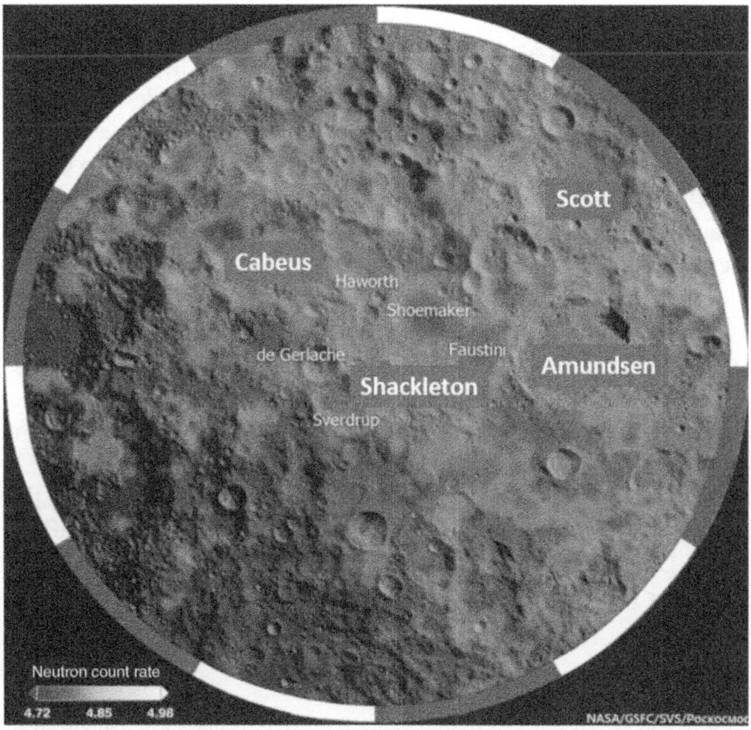

Figure 12.2 Likely locations of ice deposits on the lunar south pole using neutron counts data. Shackleton is the crater of interest. *Source*: Ref. [3] / NASA / Public domain.

pulling water from the regolith. Option a) is directly applying energy to warm the target spot and collect water in liquid or vapor form. Option b) aims to remove an amount of material from the target area and transport it to a processing facility for extraction. In both options, water is obtained by heating the ice.

Option a) amounts to providing energy to generate heat in situ. One possibility is to have a portable nuclear power source or fuel cells to energize a heat generator such as a microwave. The other possibility is to use solar power. Another possibility is to direct power from a power source to the site of operation. One idea proposed in a 2018 Nature article [4] is to mount large reflectors at high places to direct concentrated sunlight to the target site. The reflectors can be steered by a remote control. The liquid and vapor collection should be designed to minimize escape to the vacuum environment.

Option b) means that a mobile facility is used to excavate ice and soil and transport them to a processing facility. An example of the machine used to perform the excavation is the RASSOR developed by NASA (see Section 12.2.4).

12.2.2 The Shackleton Crater

A candidate crater should be accessible to early explorers. The availability of energy sources such as solar power nearby would be highly desirable. The Shackleton Crater has been chosen by NASA as the destination of the Artemis lunar missions. The Moon's rotational axis passes through the Shackleton Crater at 89.90°S, 0.00°E, i.e. the tip of the south pole.

The peak along the Shackleton rim is illuminated by sunlight 93% of the time, which makes it an ideal location for solar panels. Nearby craters are even larger. The Shackleton Crater is 21 km in diameter and at least 4 km deep with a slope around 30–35° from the rim to the floor. The crater floor is about 6.6 km in diameter and appears to be smooth. The slope gradient is a challenge to an average person; we could imagine what astronauts or machines have to do to go up and down these slopes. Using 30°, the slope itself would be 8 km long to traverse in a super cold and dark environment.

The Shackleton crater is not the only location near the south pole where the above conditions exist. There is ample room to land and set up a base and settlement. Figure 12.3 shows the 13 possible landing areas proposed by NASA for Artemis 3 (Chapter 11). These candidate areas measure at least 15 × 15 km.

12.2.3 Extracting Water from Lunar Soil

Satellites sent to the Moon have also detected a strong presence of hydrogen. It is believed that hydrogen could have come from past bombardments by comets and water-bearing asteroids or from cosmic and solar streams containing hydrogen.

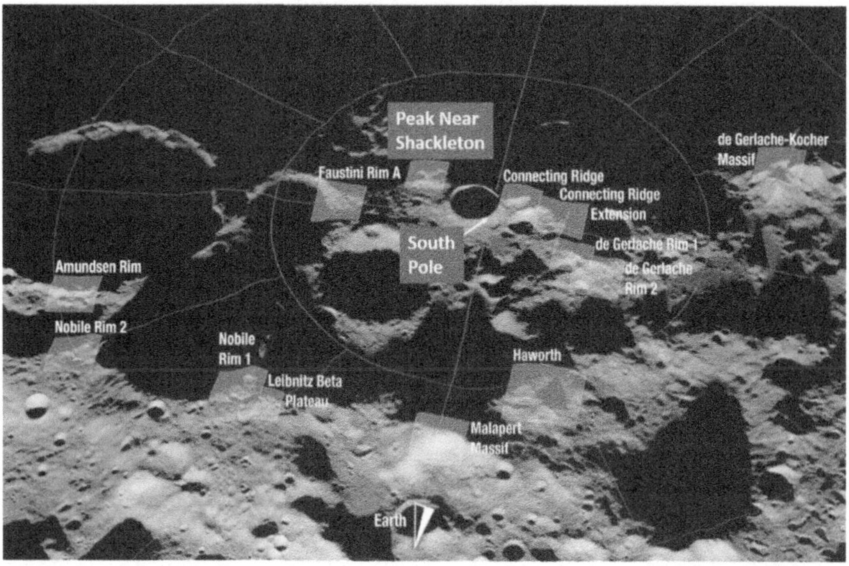

Figure 12.3 NASA proposes 13 candidate landing regions for Artemis III. Each region measures approximately 15 × 15 km. A landing site has a diameter of 200 m in each region. *Source*: Ref. [5] / NASA / Public domain.

The hydrogen ions could be present in lunar soils or interact with oxygen in lunar rocks to form hydroxyls (OH). The strong hydrogen presence can be exploited to make water from lunar soils as an alternative source. The finding also raises the possibility of extracting hydrogen and oxygen from the soil and making water and other useful materials.

In August 2024, scientists from the Chinese Academy of Sciences announced the discovery after analyzing the lunar soil collected by the Chang'e-5 mission in 2020. The mission collected 1.73-kg soil from the Oceanus Procellarum region in the northwest of the Moon's near side. High amounts of hydrogen are reportedly found in the soil minerals. The research team advanced their finding that heating the soil specimens to 1,000 °C resulted in between 51 and 76 mg of water vapor per gram, i.e. between 5% and 7.6%. If scaled up, every metric ton of soil would get 51–76 kg of water.

When heated, hydrogen reacts with the abundant iron oxides in the minerals and produces iron and water, which manifests as steam. The Chinese team proposed using solar power to provide the thermal energy to treat the lunar soil in situ, as illustrated in Figure 12.4. In this figure, solar power is collected and focused to heat up an enclosure full of lunar soil. Water will be collected as steam. The water is further dissociated by electrolysis into hydrogen and oxygen. In short, solar energy enables the extraction of hydrogen and oxygen contents in minerals with water as

Figure 12.4 A concept proposed by China to extract water by warming lunar soils. Solar power is the power source used to heat the soil. The steam produced is dissociated into hydrogen and oxygen [6–8]. *Source*: Ref. [6] / Elsevier.

an optional byproduct. Iron is also a useful byproduct as it can be used in various construction projects. This development opens the possibility of extracting water from anywhere on the Moon where there is sunlight or sufficient power sources.

12.2.4 Ice and Soil Excavation

Excavators on Earth are usually heavy, so their weights anchor to the ground and provide leverage for the blades to penetrate the soil. In low- or zero-gravity environments, however, an excavator's weight offers no leverage for the blade to remove high-density or hard soil. The soil pushes back on the blade as a reaction, and the excavator loses its footing.

Lunar missions that aim for long-term stays must include building facilities and moving lunar soil by the tons. It would be costly to send a commercial excavator to the Moon owing to its weight. A new design is required. In 2013, NASA developed the Regolith Advanced Surface Systems Operations Robot Excavator (RASSOR). The tele-operated electric excavator has counter-rotating bucket drums on its two arms that exert almost no net horizontal and vertical reaction force. Therefore, excavation does not rely on traction or the evacuator's weight in low-gravity environments. The bucket drums can load, haul, and dump their contents.

Figure 12.5 RASSOR is a machine developed by NASA designed to excavate soils in low-gravity conditions. RASSOR does not need traction as required in traditional excavators and is good for working in loose regolith. *Source*: Ref. [10] / NASA / Public domain.

According to the NASA literature, RASSOR 2.0 has a mass of 66 kg and can remove 2.7 metric tons of lunar soil per day based on 16 hours of operation and 8 hours of recharging. The power usage is 4 W per kg of soil and traverses at a velocity of 0.27 m/s (0.97 km/h). The RASSOR is shown in Figure 12.5. Scaled-up versions with a higher payload capacity are expected [9].

12.3 Electric Power Source

Electricity powers everything. The lunar outpost's power budget is preliminarily set at an average of 60 kW using a combination of solar and nuclear fission, while the rovers and mobile equipment would be powered by fuel cells. As a reference, the power consumption on board the ISS varies between 80 and 90 kW, with 120 kW as the margin. The ISS power primarily comes from solar panels with battery storage as a supplement.

A primary energy source is converted into electric power. The primary sources on Earth include fossil fuels, solar, nuclear, hydrogen fuel cells, wind, hydro, and biomass. Fossil fuels, hydro, wind, and biomass can be ruled out on the Moon's surface. That leaves nuclear, fuel cells, and solar as three viable options to supply electricity to support equipment and pioneers' activities. Rechargeable chemical batteries can serve as a backup.

Solar power is location dependent, given the Moon's long day–night cycle, while nuclear and fuel cells can provide energy continuously. The best arrangement is to combine nuclear, solar, fuel cells, and batteries to provide the power to operate the outpost, rovers, conduct experiments, and small-scale industrial processes. In the long run, the Moon's resources need to be utilized to produce power. The Moon, for example, has uranium and thorium that can be processed for local use.

12.3.1 Solar Power

Planets and their moons with a rotational axis almost perpendicular to the ecliptic plane would see their poles permanently illuminated in a tangential direction. Photovoltaic panels have been proven to work in space stations and satellites. Solar panels on orbiting stations and satellites can achieve up to 40% efficiency using advanced materials.

Solar power can also be obtained by placing large reflectors at high locations, where exposure to sunlight is guaranteed. The reflected light goes to the solar panels placed at or near the outpost or areas where work is in progress. Such high points exist at both poles.

In summary, solar power is straightforward to deploy on the Moon. The maximum power density on the surface of the Moon is also 1.37 kW/m^2 (maximum) and with no atmosphere to contend with. However, the grazing angle at the poles would reduce the power intensity substantially. Adding to the conversion efficiency, solar panels of hundreds of square meters would provide tens of kilowatts. This estimate gives an indication of the square footage of the solar panels needed.

12.3.2 Nuclear Power

Nuclear power refers to the radioactivity process. Nuclear power devices have been used in space missions since the 1960s. They are reliable and long-lasting (10 or more years) compared to chemical-based batteries. Nuclear power goes through two conversion processes to generate electricity. Radioactivity generates thermal energy, which is then converted to electric power.

There are two types of radioactivity: 1) radioisotope thermoelectric generators (RTGs) without a nuclear chain reaction and 2) fission surface power (FSP) systems through controlled chain reaction of the fission process.

12.3.2.1 Radioisotope Thermoelectric Generator

Radioactive isotopes produce heat naturally due to the decay of their heavy atoms but without any chain reaction. Plutonium-238 is commonly used in RTGs and

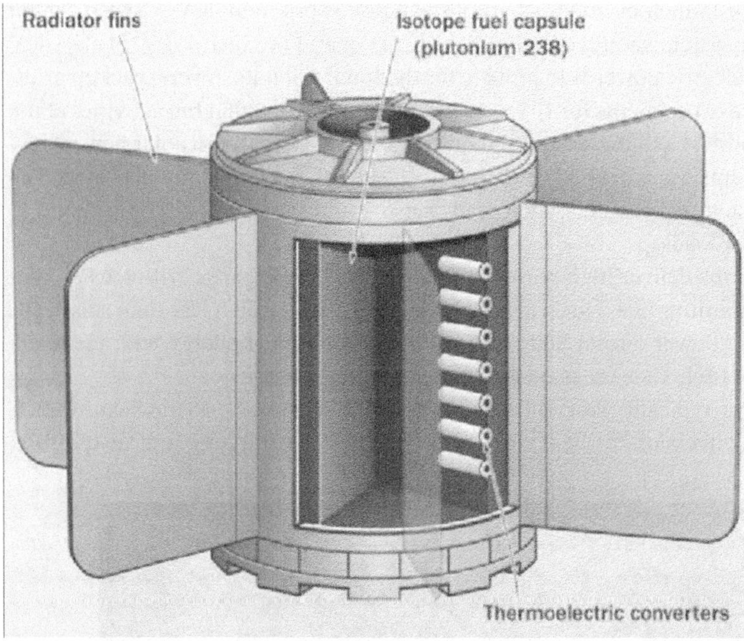

Figure 12.6 Construction of the radioisotope thermoelectric generator. Capsules of radioisotope materials are installed as shown. The excess unused heat generated by the radioisotope is dissipated by the fins. *Source*: Adapted from [11, 12].

produces 30–40 W of electric power per kilogram of fuel. The heat is converted into electricity via a thermocouple. An RTG weighs tens of kilograms. Several kilograms of nuclear fuel are used to produce tens to more than 100 W of electricity. The conversion efficiency from heat to electricity is not high. Thin shielding with lead is often needed.

RTGs provide lower electric wattage and are less cost-effective than FSPs. The power cannot be shut down. Electric power from RTGs is typically used to operate equipment in spacecraft and satellites. Figure 12.6 is a cutaway diagram of an RTG showing the assembly of isotope modules with fins as a heat sinks to radiate away excess heat.

12.3.2.2 Fission Surface Power System

Fission surface power systems have a cost advantage over RTGs. Enriched uranium-235 is usually used. Until recently, only Russia used FSP in space missions. The US is now developing FSP for deep-space missions. In 2022, NASA, in conjunction with the Department of Energy, worked with contractors to develop concepts for an FSP system suitable for use on the Moon. The demo designs are to be ready for launch by 2030 as part of the Artemis program. NASA specifies that the reactor system should remain under 6,000 kg and produce 40 kW of electrical power. Its electric power is to provide for the lunar habitats, rovers, backup grids, and science experiments for 10 years. It can be envisioned that the activities at the lunar outpost(s) will increase and require more power than just solar power.

FSP systems are launched cold, that is, the chain reaction is not activated. The reactor system is activated when ready on site and has the ability to respond to varying user loads.

FSP systems deliver high power in the tens of kilowatt range with tens of kilograms of uranium fuel. Nuclear-based systems usually weigh less than solar cells of the same power output and provide larger amounts of energy with moderate amounts of fuel. They are good for extended space missions.

There are typically four subsystems in an FSP system: 1) nuclear reactor, 2) power conversion, 3) heat sink, and 4) power conditioning and distribution

Figure 12.7 Concept of a stationary nuclear fission station for use to provide electric power to facilities on the Moon. *Source*: Ref. [13] / NASA / Public domain.

(PCAD). The chain reaction in the reactor produces heat. Heat is transferred from the reactor to the power conversion subsystem. Residual heat is managed by the heat sink. The electrical power output of the power conversion subsystem goes through the PCAD for conditioning and then to the external load.

High-kilowatt fission systems are being developed for use on lunar outposts and other destinations beyond. They are most suitable for providing power where solar power is intermittent or lacking. They can be placed in permanently shadowed regions or for industrial activities in the future as shown in Figure 12.7.

FSP systems work on the chain reaction of neutrons when activated. Shielding similar to common nuclear power plants is necessary to protect personnel nearby. The shielding materials should absorb neutrons and some gamma radiation. Since the FSP for the Moon is turned off during flight, shielding materials can be added at the installation site. Shielding materials are usually low-density materials such as water, plastic, and a list of other candidate materials. Water is the most convenient neutron absorber.

12.3.3 Hydrogen Fuel Cells

Fuel cells can generate electricity or serve as rechargeable storage devices. They work reliably for a long time. Fuel cells deliver electricity by the chemical reaction of mixing hydrogen and oxygen ions. Free electrons end up being the byproduct.

Hydrogen gas is fed to the anode and ionized by a catalyst, producing hydrogen ions (protons) and free electrons in the solution. Air or oxygen is fed to the cathode and ionized by another catalyst to form negatively charged oxygen ions. The positive hydrogen ions proceed to chemically combine with the negative oxygen ions inside the cell to form water and heat. The electrons go from the anode to the cathode and pass through an external circuit, creating a flow of electricity. This is illustrated in Figure 12.8.

A rechargeable, or regenerative, fuel cell works by dissociating water back to hydrogen and oxygen by means of electrolysis powered by an external energy input. On Earth, such external energy input usually comes from solar panels and other green sources.

NASA has used fuel cells on many missions, especially when deploying solar panels is not possible or insufficient. Fuel cells are particularly suitable for long missions such as deep-space exploration. They are currently used to power electrical systems on spacecraft.

The life expectancy of hydrogen fuel cells is between 5 and 10 years with maintenance.

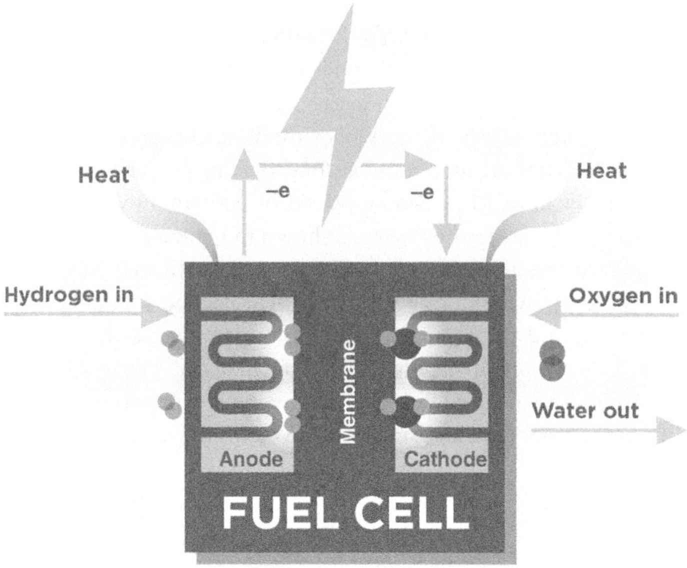

Figure 12.8 Working of a hydrogen fuel cell describing the inflow of hydrogen and oxygen. The two gases combine to produce electricity with water as a byproduct. *Source*: Ref. [15] / University of British Columbia.

Fuel cells are classified by the kind of electrolytes used, each with different performance specifications and applications [14]. Two types of fuel cells are discussed here.

1) Alkaline fuel cells (AFCs) were first used in the US space programs. The electrolyte is a water-based solution of potassium hydroxide (KOH). Hydrogen and oxygen are ionized by the catalysts in the electrodes as before. The negative OH ions migrate to the anode to combine with the hydrogen ions. The positive potassium ions migrate to the cathode to combine with oxygen ions. The electrons released from the ionizing hydrogen atoms go to the external load as current. Water and heat come out of the anode side. Most AFCs are not rechargeable.
2) Polymer electrolyte membrane fuel cells (PEMFCs): The electrolyte is a polymer membrane that is conductive to hydrogen ions. After the catalysts, the hydrogen ions migrate through the membrane and combine with oxygen ions. The electrons go to the external circuit, while water and heat come out of the cathode side. These cells are mainly used for transport and portable applications.

The advantage of the PEMFC is the high-power density produced and the low temperature in operation. The disadvantage is its high cost and limited lifetime of 5 years or less.

Fuel cells play important roles on the Moon. They can be used at the outpost (stationary) and on vehicles (mobile). In the long run, hydrogen and oxygen must be obtained in situ. Hydrogen and oxygen can be obtained by dissociating lunar water and volatiles using solar power. Stationary fuel cells are expected to provide electric power in lunar habitats when solar is unavailable or insufficient. Fuel cells can be recharged or replenished the hydrogen and oxygen supply.

To prepare for lunar missions, space agencies such as NASA and ESA have ongoing projects with commercial contractors to develop fuel cells.

12.4 Transportation and Mobility

Once landed on the lunar surface, the mode of local transportation is next for the pioneers. We are decades away from building big cities and intercity transportation networks on the Moon. Aviation on wing-type vehicles also would not work due to a lack of atmosphere. In the near term, only rovers are proven vehicles for lunar settlers.

Industrialization on the Moon means the transportation of large volumes of materials between places. Special vehicles must be developed. They would be akin to trucks and trains on Earth as the primary modes of cargo transportation on land.

12.4.1 Rovers

Apollo astronauts rode on a lunar roving vehicle (LRV), also known as the Moon buggy. Weighing 210 kg without payload, the LRV was powered by batteries; and its top speed was 9.7 km/h with a range of 92 km.

The new generation of rovers can be autonomous, semiautonomous, or operated by humans. The immediate version of a lunar vehicle should serve multiple purposes, ranging from carrying astronauts around for exploration to sample collection. It should be able to maneuver around the loose regolith and preferably have a contained environment to shield it from a range of temperatures, pressures, and radiation.

NASA's new lunar vehicle for the Artemis program is termed the lunar terrain vehicle (LTV). The specified capabilities of the LTV will include autonomous driving and being equipped with communications equipment, navigation systems, and extreme environment technologies to operate at the south pole. It will

Figure 12.9 Conceptual design of NASA's lunar terrain vehicle for astronauts to move about on the lunar surface. *Source*: Ref. [16] / NASA / Public domain.

collect samples while the astronauts safely stay inside the vehicle. When no astronauts are on the Moon, the LTV will be capable of operating remotely to perform scientific experiments.

One of the conceptual designs by NASA is shown in Figure 12.9. The actual design will come from commercial contractors. In April 2024, NASA selected three companies to develop the LTV to be completed by 2026. The LTV will not fly until Artemis 5 in 2030.

JAXA is also working with domestic manufacturers to develop lunar rovers. Figure 12.10 shows the conceptual design of the rover by the Toyota Motor Corporation. The rover has a pressurized cabin and is powered by fuel cells with deployable solar panels. Unlike the Moon buggy before, the JAXA/Toyota rover is expected to go out for 42 days at a time and 1,000 km on one tank of hydrogen. The projected time of availability is 2029.

12.4.2 Moon Train

In an Innovative Advanced Concepts program (NIAC) Phase II award meeting in May 2024, NASA announced that six breakthrough projects were selected for the Phase II study. Among the selected projects is a lunar railway system called "Flexible Levitation On A Track," or FLOAT. The function of the Moon Train is to transport, for example, regolith and payloads around the outpost to and from the landing zone and mining sites. Regolith can be processed to obtain minerals, building materials, Helium-3, and water.

Figure 12.10 Toyota's conceptual design of a manned lunar rover powered by regenerative hydrogen fuel cells. *Source*: Ref. [17] / Toyota Motor Corporation.

The FLOAT system will utilize the magnetic levitation technology using diamagnetic materials to construct the "railcars." Diamagnetic materials repel against an external magnetic field. Such a phenomenon is not related to the polarity of the magnetic field. Instead, an induced magnetic field is created in the diamagnetic material to oppose the external magnetic field. Such a repulsive force can be used to levitate an object, especially on the Moon, due to its lower gravity. Graphite is a common diamagnetic material.

Termed "Tray Robots," these trays are made of magnetized materials but otherwise have no power to facilitate motion. Instead, the tray motion is controlled by the circuits in one of the layers on the "railroad tracks" underneath. These railroad tracks require no construction but are "unrolled" on the lunar ground without any form of anchoring, as illustrated in Figure 12.11.

The track has two layers. An optional layer of thin-film solar panel can be added. The top layer is graphite (diamagnetic) to produce a repulsive force against the magnetic tray robot. The next layer generates electromagnetic thrust to move the tray robot forward. The amount of lifting force depends on the area of the magnetic materials on the tray robot. The current specification for the payload is 33 kg per square meter. The robot can move at over 0.5 m/s. When scaled up, the FLOAT system can move thousands of kilograms over several kilometers a day.

The FLOAT system can be regarded as the prototypical workhorse for transporting materials within a radius of several kilometers from a lunar base. It is part of the commercialization of the Moon as envisioned by NASA. Integrating with

Figure 12.11 Concept of the Moon railway based on diamagnetic levitation. *Source*: Ref. [18] / NASA / Public domain.

fields such as construction and logistics, we can anticipate that lunar expeditions will move into the infrastructure phase after exploration.

12.4.3 Aerobots and Ground Robots

Unmanned aerial vehicles (UAV), drones, and robots are modern products that can be applied innovatively on the lunar surface.

A helicopter has been used on the Perserverance Mars mission. On Earth, drones are useful for exploring nearby areas from 1 km to as far as several hundred km for purposes such as reconnaissance and detailed mapping of landscapes. However, aerial drones require an atmosphere for their blades to generate lift. These machines cannot fly on the Moon due to the lack of atmosphere. Therefore, thrust have to come from the principle of propulsion using the appropriate propellants. Cold gas thrusters can provide the required thrust for light-to-medium-weight crafts.

Aerobot is a term used by NASA to refer to an aerial vehicle for planetary exploration, including the Moon. Aerobots can fly low in altitude at slow speeds and are most suitable for the rapid assessment of an area over a long distance as well as taking close looks at hard-to-reach areas such as cliffs and steep slopes. Such perspectives complement satellite imaging. These vehicles are reusable and rechargeable.

Robotic devices are key tools for lunar exploration and commercialization. They can perform tasks that humans cannot do or find too risky to do. Ground robots usually move around on wheels or legs. Jumping robots are an interesting addition to the portfolio. Jumping robots utilize advanced electromechanical technology. They are suitable for exploring areas such as lava tubes and caves that wheeled robots are not suitable for.

The University of Southampton, for example, developed the Lunar Hopper Mk. It is a vertical takeoff, vertical landing (VTVL) vehicle. The thrust comes from four nitrogen cold gas thrusters mounted in the center shown in Figure 12.12. Weighing 37 kg, the hopper is capable of maintaining balance and overcoming obstacles.

Another example is the SphereX (Spherical Rocket Propelled Robots for Extreme Environment Exploration) from the University of Arizona. This family of robots weighs 3–5 kg each and can be deployed from a lunar lander or rover. These robots have three independently operated support legs that can change jumping angles. If connected with ropes, these robots can help one another in climbing. This team

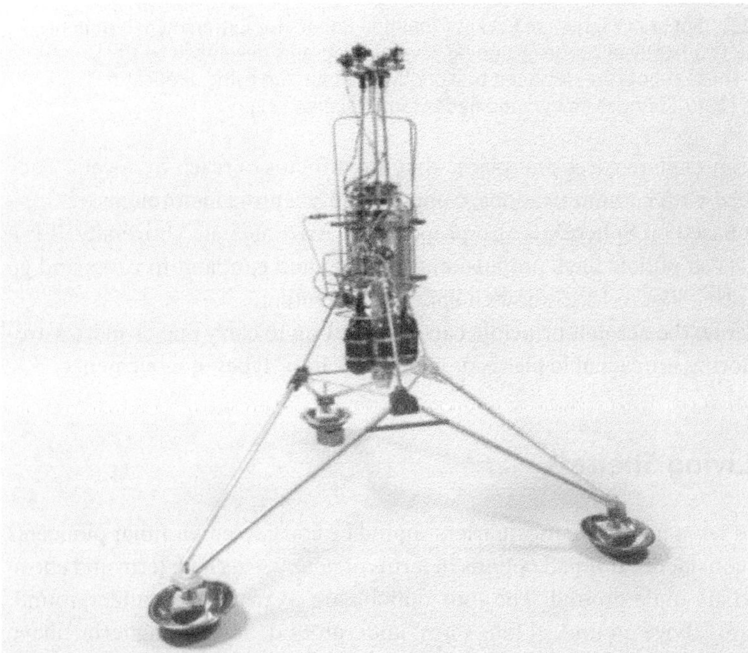

Figure 12.12 The Lunar Hopper Mk is designed by the University of Southampton. Nitrogen thrusters provide the thrust to move the hopper and maintain balance. *Source*: Ref. [19] / Lunar Hopper Enhancements / Engineering and the Environment / University of Southampton.

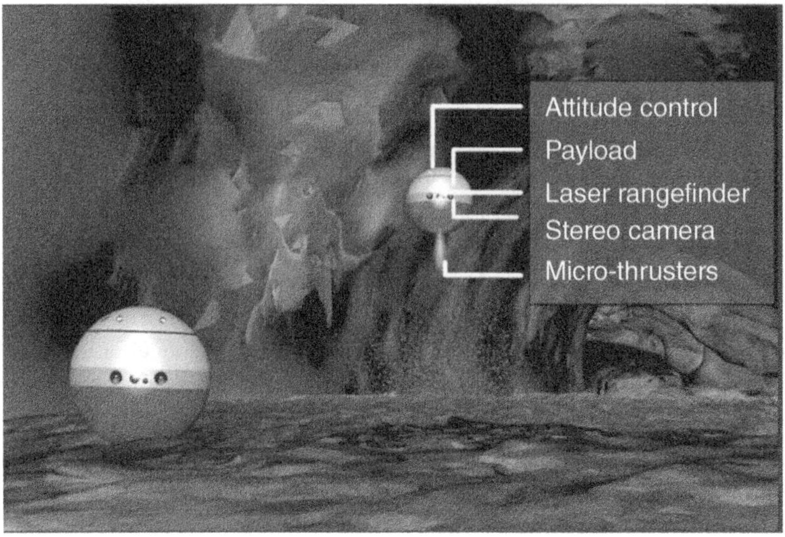

Figure 12.13 SphereX (Spherical Rocket Propelled Robots for Extreme Environment Exploration) is a family of robots weighing several kilograms developed by the University of Arizona. These robots are designed to work in caves for surveying applications. *Source:* Ref. [20] / Springer Image modified by Smithsonian [21].

of robots can explore caves and places that are difficult to reach by rovers. They are equipped with communications, cameras, and scientific instruments.

An idea based on SphereX is also proposed by Arizona State University. It is a "pitbot." These pitbots have propulsion capability and can land in caves and go into lava tubes. Figure 12.13 shows a SphereX and pitbot.

In principle, the aerobot principle can be scaled up to carry one or more astronauts exploring unreachable places or just "quick hop" types of movements.

12.5 Living Shelters

Where and what form of living quarters should be considered for lunar pioneers?

We can consider four broad options in terms of near versus long term and above ground versus underground. The four options are a) near term, underground, b) near term, above ground, c) long term, underground, and d) long term, above ground.

Working backward from the future, one can envision communities and industrial complexes in different regions of the Moon. Early Artemis missions plan for the astronauts to stay on the south pole for 6.5 days and use the lunar landers as shelters. The lander offers environmental protection. Equipment needs to be progressively accumulated to set up an outpost. By near term, therefore, we mean that pioneers are ready to reside outside the lander and stay longer than 6.5 days.

Option a): This possibility can be readily ruled out due to insufficient information and experience about these natural cavities. This is the time for the early pioneers to explore these caves and lava tubes with the aid of the robots mentioned in the previous section.
Option b): We will expand our discussion in the next section.
Options c) and d): Until proven otherwise, both options will very likely happen. The underground option will come after a thorough understanding of the caves and lava tubes, which will take time to complete.

Whether it is a single lunar outpost or multiple communities in the future, there are three criteria to consider:

1) Access to vital natural resources such as water, sunlight, and a list of useful minerals
2) Communications and transportation infrastructure
3) Serving scientific or commercial purposes or both

Criterion 1) is related to human survival and sustainability. Criterion 2) means that constant communication needs to be established among outposts, rovers, satellites, the lunar space station, and Earth. Autonomous and semiautonomous machinery (robots) are expected to perform much of the heavy lifting and will require constant communication with local and remote command centers. Transportation infrastructure requires the outpost's location to be open enough for landers to land and take off. Ground transportation and exploration on the lunar surface is currently limited to rovers.

Criterion 3) means that the outpost's location serves scientific or commercial interests within a reachable distance. They could be craters, lava tubes, caves, and sites to extract minerals. Appropriate laboratories and equipment would need to be brought in from Earth or built with local materials. Many scientists have also suggested that the lunar far side will be a perfect place to set up observatories to study the universe without electromagnetic interference coming from Earth.

12.5.1 Living Above Ground

Above-ground shelters can be broadly divided into two categories: prefabricated and built in situ. Prefabricated housing refers to modules brought in from Earth and installed on-site. Shipping prefabricated modules to the south pole makes logical sense. Such modules can be inflatable enclosures that are initially collapsed. Prefabricated modules are also expandable in that more modules can be connected easily, like the space stations.

Robots can do most of the heavy lifting. Once the outpost has a steady supply of electricity, equipment such as 3D printing will fabricate these items locally.

A base built on the lunar surface needs to be mindful of the radiation and micrometeoroids. Extra protection should be installed. The recommendation has been to use treated lunar regolith to cover these prefabricated housings.

12.5.1.1 Expandable Habitat

Future space stations covered in Chapter 4 can be prefabricated with lightweight and inflatable materials. Bigelow Aerospace [22], for example, fabricated the Bigelow expandable activity module (BEAM) and tested it favorably as an attachment to the ISS in 2016. The material proved to be able to keep the radiation and micrometeoroids to an acceptable level, as well as the pressure differential between the vacuum in space and the atmospheric pressure inside the ISS. Bigelow plans to make multipurpose expandable habitats for space stations and deep-space explorations with inflatable materials [23]. The BA-330 in Figure 12.14, for example, has a 330 m^3 pressurized volume and is equipped to support a crew of six in low earth orbit (LEO). The company is planning on bigger

Figure 12.14 Concept of the commercial space station BA-330 by Bigelow aerospace. The space station is made from inflatable materials. The internal volume is 330 m^3. *Source*: Ref. [24] / Bigelow Aerospace.

units such as BA-2100 to be delivered to the lunar surface as living quarters. Extra layers of protection against radiation can be added in situ.

12.5.1.2 3D Printing

Space agencies have been experimenting with 3D printing in space. Polymer-based 3D printers are already functioning on board the ISS, doing spare parts and simple tools instead of waiting for shipments from Earth. In January 2024, a metal 3D printer made by ESA was sent to the ISS as a technology demonstrator [25].

3D printing is additive manufacturing. Its advantage is zero waste. Not only does it minimize the need to send finished goods from Earth to the Moon at enormous cost, but it also facilitates in situ resource utilization (ISRU). Local materials can be processed, and the desired hardware can be 3D-printed on the Moon. The lunar regolith, for example, contains silica- and iron-containing compounds that can be fused when heated with microwaves. The glass-like solid is a promising material for rigid structures. Similar products can be produced for use in the construction of surface habitats.

In 2022, NASA awarded a contract to the company ICON to develop a large 3D printing system called Olympus [26] for 3D printing of buildings on the Moon and Mars.

3D printing will be a key technology for ISRU in deep-space expeditions. It is a matter of knowing the material properties of local resources. 3D printing technology has also been applied to growing foods in a space environment, as discussed later.

12.5.2 Underground

The underground is a natural environment offering protection against radiation and micrometeoroids. It can also maintain a moderate temperature. Shelters must be built in self-contained, closed structures inside the cave.

Lava tubes are the second most common type of caves. They are hollow voids formed on slightly sloped surfaces after the lava has cooled down following volcanic eruptions and drained completely or partially. Their size ranges from several meters to tens of meters wide and several meters below the surface. Lava tubes are underground passages that we only have limited knowledge of from analyzing satellite data.

In the medium term, outposts away from the poles will be set up. Some materials such as helium-3 and rare earth elements are more abundant in the lunar equator and at low latitudes. The large temperature swing between day and night in the low-latitude regions makes it challenging for both humans and machines. Different estimates suggest that the temperature inside a lava tube and lunar pit would be stable from −20 to +20 °C, which is a habitable range.

There is evidence that some lunar cavities could be connected [27]. Hence, lava tubes and caves have strong prospects as lunar habitats.

Whether the lunar poles have suitable caves or lava tubes needs to be explored and verified. Therefore, technological developments for building bases above ground must continue.

12.6 Foods

Astronauts on space stations have limited choices of food and beverages due to the effects of microgravity. Although astronauts of early Artemis missions will bring their own supply of food and water during their 6.5 days on the lunar surface, pioneers staying for longer terms need to grow their own food.

12.6.1 Plants

Plants grow on soils that contain minerals, water, and organic matter. Plant nutrients come from organic matter. Although organic matter makes up only 5%, without them, plants cannot grow sustainably. Plants also need sunlight and carbon dioxide. These elements conducive to plant growth are absent in open space on the Moon. Therefore, plants must be grown in a protected environment.

According to a NASA report in 2013 [28], plants can grow under quasipressure as low as 7% of Earth's standard atmosphere, which is 15–20 km above sea level on Earth. Oxygen is more important than nitrogen in this regard.

In 2019, the Chinese Moon Lander of the Chang'e 4 mission landed on the lunar far-side. Among other instruments, a sealed container carried Earth soil containing cotton and potato seeds, yeast, and fruit fly eggs [29]. The container had a supply of air, water, and nutrients. The cotton seeds successfully germinated, making it the first plant to sprout in protected conditions on the Moon. Loss of temperature control in the container resulted in the cotton dying on a cold lunar night. The temperature swings the lander experienced were between −173 and 100 °C.

The above information provides the precursors for growing plants in enclosed environments. Food technologies play a huge role in the lunar program. Below is a short list but certainly not exhaustive.

12.6.1.1 The Vegetable Production System

Space agencies have always had experimental projects to grow plants in space.

a) The Veggie system: The Veggie system is a space garden installed on the ISS in 2014. NASA uses this system to study plant growth under microgravity. The output is consumed by the astronauts. The system is about the size of a small

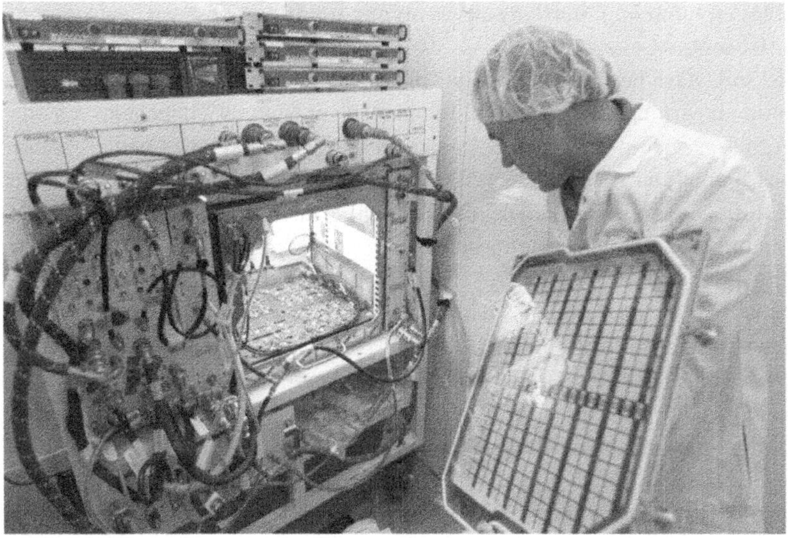

Figure 12.15 Advanced plant habitat is a NASA developmental project for growing plants in microgravity environments. Light is provided by LEDs, and the chamber is equipped with a variety of sensors. *Source*: Ref. [30] / NASA / Public domain.

suitcase and can hold six plants. As there is no gravity, plants are grown in a medium with fertilizer. The medium plays the role of distributing water, air, and nutrients around the roots. A bank of light-emitting diodes (LEDs) generates a spectrum of light suitable for plants.

To date, the Veggie system has produced flowers and plants. Part of the produce would be sent back to Earth for analysis. Harmful contamination has not been detected.

b) Advanced plant habitat (APH): The APH is an expanded version of the Veggie system with more automation and built-in control. It was first tested on the ISS in 2018. The APH is an enclosed chamber for plant growth research. Fertilizer, nutrients, oxygen, and water are delivered to a porous clay substrate. The chamber is equipped with 180 sensors and cameras. The temperature, pressure, and moisture levels are controlled and automated. The APH adds more wavelengths of light to the LEDs. Figure 12.15 shows the APH setup at the Kennedy Space Center.

12.6.1.2 Growing Plants on Lunar Soil

Growing plants and vegetables in lunar soils is not only of scientific interest but is also critical from the ISRU perspective. Space agencies have long regarded growing vegetables on lunar and Martian soils as an important part of space development in the 21st century.

Lunar soils have been considered for plant growth in lunar greenhouses. The issue is about turning the lunar regolith to be "plant friendly." Between 1969 and 1972, Apollo astronauts brought back a total of 382 kg of lunar rocks, pebbles, sand, and dust. Lunar soil was found to contain various mineral nutrients but lacked carbon, nitrogen, and microorganisms for plant growth. Investigators have been experimenting with small amounts of lunar soil, while others use simulated lunar regolith.

Growing plants and vegetables with lunar soil is at the research stage. We are not anywhere near mass-producing different vegetables grown on lunar soil. Here we provide some of the current development work and results.

Funded by NASA, a team at the University of Florida used one gram of Apollo lunar soil to grow Arabidopsis thaliana in the laboratory. Arabidopsis thaliana is a relative of the Mustard Family that includes broccoli, cauliflower, kale, etc. The seeds do not germinate without some deliberate actions by the team and are slow to develop.

A team from the China Agricultural University in Beijing used volcanic materials to simulate lunar soil. The team tested the idea of adding microbes to the simulated soil and found that the soil became more acidic, thereby changing its pH. In the process, the amount of phosphate dissolved in the soil increased. Phosphate is a major plant nutrient and is vital for seedlings and young plants. The results suggest a promising possibility of turning lunar regolith into fertile soil to grow plants [31].

Artemis-3 Mission One of the three scientific experiments in this mission is that astronauts will set up experiments to grow plants in lunar soil. The project is called LEAF, which stands for Lunar Effects on Agricultural Flora and will observe plant biology, including photosynthesis, growth, and any stress induced by space radiation and lower gravity. Red and green plants will be housed inside a protective chamber to avoid excessive temperature swings and radiation. Samples of the seedlings will be brought back for analyses at the molecular level at the end of the mission [32, 33].

12.6.1.3 Sources of Meat

Lunar pioneers cannot just survive on a diet of freeze-dried food and vegetables during their extended stay. Various meats will do both physical and psychological good. The question is which animal groups can be raised in the lunar environment. To be clear, there is no chance to raise any animal species in the open on the Moon. Vertebrates take up physical space and oxygen. Regular shipping of frozen meats from Earth to the Moon is neither practical nor economical. Here we focus on raising suitable animal species locally.

There are two challenges to overcome in bringing animals to the Moon: adaptation to the Moon's low gravity and surviving the launch.

Fish Fish have the best chance to adapt to low gravity since they already live in a buoyant environment. Hence, farming fish on the Moon using eggs originating from Earth is a viable idea. In 2019, the Lunar Hatch mission was launched with support from the Center National D'études Spatiales and ESA. The project was undertaken by the Montpellier University Space Center and the French Research Institute for Exploitation of the Sea (IFREMER). They selected seabass and meager eggs.

The team subjected 200 eggs to simulated vibrations, weightlessness, and hypergravity (1 to 5g) during the rocket launch. Seventy-six percent (76%) of the seabass eggs still hatched, relative to 82% for the control batch. Ninety-five percent (95%) of the meager eggs also hatched versus 92% for the control batch. Researchers from IFREMER stated that the eggs hatching date could be programed by controlling the water temperature: around 4 days at 14 °C and 8 days at 10 °C. These durations work out perfectly with the travel time from Earth to the Moon. The seabass is also tolerant to different levels of water salinity.

In short, conditions for fish candidates suitable for the Moon include a) surviving the various conditions from launch to travel and some level of radiation, b) a relatively short hatching time, c) low to modest oxygen requirement, and d) low carbon dioxide output. In building up an aquaculture on the Moon, invertebrates such as shrimp and insects such as crickets and mussels are possibilities.

Cultured Meat Cultured meat (also known as lab-grown meat) is grown in a laboratory bioreactor using an animal's stem cells to cultivate different cells for various tissues. It is genuine meat. This method bypasses the need to raise and slaughter animals and is applicable to both mammals and seafood. Figure 12.16 describes this process.

The cultivation process is currently reported to be 2–8 weeks, depending on the meat. Several startups in cultured meats have emerged in the past decade.

Space agencies are also targeting the development of 3D-printed foods for medium- to long-term missions. 3D-printed meat is made of cultured meat. It can be made of animal cells or using plant-based ingredients.

12.6.1.4 Food Technology Challenge

Provision of food and relevant food technology is high on space agencies' priority list. Necessary facilities and infrastructure to become self-sufficient on food are needed. Innovative food technologies can be solicited from interested parties.

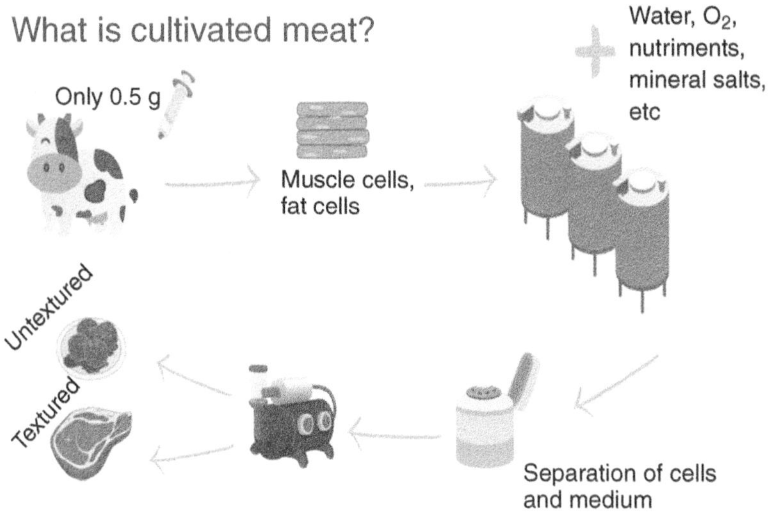

Figure 12.16 Process flow showing how cultivated meats are produced. An animal's stem cells are collected and cultivated into different cells and then meat in a bioreactor. *Source*: Ref. [34] / FUDZS.

NASA's Deep Space Food Challenge is an annual competition-based, multiphased event. The competition first started in 2021 in coordination with the Canadian Space Agency. According to NASA, participating teams are challenged to come up with food technologies that can "fill food gaps for a crew of 4 for a 3-year round-trip mission with no resupply.... This Challenge seeks to incentivize teams to develop novel technologies and/or systems for food production that need not meet the full nutritional requirements of future crews but can contribute significantly to and be integrated into a comprehensive food system."

Competitions like these open the door for more participants to innovate on food technologies that can benefit both Earth and space explorations [35].

12.7 Helium-3

Helium-3 (He-3) is currently the most sought-after material extractable from the lunar surface owing to its promise in nuclear fusion. Helium-3 is the radioisotope of helium-4 (He-4). It has two protons and one neutron and is stable and not radioactive. Helium is about 5 parts per million by volume on Earth and is essentially all He-4. In the Earth's atmosphere, He-3 is about 1.4 parts per million of He-4, that is, 7 parts per trillion in the atmosphere. The world consumes several kilograms of He-3 per year for various research and industrial applications not related to fusion energy.

He-3 is a key byproduct of fusion in the core of the Sun. The Sun is a huge fusion reactor fueled by hydrogen. Ionized He-3 is among the materials ejected by the Sun as part of the solar wind. The Earth's magnetic field deflects most charged particles away, making He-3 rare on Earth. Magnetic field and atmosphere are absent on the Moon. Therefore, He-3 and other ionized materials have been directly bombarding the lunar surface for billions of years. He-3 is embedded in the lunar regolith all over the Moon's surface. Its abundance is latitude dependent. He-3 is believed to be found within 3 m of the surface of the lunar regolith, making it relatively easy to excavate.

He-3 was found in the lunar soil samples collected by Apollo astronauts. The University of Wisconsin estimates that about 25 tons of He-3 would be enough to power the United States for a year. From an economic point of view, He-3 is highly valuable and can disrupt the entire power generation industry. It is impractical for Earth to have the amount of He-3 needed by the energy industry. It is fair to state that countries going to the Moon have He-3 on their minds.

Data from China's Chang'e-1 lunar satellite in 2010 suggests the distribution of He-3, as shown in Figure 12.22. The total mass of He-3 is estimated to be a little over 1 billion kg (1 million tons), with 660 million kg on the near side and 370 million kg on the far side. In Figure 12.17, the concentration near the equatorial

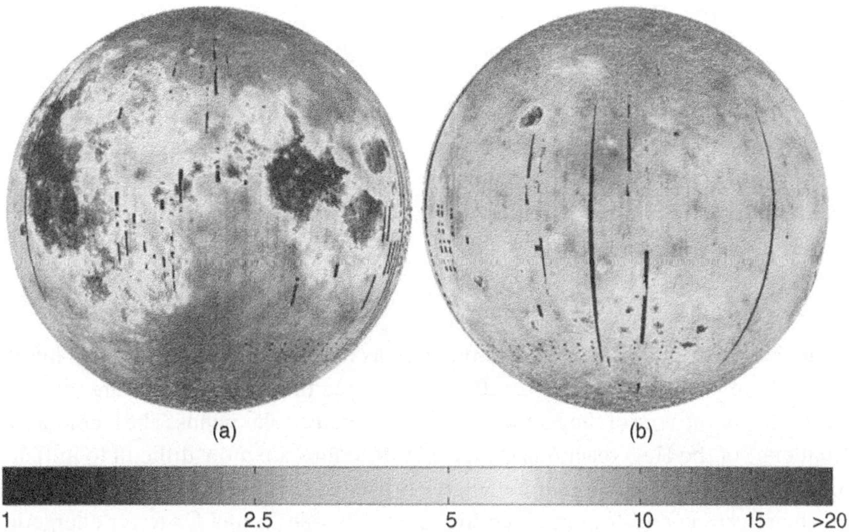

Figure 12.17 The distribution of helium-3 on the near side (a) and the far side (b) of the Moon. The centration in parts per billion (ppb) is represented by different colors. The equator on the nearside shows a higher concentration than the polar regions. *Source*: Adapted from Refs. [36–38].

region reaches 20 ppb and becomes considerably less near the polar regions. Ppb is part per billion.

12.7.1 He-3 Nuclear Fusion

Nuclear fusion has been under development and experimentation for several decades. Fusion is the holy grail of the energy industry. While nuclear fission is breaking up heavy atoms, fusion energy comes from forcing two light atoms together. Fusion is initiated by an input energy, and the goal is to achieve an output energy larger than the input, that is, a net positive.

Current work on nuclear fusion primarily focuses on using Deuterium (D) and Tritium (T), both of which are heavier isotopes of the hydrogen atom. Possible fusion fuels include He-3, Deuterium (D), Tritium (T), and Boron (B). Only He-3 is not available on Earth in quantity. To fuse atoms together, these atoms need to be ionized and accelerated to super-high speeds close to the speed of most of the. Therefore, it takes a lot more energy to ionize and accelerate heavy ions.

12.7.1.1 Deuterium-Tritium (D-T) Reaction

Tritium is radioactive (half-life 12 years) and requires care in handling it. The D-T interaction produces highly energetic neutrons and He-4 atoms while releasing thermal energy. This thermal energy is used to produce steam to drive turbines similar to traditional power plants. Such thermal conversion efficiency is usually limited to 50% or less.

Additionally, the energetic neutrons can only be slowed down but not fully contained. Energetic neutrons cause material damage in the containment structure and even turn other atoms into radioactive by transmutation. They also pose severe health risks. Contaminated materials become radiation hazards and need to be replaced every few years [39].

12.7.1.2 He-3 Reactions

Deuterium-He3 and He3-He3 fusion reactions have been considered. He-3 interactions are promising on a theoretical basis since these interactions are already occurring in the core of the Sun and are present in the solar winds. The technology challenge of the He-3 reactions is that such reactions are more difficult to initiate than the D-T reaction. More input energy is needed to cause the He-3 ions to fuse.

The reward comes from a much higher energy release and far fewer energetic neutrons. Neutron damage would be much reduced. The most attractive byproduct of the He-3 reactions are the protons. The He3-He3 reaction produces twice as many protons as D-He3. Protons can be directly used as electric current without the thermal conversion process, thereby increasing the conversion efficiency.

The extraction of He-3 on the Moon will involve digging up and transporting large quantities of regolith to a processing facility to extract the He-3 and other useful resources. A mobile facility is also possible. If we use 10 ppb as the concentration from Figure 12.17, we need 10 billion-kg regolith to yield 100 kg of He-3. We use 100 kg of He-3 to do the estimate because 100 kg of He-3 and 67 kg of deuterium would produce approximately 1 gigawatt-year of electric energy, i.e. like a 1 GW power plant.

What volume of regolith would yield 10 billion kg? The regolith density is typically 1,500 kg/m^3. One would need 7 million cubic meters of regolith to obtain 10 billion kg. If we excavate the regolith to 3 m deep, the area would be about 2.5 km^2 (assuming even distribution of He-3). We have an idea of the size of the dig.

Since He-3 is theoretically promising, a substantial quantity of He-3 will be needed to support scientific investigations on Earth [40]. Extracting He-3 from lunar regolith is the answer since it would be too expensive to transport billions of tons of raw lunar soil to Earth for processing.

Since the He-3 concentration is relatively low in the ppb range, extraction requires heating a large volume of regolith. Fortunately, the process of extracting 1 kg of He-3 also yields thousands of kilograms of hydrogen, He-4, water, carbon dioxide and carbon monoxide, and methane as byproducts. These gases will then need to be separated. To achieve industrial scale, a fission reactor in the megawatt range would be needed for this operation.

12.8 Lunar Minerals

This chapter only focuses on identifying key materials and their potential applications that can benefit the Earth and/or help develop the Moon. The processes of extracting these resources will not be covered.

By concentration, we can expect the Moon to have more materials in some categories and lack in others. We can further expect that the extraction processes that work on Earth may need either modification or to be redeveloped altogether to suit local conditions.

Many industrial processes on Earth work by a) using 1-g gravity, b) using plenty of water, c) using plenty of electricity, and d) generating wastewater and toxic chemicals like liquids and gases. All these factors are either not possible or extremely uneconomical on the Moon.

The south pole is endowed with sunlight and a rich supply of oxygen locked in useful mineral resources such as iron, aluminum, silicon, and titanium, to name a few. In 2020, the US Geological Survey (USGS) published a complete geologic map of the Moon, both near side and far side. The map, called the "United Geologic Map of the Moon," is shown in Figure 12.18. Geologic maps benefit interested

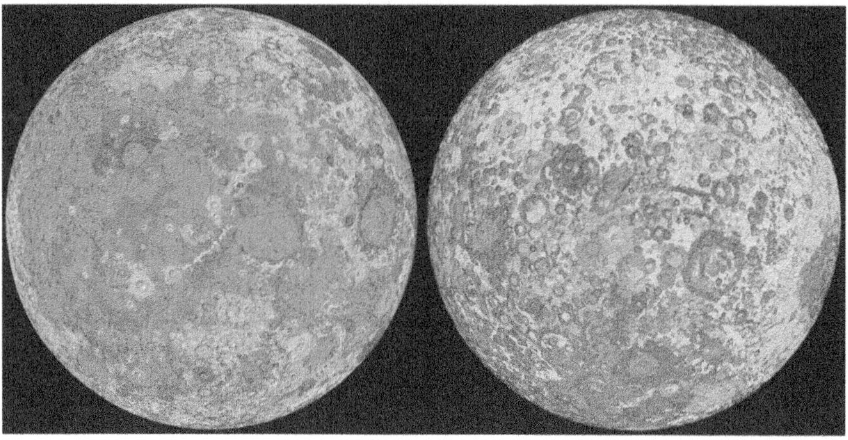

Figure 12.18 Map produced by USGS showing the distribution of minerals. The left is near side, and the right is far side. Readers are referred to the link to identify the materials. *Source*: Ref. [41] / NASA / Public domain.

parties, including geologists, mission planners, and the mineral and construction industries alike. Readers are referred to the USGS website [41] for details of the geologic map.

A 2019 Popular Science article provides a short list of seven elements that can be mined on the Moon [42]. These elements are either vital for human survival on the Moon or serve Earth's needs. Some materials may appear abundant on Earth, but it would be uneconomical to transport these resources to the Moon.

The 7 listed elements include: silicon, rare earth elements (REE), titanium, aluminum, precious metals, water, and helium-3. Water and He-3 have been covered earlier; we will focus on the first five elements and add in other elements to set up a portfolio of important materials needed on the Moon.

Figure 12.19 shows the concentration by percentage of key elements based on satellite data on the highlands and lowlands relative to Earth. Figure 12.20 shows the percent concentration of these vital materials found on lunar soil.

Figure 12.19 shows that titanium, magnesium, calcium, aluminum, and iron on the Moon have a higher concentration (but not necessarily by total weight) than those on Earth. But potassium, sodium, and nitrogen are grossly lacking.

We will list the key applications of some of these vital materials.

1) Silicon: Silicon is known on Earth for the fabrication of integrated circuits. The majority of the photovoltaic (PV) cells are made from silicon. Compared to integrated circuits, the purity requirement of silicon is not as stringent in making solar panels. Therefore, solar panels are the first lunar application as the demand for solar power on the Moon obviously takes priority.

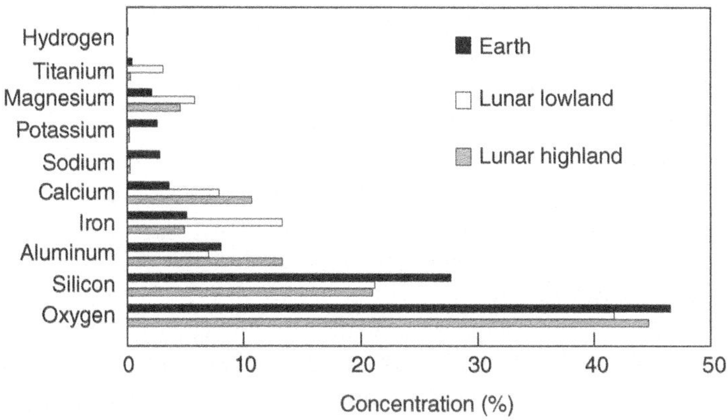

Figure 12.19 Histograms of key elements on the lowlands and highlands on the Moon. Earth data (black) is included for comparison. *Source*: Ref. [43] / Roger wilco / wikimedia Commons / Public Domain.

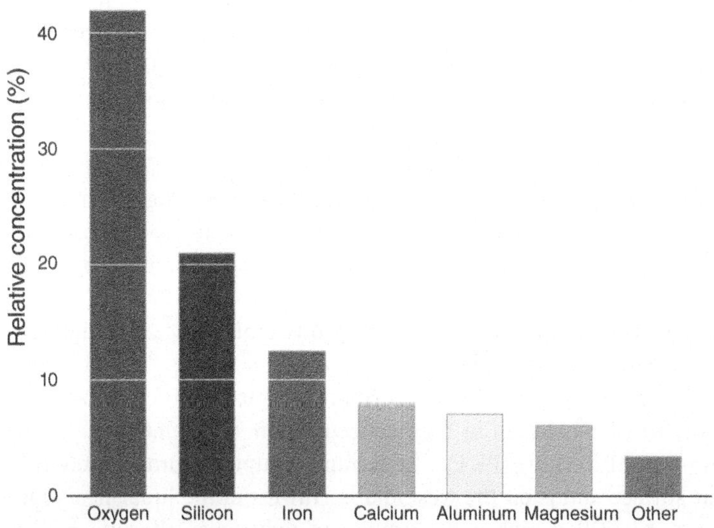

Figure 12.20 Composition of elements in the lunar soil. The oxygen concentration is high and is locked in as oxides in minerals. *Source*: Ref. [44] / Hateras / wikimedia Commons / Public Domain.

2) Iron: More iron is found in the lowlands. It exists in the form of free iron and oxides. Free iron tends to alloy with nickel and cobalt, making it easy to sort out by magnets. Iron is an important building material. In powder form, it can be used in 3D printing.

3) Aluminum: A higher concentration of aluminum is found in the highlands. Aluminum is used in many products due to its lightweight and good electrical conductivity. While the Earth does not need to import aluminum from the Moon, it is useful in ISRU to build up on the Moon.
4) Titanium: Most titanium is found in the lowlands with a richer concentration than on Earth. Known for its lightweight, resistance to corrosion, and tolerance to high temperatures, it can be used in the construction of spacecraft and aerospace devices. The material can benefit Earth if the economics of transporting them is justified.
5) Precious metals: The quantity of precious metals is unknown but likely lower in concentration than Earth. Precious metals usually refer to gold, silver, platinum, palladium, rhodium, ruthenium, iridium, and osmium. Gold, silver, and platinum are often used in jewelry and traded as investment commodities. They all have important applications in the high technology and medical industries. Hence, precious metals are highly valuable to the Earth.
6) Rare earth elements (REE): The amount of REE on the Moon has yet to be determined accurately, although the current estimated concentration is lower than that on Earth. Their distribution also varies depending on the location. The Earth has a relatively high abundance of REE in the 100–200 ppm range, but extracting REE consumes plenty of energy and water, not to mention industrial pollution.

   Space agencies such as NASA place emphasis on finding REE on the lunar surface due to the materials' important applications in the high-technology industries. There are 17 rare earth elements. Many REEs are used in the manufacture of new products such as electric vehicles, magnets, and electronic devices.

   Interestingly, REE is found on the Moon with two otherwise incompatible elements, potassium and phosphorous, from Apollo samples. The acronym KREEP stands for potassium (chemical symbol K), rare earth elements, and phosphorous. KREEP is found in high concentration, about 4.8 ppm, in the Procellarum KREEP Terrane (PKT). The terrane occupies approximately 10% of the lunar surface and is on the west and northwest of the lunar near side. In addition to KREEP, a high concentration of radioactive thorium is also found. These materials are believed to have been related to mare volcanism in the past.

   Potassium and phosphorous are important elements in plant growth. Potassium varies between 0.05% and 0.4%, depending on the location on the Moon, while its concentration in Earth soil is up to 2%. The concentration of phosphorous on the lunar surface is low but is higher in KREEP, ranging from 0.1% to 0.3%. Lunar soil has some phosphorous compounds, but the alkalinity prevents the phosphorous compounds from dissolving as nutrients.

7) Calcium: Calcium is abundant and more so in the lunar highlands. The calcium metal is flexible and makes a good conductor when oxygen is absent. Calcium is also an important component in the manufacture of silicon solar cells. Existing in different compounds, calcium compounds can be used to make ceramics.
8) Carbon and nitrogen: The Moon lacks carbon and nitrogen. The concentration of carbon in the Earth's soil is 25,000 ppm compared to 100 ppm on the Moon. On Earth, the nitrogen in the soil is about 50 ppm primarily in the form of organic matter. But 78% of the Earth's atmosphere is nitrogen. The nitrogen concentration in lunar soil is less than 5 ppm and was found to be the case in Apollo soil samples.

Carbon and nitrogen are important elements for plant growth. The question is how to source them. Carbon exists in the form of carbon-bearing ice found in permanently shadowed regions. The temperature is cold enough that solid carbon dioxide may exist, but verification is needed. Carbon-bearing ice is about 20% or less by weight of all the polar ice, but carbon's concentration is only between 0% and 3%.

Finding nitrogen on the Moon remains to be a challenge. Nitrogen compounds such as ammonia may need to be brought in from Earth for lunar plant growth.

12.9 Lunar Communications Infrastructure

The Earth–Moon distance is 384,000 km. It takes between 1.2 and 1.35 seconds for radio signals to travel for one-way communication. The Apollo lunar landers had transponders to directly communicate with the mission control. Not only is there a signal delay, but the radio frequency (RF) power also spreads by the time it reaches Earth. A large dish antenna is needed to collect the weak signals. This issue goes in both directions.

As of 2024, 5 interstellar space probes are still operating, and since the 1950s, over 50 probes and satellites have been launched to investigate the Moon. NASA has been operating a Deep Space Network (DSN) since 1958. The DSN is designed to track spacecraft starting from 16,000 km away from Earth and going beyond the solar system. DSN has three sites around the world in California, Australia, and Spain to track deep space probes that send signals to Earth. These sites have numerous antennas. The locations were chosen to ensure constant communications to compensate for the Earth's rotation.

As the number of human and machine lunar missions has been increasing, large volumes of data are generated constantly from videos to scientific data. A new lunar communications network has been called for. NASA has been working on the framework of a new network called LunaNet. A lunar communications network will aim to support the exploration of the Moon and ready us to go on to Mars.

One issue is navigation on the Moon. On Earth, we are accustomed to using the global positioning system (GPS) or similar navigation products. Our positions are accurately defined by orbiting satellites. Currently, all activities that occur on the Moon are referenced back to mission control on Earth. Is there a need for a lunar version of the GPS? Ground control on Earth would need to be on the same page as lunar pioneers and mobile equipment in terms of their navigation. This issue is particularly important for the far side of the Moon.

Establishing a lunar GPS will require a network of satellites orbiting the Moon.

12.9.1 Lunar Communications Relay and Navigation Systems (LCRNS)

The LCRNS is a communications network with Artemis missions in mind, but it is also the foundation for future growth. Figure 12.21 describes the picture.

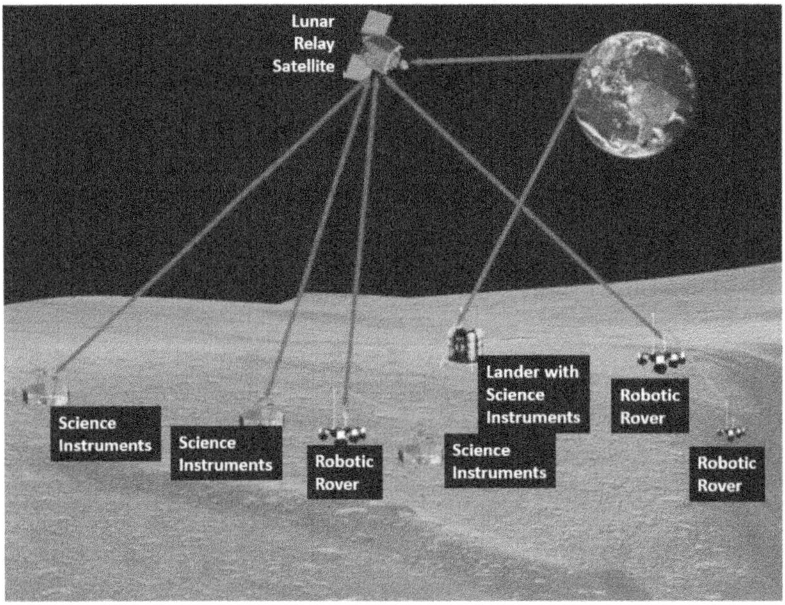

Figure 12.21 Lunar Communications Relay and Navigation Systems by NASA connect the Earth with various activities on the lunar surface. Instead of the Earth communicating directly with each activity with large antennas, the relay satellite takes care of the transmission and reception with the Earth and relays with the lunar surface. *Source*: Ref. [45] / NASA / Public domain.

Essentially, a series of lunar satellites, including the Gateway, will serve as relays between the lunar surface and Earth. The satellites will be strategically placed in lunar orbit to ensure constant communications with all the activities around the outpost on the south pole and later the entire lunar surface. Lunar activities include scientific experiments, data collection, operating equipment, and crew activities. Continuous connectivity and interconnectedness between the Earth and multiple "lunar assets" are the key goals for this network. These foundational concepts will be expanded to beyond the Moon to Mars.

In practice, a central antenna at the outpost functioning like a "cell tower" may work better and ease the transmission power requirement of the local activities. The LCRNS is a collaboration between government agencies and the private sector to ensure that the most advanced systems from hardware to software are used.

12.9.2 LunaNet

LunaNet is a system architecture for communications and navigation akin to the popular internet we use every day. The LunaNet architecture comes from collaborations among governments, industries, and science communities worldwide to build up the LunaNet nodes. The system is open, expandable, and interoperable. The idea is to avoid "duplication of efforts" as more nations plan human and machine missions to the Moon.

The LunaNet nodes will offer four services to lunar missions: networking, navigation and positioning, detection and information, and radio/optical science services.

12.9.2.1 Network

To put simply, the LunaNet protocol allows lunar users (probes, local machineries, and pioneers) and parties with vested interests to just "plug and play." State-of-the-art error correction and detection technologies will be incorporated, given the long distance between Earth and the Moon and the possibility of signal disruption. LunaNet is characterized as a delay/disruption-tolerant network (DTN). Both laser and RF communications are used as the laser links carry much higher data volumes.

12.9.2.2 Positioning, Navigation, and Timing (PNT)

PNT is what GPS is about. PNT not only provides positioning data (longitude, latitude, and altitude) but also adds the dimension of time. Each GPS satellite is equipped with several precise timing devices known as atomic clocks. Riding on the lunar relay and navigation system, LunaNet users on the Moon will have a positioning and navigation system that will not need to wait for the Earth's response. Lunar users can also synchronize their time and frequency reference for their equipment. A lunar time standard is now established.

12.9.2.3 Detection and Information System

This situation-awareness function is similar to an emergency alert, distress, and search and rescue. Instead of waiting for the Earth to inform users of impending danger or emergency and provide directions, this system communicates with lunar users directly.

12.9.2.4 Science Services

A LunaNet node can be a scientific instrument making global or regional observations of the Moon. Through the Science Services, a researcher on Earth can seek the help of the node to make observations for the benefit of his or her research.

12.9.3 Queqiao System (China)

Queqiao is a Chinese legion with the direct translation "Magpie Bridge" that links across the galaxy. The immediate goal of this system is to build a communications and navigation system to cover the lunar surface to facilitate communication on both sides of the Moon. The long-term goal is the Cislunar space and beyond. This system will support China's ambitious programs on the Moon, including a future research station on the south pole.

The system will take three phases to build:

Phase One: A pilot constellation by 2030 to support the lunar missions as planned
Phase Two: A working basic constellation by 2040 to cover regional communications and navigation for China and other international missions
Phase Three: Expand the system to facilitate further exploration of Mars by 2050

12.9.4 Andromeda Satellite Constellation

This constellation is proposed by Argotec, an aerospace company in Italy, in collaboration with the Jet Propulsion Laboratory (JPL) [46]. The constellation is composed of 24 satellites evenly divided among 4 orbits. The constellation aims to cover the entire lunar surface and be the communication link between the Earth and the activities on the lunar surface.

The proposal is motivated by the anticipation of increased lunar missions and therefore the need for communications between Earth and humans and machines operating on the Moon as well as the space station Gateway. Each satellite measures $44 \times 40 \times 37$ cm before its solar panels and antennas are deployed. The goal is similar to that of the LCRNS above.

The chosen orbits are known as frozen orbits and are stable. A frozen orbit is a special class of elliptical orbits that are not sensitive to the irregularities of the Moon and therefore do not drift over an extended period of time. Fuel for orbit adjustment is kept to a minimum. The Andromedia orbits have 57° inclination relative to the lunar equator with apoapsis at about 8,000 km and periapsis at 720 km. A satellite at the apoapsis is the slowest and spends a lot of time there and can therefore take a long look at the target on the ground. The orbits are designed to have the maximum coverage of the lunar surface.

12.9.5 Data Centers on the Moon

Data centers are integral parts of the space's communications infrastructure. Like the many data centers now located in terrestrial cities, they serve as the repository of information.

The lunar missions of this century are expected to generate enormous volumes of data. The operation of autonomous and semiautonomous machines on the Moon requires real-time data exchange with a local data depot instead of always relying on the Earth side. For the near term, a bank of data servers on the Moon would suffice and could also be a communication node linking to relay satellites. A similar case has been tested on the ISS where artificial intelligence (AI)-enabled servers were installed to assist astronauts to expedite experiments instead of transmitting all the data back to the ground control to be analyzed.

Large data centers on Earth consume plenty of electric power—typically around 100 MW, about 40% of which goes to cooling. This much power is not available on the Moon, nor is it needed initially. Operating a data center on the Moon would have the advantage in thermal management. The data center will have to be inside a shielded structure due to high solar and galactic radiation on the lunar surface, which can cause reliability issues with semiconductors. Protecting the data servers from radiation should be a high priority.

Computing power is also a capability accompanying a data center. Edge computing capability in space is another topic under research and development to expedite the processing of large volumes of data collected by the lunar activities.

12.10 Physical Hazards

There are all kinds of physical hazards on the Moon due to its harsh environment. We can generally classify these hazards into five groups: low gravity, pressure and temperature, radiation, electrostatic charge, and Moon dust. All of these roadblocks are known but cannot be physically removed. Technologies will solve some of these challenges over time.

12.10.1 Low Gravity

Humans have limited first-hand experience of living on the lunar surface. Apollo astronauts spent several days on the Moon. And they wore protective spacesuits while outside of their landing capsules.

By now, astronauts have plenty of experience functioning in microgravity (weightless) environments in LEO space stations but not under the Moon's low gravity. Some of the known health issues under microgravity conditions may or may not apply to the low gravity on the Moon.

Whether humans will experience similar health effects in 1/6 of Earth's gravity for days and months needs supporting data, which Artemis astronauts will provide.

12.10.2 Electrostatic Charge

Due to bombardment by solar and cosmic radiation without the protection of an atmosphere, lunar dust and regolith are negatively charged to hundreds to thousands of volts. Charged lunar dust repels each other and therefore stays suspended once it is kicked up. The charge potential is location dependent.

Static charge can cling onto space suits, rover wheels, and other equipment, resulting in potential arc damage to electronic circuits such as short-circuiting. Humans' susceptibility to high voltage ranges from a tinkling feeling to potentially lethal damage, depending on the amount of electric current passing through the body.

Unlike on Earth, the term "electric ground" has no meaning on the Moon. The lunar regolith is electrically nonconductive so "ground" cannot be established that way. Rather, "ground" should be understood as sharing a common electric potential reference such as using connected conductive strips.

12.10.3 Radiation

Radiation on the lunar surface is about 200 times more than on Earth due to the lack of a magnetic field and atmosphere. This ionizing radiation can move through materials before being stopped and often alters the electronic structure of the materials along the way. Electronic components and devices are the most susceptible.

Radiation can only be minimized by choosing the right shielding for shelters, mobile equipment, and spacesuits when working in the open. Radiation protection is a main topic to be resolved before humans can stay on the Moon for a long time [47].

12.10.4 Moon Dust

Moon dust is part of the lunar regolith. It was first noticed on the suits worn by Apollo astronauts when they returned to their capsules after their Moonwalks. The dust reportedly caused problems with some equipment. Several astronauts who had stepped on the Moon reported hay fever symptoms.

Moon dust is a mixture of compounds, with silicate being the dominant component. The micrometer-sized fine powder is razor sharp and abrasive. Silicate dust is known to cause severe scars and damage to miners' lungs on Earth. The longer the lunar dust stays in the lungs, the greater the chance of toxic effects. Therefore, lunar dust is a threat to long-term lunar missions.

Moon dust is believed to have formed over the age of the Moon from rocks and meteorites bombarding the lunar surface, turning rocks into shards and glassy sands. Moon dust has never gone through weathering processes by the forces of nature like the sands on Earth. The dust's fine size makes it easy to be electrostatically charged when bombarded by solar winds. Spacesuits and equipment easily pick up lunar dust. With an increased number of lunar missions and landings and takeoffs by crewed and noncrewed spacecraft, plumes by rocket engines and thrusters will kick up the Moon dust. Once kicked up, the dust will stay suspended for a long time. Technological means must be applied to reduce the hazards to personnel and equipment working inside and outside their enclosed quarters.

12.10.5 Space Suits

Humans can stand on the lunar ground. Lunar space suits are similar to those worn by astronauts doing extravehicular activities (EVA) outside space stations. They are equipped with many systems vital to an astronaut's survival outside the space station. The environment on the lunar surface is very similar in terms of pressure, temperature, and radiation exposure. The difference is that the lunar surface is a solid ground with some gravity.

The bulky EVA suit plus all the attached equipment caused the center of gravity to shift, making Apollo astronauts difficult to walk and easy to tumble. Called the Extravehicular Mobile Unit (EMU), the suit used in the Space Shuttle program weighs 125 kg on Earth. The outer layer of these EMUs is made from a stiff fabric called Ortho-Fabric. The EMU has not changed much since the Shuttle program. It would be highly inconvenient for lunar settlers to bunny-hop around in the same EVA spacesuits on the Moon during their extended stays.

The lunar spacesuit design combines material properties with mechanical and electronic gear. Artemis astronauts will have more tools and autonomous machines, dexterity, and flexibility at their disposal. For example, the ESA funded a Pextex (Planetary EXploration TEXtiles) project to identify new materials and textiles capable of withstanding the hostile environment of the Moon and Mars for use in future space suits.

Candidate materials should meet the following requirements: a) thermal insulation with 17–25 °C as design range, b) tolerance of the temperature swing from −170 to 120 °C, c) radiation and micrometeoroid resistant, d) withstand vacuum and pressure cycles from 6 psi (about 0.4 atmosphere) over a specified number of cycles, e) abrasion resistant to lunar soils including Moon dust, f) nontoxic and nonflammable, and g) impermeable to fluids and 180° foldable.

The Pertex project requires the duration to be at least 2,500 hours (about 3.5 months). In 2023, fashion design house Prada announced a partnership with

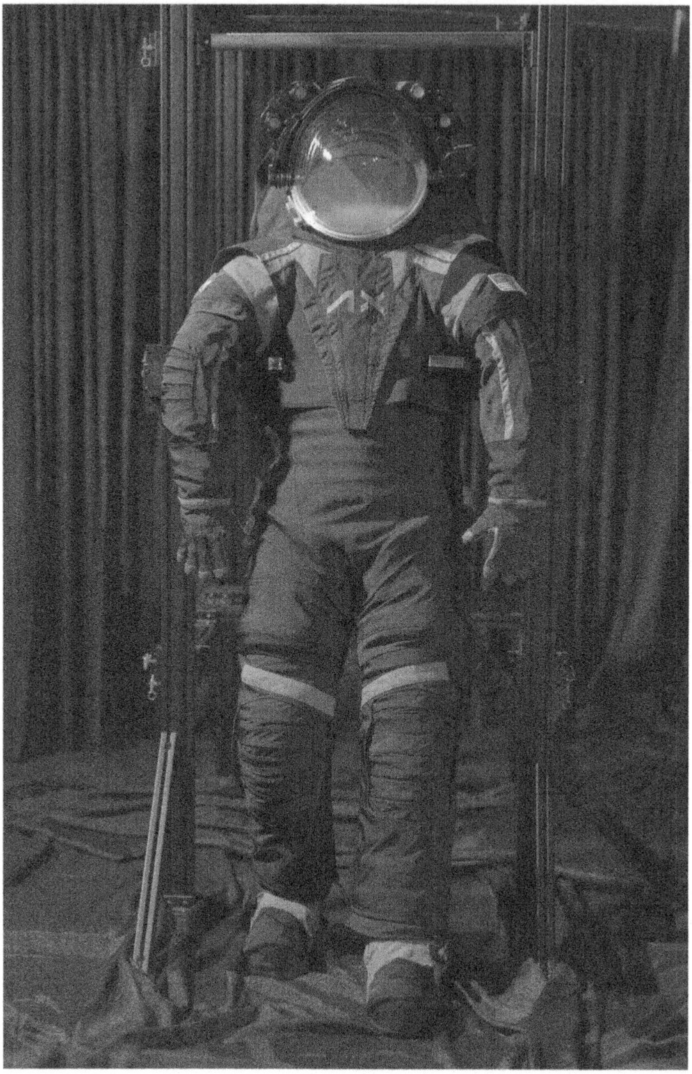

Figure 12.22 Prototype of a lunar suit proposed by Prada and Axiom Space with funding from NASA in 2023. *Source*: Ref. [48] / Axiom Space, Inc.

Axiom Space to design the next-generation lunar suit for the Artemis Program under funding from NASA. The prototype AxEMU is shown in Figure 12.22. The suit has a preliminary mass of 55 kg. In 2024, Axiom and NASA showed the next-generation EMUs that are destined for Artemis astronauts, as illustrated in Figure 12.23. These suits are to be put through extreme testing.

Figure 12.23 Axiom and NASA demonstrate the extravehicular mobile unit (EMU) in 2024. The suit is intended for use in the open lunar environment. *Source*: Ref. [49] / NASA / Public domain.

12.10.6 Moonquakes

One may wonder if quakes occur on the Moon and are hazardous. The seismic instruments installed by the Apollo astronauts recorded nearly 22,000 seismic tremors between 1969 and 1977. They are moonquakes.

The Moon has no active volcanoes and no tectonic plate motions. Moonquakes are believed to be the result of several factors, including cooling and shrinking of the interior, tidal forces from Earth acting on the still-hot lunar magma, and impacts by meteors. Even the day–night cycle every two weeks can cause the local lunar crust to expand and contract, thereby causing moonquakes.

Moonquakes are not big by Earth's standards. Most are less than 3 on the Richter scale and the largest between 5 and 5.7. Many are found to recur following the day–night cycles.

12.11 Conclusion

The Moon is the first celestial body we set foot on beyond Earth's gravity well. Nations with active programs going to the Moon share the same goal: the Moon is a must-go place this century, and the next stop is Mars.

The environments on the Moon are harsh. That did not stop humans from stepping on the Moon's surface decades ago, and now we are ready to do that again and plan to stay. Satellite data suggest that the lunar surface contains almost all the resources humans need to form settlements. Key resources such as water, oxygen, and hydrogen can be extracted from the water. Sources of energy and material resources have been identified, but a few will likely need to be brought in.

The Moon also has resources of high value to Earth. Helium-3 is the superstar material waiting to be mined on the Moon and tested on Earth. Conversely, key elements such as nitrogen will likely have to be brought in to facilitate plant growth.

Infrastructures for energy, transportation, and communications will be built as the lunar missions move beyond the exploration phase.

While going to the lunar south pole first makes sense owing to the abundance of sunlight and access to water, materials of high interest such as He-3 and KREEP are found in low latitudes. Humans will have to set up shops at low latitudes and overcome the various harsh conditions.

Reading List

Barnatt C. *The Next Big Thing: From 3D Printing to Mining the Moon.* ExplainingTheFuture.com; 2015.

David L. *Moon Rush: The New Space Race.* National Geographic; 2019.

Dietzler D. *Mining the Moon: Bootstrapping Space Industry* (3rd Edition). Amazon Digital Services; 2020.

Poliskie M. *The Planning and Execution of Human Missions to the Moon and Mars (Progress in Astronautics and Aeronautics, 266)*. American Institute of Aeronautics Ast.; 2023.

Pyle R. *Missions to the Moon*. Andre Deutsch; 2018.

Reeves DL. *Mining the Moon*. Lerner Publications; 2024.

References

1 Foust J. DARPA picks 14 companies for lunar architecture study. Space News [cited 2025 Feb 21] [online]. Available from: https://spacenews.com/darpa-picks-14-companies-for-lunar-architecture-study/.

2 Ice Confirmed at the Moon's Poles. NASA Jet Propulsion Laboratory [cited 2025 Feb 21] [online]. Available from: https://www.jpl.nasa.gov/news/ice-confirmed-at-the-moons-poles/.

3 A map of the most likely locations of water ice deposits on the Moon's south polar regions. NASA/Goddard Space Flight Center. Cited by Papadopoulos L. Buried in the Deep: Could Water Ice Deposits on the Moon Generate Their Own Organic Compounds?. America Space [cited 2025 Feb 21] [online]. Available from: https://www.americaspace.com/2014/05/12/buried-in-the-deep-could-water-ice-deposits-on-the-moon-generate-their-own-organic-compounds/

4 Gibney E. How to build a Moon base. Nature [cited 2025 Feb 21] [online]. Available from: https://www.nature.com/articles/d41586-018-07107-4.

5 Wilcox K. The Moon's rugged, intriguing south pole. NASA [cited 2025 Feb 21] [online]. Available from: https://appel.nasa.gov/2023/08/25/the-moons-rugged-intriguing-south-pole/.

6 Chen X, Yang S, Chen G, Xu W, Song L, Li A, et al. Massive water production from lunar ilmenite through reaction with endogenous hydrogen. The Innovation. 2024 [cited 2025 Feb 21]. Available from: https://www.cell.com/the-innovation/pdf/S2666-6758(24)00128-0.pdf.

7 Depond C. Chinese scientists succeed in making water from lunar soil: here's how. Techno-science [cited 2025 Feb 21] [online]. Available from: https://www.techno-science.net/en/news/chinese-scientists-succeed-in-making-water-from-lunar-soil-here-how-N25519.html.

8 Korotev RL. The chemical composition of lunar soil. Department of Earth, Environmental, and Planetary Sciences Washington University in St. Louis [cited 2025 Feb 21] [online]. Available from: https://sites.wustl.edu/meteoritesite/items/the-chemical-composition-of-lunar-soil/.

9 Schuler J, Smith D, Nick A, Mueller R, Langton A, Buckles B. Capabilities and scalability of the regolith advanced surface systems operations robot (RASSOR). AIAA ASCEND 2020; 2020 [cited 2025 Feb 21]. Available from: https://ntrs.nasa.gov/api/citations/20205008904/downloads/ASCEND%202020%20Presentation_ID%233411599.pdf.

10 Regolith advanced surface systems operations robot (RASSOR) excavator. NASA [cited 2025 Feb 21] [online]. Available from: https://technology.nasa.gov/patent/KSC-TOPS-7.

11 Cataldo, R, Bennett. (2011). *U.S. Space Radioisotope Power Systems and Applications: Past, Present and Future.* 10.5772/23914.

12 Salh H. Improving the overall efficiency of radioisotope thermoelectric generators. Advances in Energy and Power. 2014;*01*(2):21–26.

13 Fission surface power. NASA [cited 2025 Feb 21] [online]. Available from: https://www.nasa.gov/space-technology-mission-directorate/tdm/fission-surface-power/.

14 Types of fuel cells. U.S. Department of Energy [cited 2025 Feb 21] [online]. Available from: https://www.energy.gov/eere/fuelcells/types-fuel-cells.

15 Antweiler W. What role does hydrogen have in the future of electric mobility? Werner's Blog [cited 2025 Feb 21] [online]. Available from: https://wernerantweiler.ca/blog.php?item=2020-09-28.

16 Hambleton K, Ugalde V, Ramji N. NASA selects companies to advance moon mobility for artemis missions. Fuel Cell & Hydrogen Energy Association [cited 2025 Feb 21] [online]. Available from: https://www.nasa.gov/news-release/nasa-selects-companies-to-advance-moon-mobility-for-artemis-missions/.

17 JAXA and Toyota announce lunar cruiser as the name for their manned pressurised rover. Toyota [cited 2025 Feb 21] [online]. Available from: https://media.toyota.co.uk/jaxa-and-toyota-announce-lunar-cruiser-as-the-name-for-their-manned-pressuri.

18 Schaler E. FLOAT — Flexible levitation on a track. NASA; 2021 [updated 2021 Feb 15; cited 2025 May 1] [online]. Available from: https://www.nasa.gov/general/float-flexible-levitation-on-a-track/.

19 Elkins A, Daogaru M, Dufresne A, Studebaker A, Musker A, Roberts G, et al. Lunar hopper enhancements. Engineering and the Environment, University of Southampton [cited 2025 Feb 21] [online]. Available from: https://uosdesign.org/designshow2016/group-design-project/lunar-hopper-enhancements/.

20 Kalita H, Thangavelautham J. Exploration of extreme environments with currentand emerging robot systems. Current Robotics Reports [cited 2025 Feb 21]. Available from: https://link.springer.com/article/10.1007/s43154-020-00016-3.

21 Spudis PD. Drones on the Moon. Smithsonian Magazine [cited 2025 Feb 21] [online]. Available from: https://www.smithsonianmag.com/air-space-magazine/drones-moon-180956187/.

22 Meet "First Base" A lunar landers best friend. Bigelow Aerospace [cited 2025 Feb 21] [online]. Available from: https://bigelowaerospace.com/pages/firstbase/.

23 Grondin YA. Affordable habitats means more Buck Rogers for less money says Bigelow. NASA Space Flight [cited 2025 Feb 21] [online]. Available from: https://www.nasaspaceflight.com/2014/02/affordable-habitats-more-buck-rogers-less-money-bigelow/.

24 An artist's conception shows the B330 space module in Earth orbit. Bigelow Aerospace. Cited by Boyle A. Bigelow Aerospace and ULA plan to launch supersized space station module in 2020. GeekWire [cited 2025 Feb 21] [online]. Available from: https://www.geekwire.com/2016/bigelow-aerospace-ula-plan-expandable-b330-space-station-module-2020/.

25 The world's first metal 3D printer for space is on its way to the ISS. Airbus [cited 2025 Feb 21] [online]. Available from: https://www.airbus.com/en/newsroom/stories/2024-01-the-worlds-first-metal-3d-printer-for-space-is-on-its-way-to-the-iss.

26 Page T. A 3D-printing company is preparing to build on the lunar surface. But first, a moonshot at home. Cable News Network [cited 2025 Feb 21] [online]. Available from: https://edition.cnn.com/style/article/icon-3d-printing-architecture-initiative-99-project-olympus-spc-intl-scn/index.html.

27 Shekhtman L. New evidence adds to findings hinting at network of caves on moon. NASA [cited 2025 Feb 21] [online]. Available from: https://science.nasa.gov/solar-system/new-evidence-adds-to-findings-hinting-at-network-of-caves-on-moon/.

28 Can plants grow in quasi-vacuum? NTRS - NASA Technical Reports Server [cited 2025 Feb 21] [online]. Available from: https://ntrs.nasa.gov/citations/19860010460.

29 China's Moon mission sees first seeds sprout. BBC [cited 2025 Feb 21] [online]. Available from: https://www.bbc.com/news/world-asia-china-46873526.

30 Huston C. Growing plants in space. NASA [cited 2025 Feb 21] [online]. Available from: https://www.nasa.gov/exploration-research-and-technology/growing-plants-in-space/.

31 Yirka B. Using bacteria to make lunar soil more fertile. Science X Network [cited 2025 Feb 21] [online]. Available from: https://phys.org/news/2023-11-bacteria-lunar-soil-fertile.html#google_vignette.

32 Space lab LEAF experiment selected for Artemis III lunar mission. Space Lab [cited 2025 Feb 21] [online]. Available from: https://www.spacelabtech.com/lunar-payload-leaf---lunar-effects-on-agricultural-flora.html.

33 Wall M. Artemis astronauts will carry plants to the moon in 2026. Space [cited 2025 Feb 21] [online]. Available from: https://www.space.com/nasa-artemis-3-plant-growth-experiment-moon.

34 Why cultivated meat? FUDZS [cited 2025 Feb 21] [online]. Available from: https://fudzs.com/why-cultivated-meat/.

35 Deep space food challenge. Methuselah Foundation [cited 2025 Feb 21] [online]. Available from: https://www.deepspacefoodchallenge.org/.

36 Fa W, Jin Y. Global inventory of Helium-3 in lunar regoliths estimated by a multi-channel microwave radiometer on the Chang-E 1 lunar satellite. Chinese Science Bulletin. 2010;*12*(55):4005–4009.

37 Whittington M. Solving the climate and energy crises: Mine the Moon's helium-3? The Hill [cited 2025 Feb 21] [online]. Available from: https://thehill.com/opinion/technology/540856-solving-the-climate-and-energy-crises-mine-the-moons-helium-3/.

38 Olson ADS. Lunar helium-3: Mining concepts, extraction research, and potential ISRU synergies. AIAA ASCEND 2021; 2021 [cited 2025 Feb 21]. Available from: https://ntrs.nasa.gov/api/citations/20210022801/downloads/AIAA%20ASCEND%202021%20Paper_211018.pdf.

39 DOE explains...deuterium-tritium fusion fuel. U.S. Department of Energy [cited 2025 Feb 21] [online]. Available from: https://www.energy.gov/science/doe-explainsdeuterium-tritium-fusion-fuel.

40 Bogossian J. Helium-3 mining on the moon: houston, we have a problem. Geology for Investors [cited 2025 Feb 21] [online]. Available from: https://www.geologyforinvestors.com/helium-3-mining-on-the-moon-houston-we-have-a-problem/.

41 Sobieszczyk S. New comprehensive geologic map of the moon released. U.S. Geological Survey [cited 2025 Feb 21] [online]. Available from: https://www.usgs.gov/news/national-news-release/usgs-releases-first-ever-comprehensive-geologic-map-moon.

42 Boyle R. The elements we might mine on the moon. Popular Science [cited 2025 Feb 21] [online]. Available from: https://www.popsci.com/elements-mine-on-the-moon/.

43 Wilco R. Moon vs earth composition. Wikimedia [cited 2025 Feb 21] [online]. Available from: https://commons.wikimedia.org/wiki/File:Moon_vs_earth_composition.svg.

44 Hateras. Composition of lunar soil. Wikimedia [cited 2025 Feb 21] [online]. Available from: https://commons.wikimedia.org/wiki/File:Composition_of_lunar_soil.svg.

45 Bhasin KB, Warner JD, Anderson LM. Lunar Communication Terminals for NASA Exploration Missions: Needs, Operations Concepts and Architectures. In: International Communications Satellite Systems Conference (ICSSC); 2008 [cited 2025 Feb 21]. Available from: https://ntrs.nasa.gov/citations/20080033045.

46 Balossino A, Davarian F. The plan to give the Moon decent wireless coverage. IEEE Spectrum [cited 2025 Feb 21] [online]. Available from: https://spectrum.ieee.org/lunar-communications.

47 Mrigakshi Dixit S. How space radiation threatens lunar exploration. Smithsonian magazine [cited 2025 Apr 28] [online]. Available from: https://www.smithsonianmag.com/science-nature/how-space-radiation-threatens-lunar-exploration-180981415/.

48 The next-generation spacesuit. Axiom Suit [cited 2025 Feb 21] [online]. Available from: https://www.axiomspace.com/axiom-suit.

49 Ugalde V. Axiom space tests lunar spacesuit at NASA's Johnson Space Center. NASA [cited 2025 Feb 21] [online]. Available from: https://www.nasa.gov/image-article/axiom-space-tests-lunar-spacesuit-at-nasas-johnson-space-center/.

13

Mars

Xue Xiaopeng[1] and Stephen W. Cheung[2]

[1]*School of Aeronautics and Astronautics, Central South University, The People's Republic of China*
[2]*Former Adjunct Professor, Department of Physics, The University of Hong Kong, Hong Kong, China*

13.1 Introduction

Mars is part of Cislunar space and an integral part of the future space economy. Going to Mars has become part of the national space agenda for superpowers since the turn of the 21st century.

To some, Mars is a better alternative to settle than the Moon as Plan B for our civilization. Mars has the materials needed to build a self-sustained industrial society, although it also has obvious shortages and limitations. The long distance from Earth makes the regular shipment of supplies to the Red Planet impractical and uneconomical. Hence, it is not about tradable goods of immediate economic value but more about self-reliance. We should keep in mind that Mars is much closer to the Asteroid Belt if we ever want to go there.

There are plenty of reasons why we go to Mars. The immediate reason is the scientific understanding of our neighboring planet. Searching for signs of life on other planets is always on the public's mind. Such findings support the idea that Mars in the past could have sustained life just like Earth. It should be emphasized that upcoming Martian missions are not intended to look for currently living organisms but for signs of past life when the conditions on Mars were conducive for life.

Although Mars is considered Earth's "fraternal brother" in terms of material makeup, its evolution has gone down a different path. Mars might have had a very similar climate to Earth in the past, but it is no longer the case. Scientists wonder if tomorrow's Earth will look like Mars today.

If experience with the Moon is any reference, humans will not be satisfied with short exploratory visits by astronauts on Mars. It will be just a matter of time before humans settle on Mars, the same as what is happening on the Moon now.

New Space: From Low Earth Orbit to the Moon and Beyond, First Edition.
Edited by Stephen W. Cheung and Chih-Yung Wen.

Developing technologies to travel to Mars and settle there will benefit Earth in many ways.

This chapter will provide an overview of Mars and its physical conditions for humans to live in. The resources available on Mars will be discussed. Other than rovers, governments have not officially announced any human missions to Mars. Hence, we use the flight paths of the rover missions to illustrate the flight aspect. After exploration, a settlement is built around three infrastructures, viz. energy, transportation, and communications. We will look at these three aspects. Finally, while the lunar south pole is our first destination in the Artemis program, we will look at where the good initial destinations are on Mars.

13.2 About Mars

Mars was formed approximately 4.6 billion years ago. Its mass is 6.4×10^{23} kg, approximately 10.7% of Earth's mass (see Figure 13.1). The equatorial radius of Mars is 3,420 km, which is about 53% of Earth's size. Mars's surface area is slightly less than the land area of Earth, while its volume is about 15% of Earth's. The average density of Mars is 3,950 kg/m^3, lower than Earth's 5,520 kg/m^3. The local gravity on the surface of Mars is only 38% of Earth's gravity, i.e. 3.7 m/s^2. Therefore, Mar's local gravity is about 1/3 of Earth's.

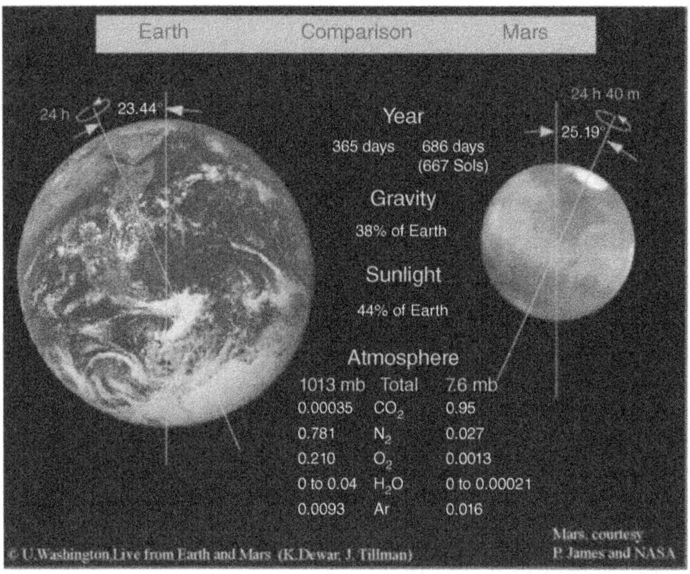

Figure 13.1 Comparison of Earth and Mars. *Source:* Ref. [1] / NASA / Public domain.

13.2.1 Physical Conditions

13.2.1.1 Celestial Motion

Mars orbits the Sun in an elliptical path with an orbital eccentricity of 0.093. Its closest and farthest points from the Sun are 207 million km and 249 million km, respectively. The average distance is 229 million km or about 1.5 AU. Mars's orbit differs from Earth's, resulting in the distance between the two planets from as close as 60 million km (0.4 AU) to as far as 400 million km (2.7 AU), depending on their respective positions.

Mars takes 687 Earth days (1.88 Earth years) to complete one revolution around the Sun. Expressed in "sol"—short for solar day—a Martian year is 668.6 sols. A sol averages 24 hours, 39 minutes, and 35 seconds, which is very close to the duration of a day on Earth. Therefore, we can say that a day on Mars feels about the same as a day on Earth.

Mars has an axial tilt of 25.19°, slightly higher than Earth's 23.44°. Consequently, Mars also experiences four seasons (see Figure 13.2). Since a Martian year is 1.88 times longer than an Earth year, the seasons are correspondingly extended. Due to "Mars" elliptical orbit, the seasons are uneven.

13.2.1.2 Atmospheric Environment

Mars has an extremely thin atmosphere, primarily composed of carbon dioxide (95.3%), nitrogen (2.7%), argon (1.6%), trace amounts of oxygen (0.15%), and water vapor (0.03%) [3]. The surface atmospheric pressure ranges from 500 to 700 Pa, which is roughly 0.5% to 0.7% of the Earth's atmospheric pressure at sea level. Hence, there is neither enough air pressure nor the right gas composition for humans to breathe freely on Mars.

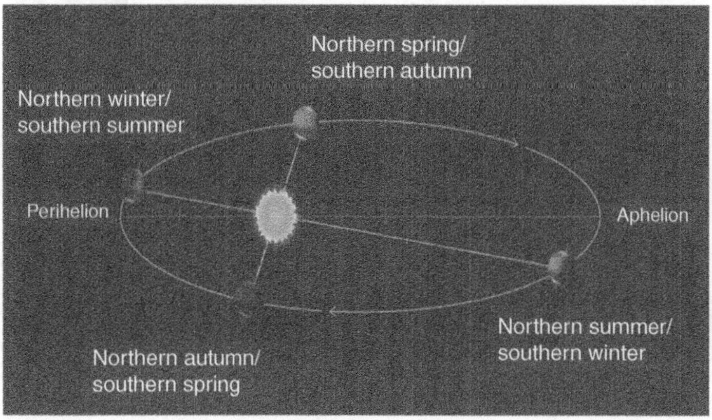

Figure 13.2 Four seasons on Mars. Since Mars describes an elliptical orbit, the four seasons are of different lengths. *Source:* Ref. [2] / Carpenter Publications.

Mars's temperature and solar radiation level vary with its orbital position relative to the Sun and its atmospheric composition [4]. The temperature on Mars is 20 °C (highest) and −153 °C (lowest), with average temperature of −62 °C, and with significant diurnal temperature variations.

Therefore, Mars is more "habitable" than the Moon in terms of temperature and day–night cycles. In this regard, the equator and low latitudes on Mars are habitable, which is the opposite to the situation on the Moon.

13.2.1.3 Terrain and Landforms

The terrain features of Mars show distinct differences between the northern and southern hemispheres [4]. The northern hemisphere is primarily composed of plains, while the southern hemisphere is dominated by highlands (see Figure 13.3). Since Mars currently has no oceans, there is no reference for zero altitude. The zero altitude is arbitrarily defined based on a certain atmospheric pressure.

The volcanoes on Mars are mainly distributed in the Tharsis Bulge, Elysium Planitia, and Syrtis Major Planum [3]. The Tharsis Bulge covers approximately 25% of the Martian surface and is home to 12 large volcanic features, including Olympus Mons, the highest peak in the solar system (see Figure 13.4). These volcanoes are predominantly effusive shield volcanoes [7].

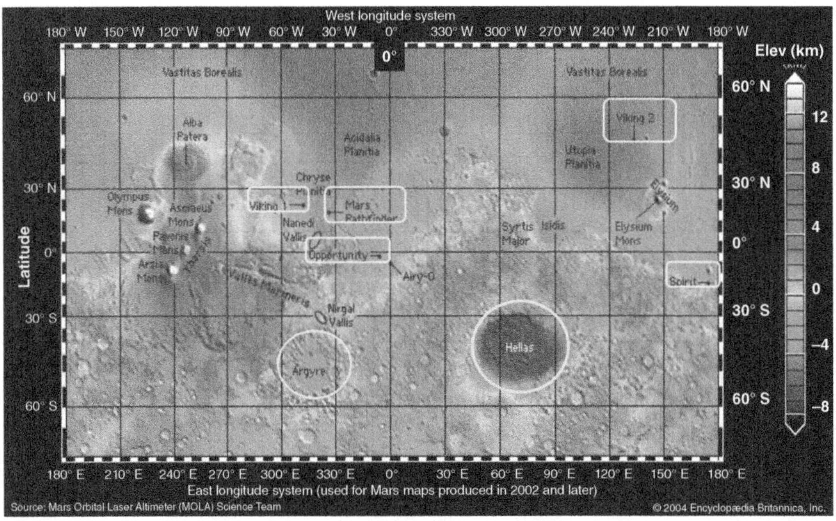

Figure 13.3 Laser altimeter data from the Mars global map. The southern hemisphere has mostly highlands, and the northern hemisphere has mostly lowlands. Several rover missions are marked in rectangular frames. The two impact basins, Hellas and Argyre, are marked in circles [5].

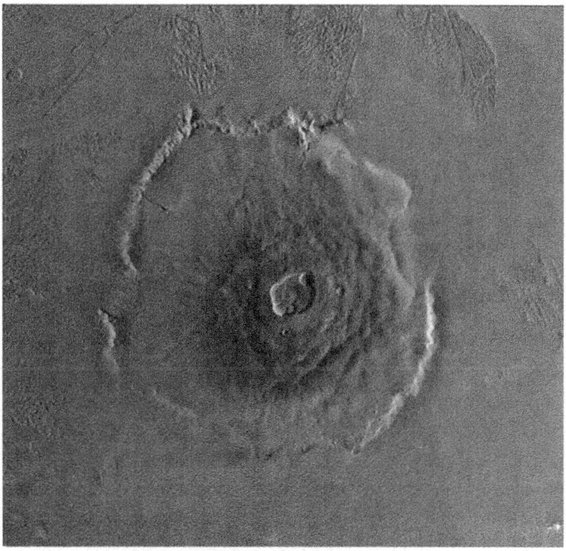

Figure 13.4 Olympus Mons is the highest peak in the entire solar system. *Source:* Ref. [6] / United States Geological Survey / Public domain.

The formation of Martian volcanoes differs from that of Earth, as they are not a result of plate tectonic movements but are primarily associated with hotspot activity [8]. Heat from the Martian interior is released from regions where hotspots are distributed, leading to the continuous accumulation of lava and the formation of vast volcanic structures.

Figure 13.3 suggests that the Martian south pole has more highlands measuring several kilometers high as well as two large impact basins. Impact craters and basins on Mars are created by meteorites hitting the planet at high speeds. Because Mars has a thin atmosphere, meteoroids do not burn as much as they do in Earth's atmosphere, so the craters on Mars are clearer and more visible [4, 7]. The two impact basins are the Hellas Planitia and the Argyre Planitia in the southern hemisphere, they are marked in circles in Figure 13.3) [7]. Hellas is as much as 8 km deep, while the Argyre is about 4 km deep.

13.2.1.4 Hydrological Characteristics

Two different satellite images help identify the presence of water. The first is the neutron count, and the second is the surface temperature.

High-resolution imagery returned by Martian orbiters reveals vast alluvial plains and potential marine traces [9]. Multiple Mars rovers have also discovered evidence of water-saturated rocks such as hematite and clay minerals [10]. Figure 13.5 shows the global distribution map of neutrons on Mars based on data collected by the Mars Odyssey orbiter in 2001 [11]. These neutrons indicate an abundance of hydrogen and therefore underground water and ice deposits. In the high-latitude

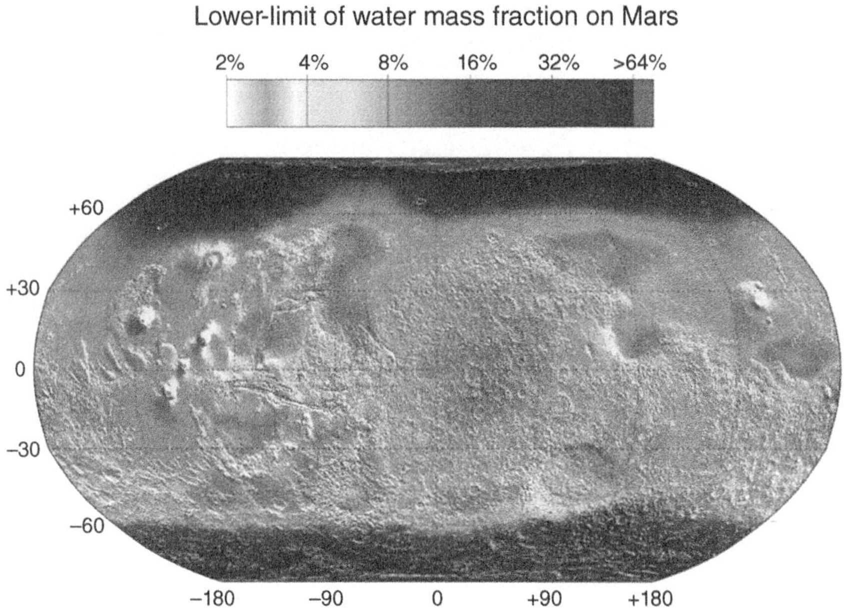

Figure 13.5 Satellite data collected in 2001. Dark represents high hydrogen and therefore reservoirs of water [11].

regions, the dark areas correspond to a rich abundance of hydrogen. These findings suggest that small-scale oceans and extensive river systems may have once dominated the Martian landscape [8, 12]. At present, the low air pressure on Mars means that liquid water exposed to the atmosphere evaporates directly. Hence, liquid water no longer exists on the Martian surface but should still be on the sub-surface.

At the two Martian poles, there are stacks of finely layered, water-ice-rich sediments that are about 3 km thick and only a few tens of millions of years old [13]. The sediments are exposed in the summer. In the winter months, the sediments are covered with carbon dioxide frost. At the north pole, the sediments extend southward to 80° latitude, while at the south pole, their extent is less clearly defined, appearing to reach farther from the pole [13].

Figure 13.6 illustrates the presence of underground water ice on Mars. The local soil surface temperature is affected by how deep the ice is buried. The dark gray color on high latitudes of the poles in Figure 13.6 indicate that the buried ice is shallow (0.2 m or less).

Subterranean aquifers of liquid water are believed to exist several kilometers below the surface. Subterranean aquifers could also exist in deep caves. Once these

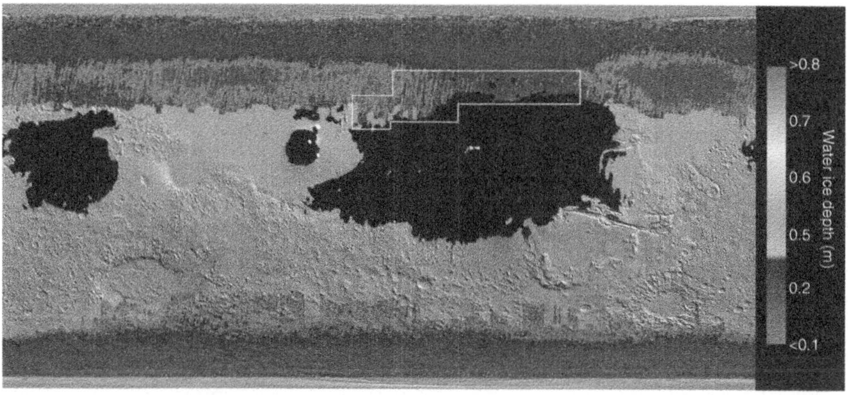

Figure 13.6 Underground water ice on Mars. Dark gray indicates that the underground water ice is close to the surface [14].

caves trap liquid water, they could create a stable environment conducive to the evolution of life.

The total estimate of ice and subsurface water on Mars is 5 million cubic kilometers. It would be equivalent to more than 30 m of water covering the entire planet Mars.

In identifying potential areas for human missions to land, ground conditions that facilitate landing and access to sub-surface water are important factors to consider. The boxed area in Figure 13.6 is believed to be an optimal site for astronauts to excavate water ice. Areas depicted in black are not conducive for landing due to the thick powder-like soil.

Water is also available in water-bearing soil. This will be elaborated in Section 13.3.3.

13.2.1.5 Mineral Resources

The mineral resources on Mars share similarities with those on Earth. The predominant minerals on the Martian surface include silicate minerals such as silicon dioxide (SiO_2), iron oxide (Fe_2O_3), and magnesium oxide (MgO), which are also widely present on Earth [15]. Hydrated minerals have also been identified on Mars, including phyllosilicate, hydrated silica, sulfates, and carbonates, which are similarly ubiquitous in 'Earth's sedimentary environments.

The Martian mineral resources also exhibit differences compared to Earth's. While nearly 6,000 different minerals are known to exist on Earth to date, only 161 minerals have been identified on Mars [16]. The absence of plate tectonic activity on Mars may be the reason [16]. Also, some minerals on Earth are the result of biological processes, which Mars currently lacks.

13.2.2 Hazards on Mars

13.2.2.1 Radiation

There are two types of radiation when traveling to Mars: solar radiation (ionizing and nonionizing) and galactic cosmic radiation (GCR).

The Sun is a source of solar radiation. Nonionizing radiation ranges from ultraviolet to microwaves. Ionizing radiation includes alpha particles, beta particles, and gamma rays. GCRs are streams of a variety of nuclei (atoms with their electrons stripped away) and come from within the Milky Way Galaxy. These nuclei are protons and as heavy as zinc. These nuclei travel near the speed of light and are highly penetrative as well as interactive with the atoms in the materials they pass through. Therefore, GCR can penetrate the spacecraft.

A journey to Mars exposes passengers to an average 1.8 milli-Sievert (mSv) per day. By comparison, the background radiation on Earth is about 2 mSv per year. Therefore, the radiation exposure during the journey to Mars is over 300 times that of Earth. Excluding the stay on Mars, a mission of 800-day round-trip travel will expose astronauts to about 1,400 mSv, that is, the same as 4,200 chest X-rays.

Cosmic radiation presents significant health risks to astronauts. Such risks include cancer, radiation sickness, problems with the cardiovascular and central nervous systems, and other diseases, including death.

Mars's weak magnetic field and thin atmosphere result in the planet's ineffectiveness in resisting radiation, thereby making the radiation environment on the Martian surface more severe than that on Earth [17]. When radiation penetrates the Martian atmosphere and soil, it undergoes a series of complex interactions with the soil atoms and molecules present. These interactions result in the production of secondary particles such as gamma photons, neutrons, and electrons, thereby creating a radiation field composed of both direct cosmic rays and secondary particles [18]. Hence, radiation on Mars's ground is about 40–50 times higher than that on Earth.

13.2.2.2 Sandstorm

Sandstorms on Mars are atmospheric vortex phenomena. They typically occur when Mars is close to the Sun [19] and are therefore thermally driven. During this period, the Sun's heating effect on the Martian surface is intensified, causing hot air to rise and lift a large amount of dust, leading to sandstorms (see Figure 13.7 [21]). Hundreds of regional dust storms occur each Martian year. These regional sandstorms sometimes merge to form global sandstorms that last for weeks. Sandstorms can affect the whole planet, but the most intense storms originate from low latitudes.

Sandstorms on Mars can severely impact Martian exploration missions. The storms can significantly reduce the transparency of the thin atmosphere and

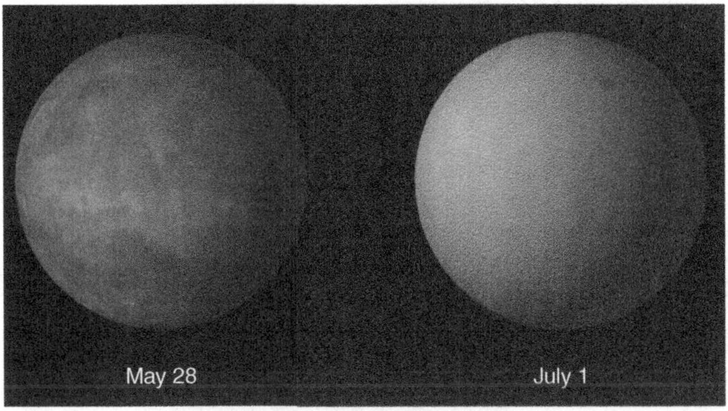

Figure 13.7 Side-by-side images show the before-and-after of a global dust storm enveloping Mars in July 2018. Images were taken by the Mars Color Imager (MARCI) camera on board NASA's Mars Reconnaissance Orbiter (MRO). The dust storm reportedly caused the loss of contact between the NASA Opportunity rover and ground control. *Source:* Ref. [20] / NASA / Public domain.

decrease sunlight from reaching the surface, thereby affecting the efficiency of the solar panels on the probes. The reduction in visibility also severely hinders the observation and data collection work of the orbiting satellites [22].

13.2.2.3 Meteorite Impact

Meteorites frequently impact Mars. First, Mars is situated next to the inner edge of the Asteroid Belt (about 0.5 AU away). Its relatively high orbital inclination of 1.85° (versus Earth's 0°) means a higher likelihood of collisions with asteroids or meteoroids [23]. Second, Mars's thin atmosphere is insufficient to fully incinerate incoming objects. Third, the absence of a magnetic field to shield against the solar wind leads to the continual thinning of its atmosphere.

Meteorite impacts physically alter the topography of the Martian surface, presenting challenges for the accurate mapping of geological features. The dust and debris produced by impacts can damage the mechanical systems of rovers roaming on Mars, affecting their normal operation.

13.2.3 Martian Moons: Phobos and Deimos

Mars has two moons: Phobos and Deimos. Deimos is the smaller one. The moons are shown in Figure 13.8.

Phobos is the innermost moon, measuring 28×10 km in size and weighing 10^{16} kg. The moon is heavily cratered. It follows a nearly circular orbit around Mars's equator at 6,000 km above. Its eccentricity is 0.015 and inclination 1.09°.

Figure 13.8 Martian moons Phobos (left) and Deimos (right). Phobos is 6,000 km from Mars's surface and Deimos is 23,000 km. Both were previously asteroids [24]. *Source:* Giuseppe Donatiello/Wikimedia Commons/Public domain.

Phobos goes around Mars three revolutions a day. Its local gravity is 0.001g. Phobos is not solid due to its low density. As a C-type asteroid, its surface is covered with a dark gray regolith.

Deimos is about half the size of Phobos, measuring 15×11.6 km in size and weighing 1.5×10^{15} kg. Its orbit is circular at 1.8° inclination and is more than 23,000 km above the Martian surface. Its surface is covered with loose dirt.

Both Phobos and Deimos are tidally locked to Mars. They orbit in the same direction and always show the same sides facing Mars. Phobos, being closer, also exerts a tidal force on Mars and deforms the planet in a small way.

The Martian moons will not stay in their present orbits forever. Deimos's orbit is getting bigger and moving further away. It will eventually fly away into space. On the other hand, Phobos is found to be slowly moving toward Mars at 1.8 m per century. In about 50 million years, Phobos will either spiral down and eventually crash into Mars or break apart at around 5,000 km altitude when the moon crosses the Roche radius, which is the boundary when the tidal force exceeds the self-gravitation of the moon's materials.

13.3 Settling on Mars

13.3.1 In Situ Resource Utilization

In situ resource utilization (ISRU) is mandatory for human missions to the Moon and much more so on Mars after the initial phase of landing and exploration.

It would be impractical to bring full supplies from Earth for the round trip to Mars plus the stay.

As will be seen in later sections, the optimized one-way travel time from Earth to Mars is 8–9 months. The return journey is much longer, depending on Mars's position relative to Earth at the time of departure. Early exploration missions may stay on Mars for a few days or weeks; longer missions must be self-sufficient.

Chapter 12 covers the ISRU tools developed on Earth for use on the Moon. Examples include using regolith to fabricate bricks as a radiation cover over prefabricated housing modules and growing foods. The hope is that these ideas will work on Mars with minor modifications and localization.

Scientists will have a better idea using actual samples from the Moon to study the contents of lunar soil. Therefore, we can only rely on the interpretation of the geological data for now. For growing plants in the Martian soil, issues are similar to the lunar situation—lack of organic matter and low in nutrients.

13.3.2 Energy

To achieve the long-term autonomous operation of a Martian base, a continuous and stable supply of power is a must. The current energy solutions under consideration mainly include nuclear, solar, fuel cells, batteries, geothermal, and wind energy.

Additionally, all ISRU-related work will consume different amounts of energy. For example, water is not readily available on the Martian surface. Therefore, it must be extracted by drilling and heating equipment, depending on the "hidden" water in a particular medium. This issue will become more severe as we move from exploration to settlement phase.

13.3.2.1 Nuclear Energy

Nuclear energy provides thermal energy for conversion into electricity. Thermal energy can come from radioisotopes or nuclear fission. The two options are covered in Chapter 12.

Radioisotope thermal generators (RTGs) are suitable for rovers owing to their relatively light weight and reliability. The Curiosity and Perseverance rovers in NASA's Mars missions have RTGs on board. The Next-Gen RTG, known as the Multi-Hundred Watt RTG (MHW RTG), for example, is in Perseverance. It weighs 55.7 kg and produces 245 W of electric power from 4,000 W of thermal energy.

Controlled nuclear fission delivers much higher power in the tens of kilowatt range and can be activated on demand. Space agencies like NASA are funding the development of stationary high-power nuclear reactors for use on the Moon and which can be used on Mars. The reactor unit must be light enough to be carried by Martian landers.

Radioactive ores are products of volcanic actions, so it is reasonable to expect to find radioactive materials on Mars. Uranium and thorium (both radioactive) are believed to be available beneath the crust.

13.3.2.2 Solar Power

Mars is on average about 1.5 AU from the Sun. The solar power density in Mars's vicinity can be scaled with the inverse square law. That works out to be about 600 W/m^2, which is a 56% reduction from Earth's orbit. Therefore, solar panels require more than twice as much area on Mars orbits than those orbiting Earth and the Moon.

The day–night cycle, the angle of incidence, and seasonal variations all reduce the intensity by more than 70%. The latitude of the outpost is a factor. The higher the latitude, the lower the power intensity. At the poles, the Sun is absent for almost one Earth year. The Martian atmosphere also contains a significant amount of suspended dust. Such dust deposition on the panels can result in a steady degradation of the solar panel performance at around 5% every 10 years. Hence, battery storage needs to be part of the energy equation.

13.3.2.3 Hydrogen Fuel Cells

Fuel cells are important components of the lunar energy infrastructure, as discussed in Chapter 12. Fuel cells are most suitable for use in rovers and mobile applications on the ground of a celestial body. They are compact and reliable, often lasting over 50,000 hours (about 6 years of continuous use) with efficiency ranging from 40% to 60%.

Some cells are rechargeable and replenishable by the electrolysis of water into hydrogen and oxygen gas. Such gases would need to come from the water supply on Mars.

13.3.2.4 Wind Energy

Studies suggest that winds can be strong enough to generate power, but extraordinarily large wind turbines would be needed. Wind power is available at night and works in high elevations and during local and global dust storms when wind speeds are high. Wind power should only be a supplement to mainstream sources due to its inconsistent availability.

13.3.2.5 Geothermal Energy

Presently, there are no active volcanic or tectonic activities on Mars. Energy can be tapped from geothermal reservoirs and requires drilling down to several kilometers. The hot subsurface water being pumped to the surface will quickly expand and become steam due to Mars' low atmospheric pressure. The steam can be piped to produce electric power [25, 26]. For geothermal to be a primary energy supply, the user community must be located near these hotspots.

13.3.3 Water-Bearing Soil

Following Section 13.2.1.4, most of the water resources on Mars are stored as ice beneath the surface and around the poles. A smaller portion exists in the form of water-bearing minerals within the Martian soil, and an extremely small amount is present as water vapor (0.03%) in the already thin Martian atmosphere [27]. It is technologically possible to extract water from the atmosphere, but the tradeoff is that these methods are energy intensive.

The substantial reserves of water ice at the poles and underground on Mars are better sources from which to extract water. For water ice in the shallow subsurface, the "Mobile In-Situ Water Extraction Device" developed by the United States is a candidate [28]. As shown in Figure 13.9, the spiral drilling technology is used to collect the Martian soil. The soil is then heated in a volatile extraction, which is a distillation method. Alternatively, the ice in the soil can be melted in the capture system to collect water, and then the liquid water is pumped through a pipeline system.

Water-bearing minerals in the Martian surface soil can release most of their water content when heated to 300 °C. The "Regolith Advanced Surface Systems Operations Robot" (RASSOR), developed by NASA in Chapter 11, is capable of excavating Martian soil and transporting it to the "Mars Atmospheric and Regolith Collection and Processing Unit." This machine releases water vapor by mixing and

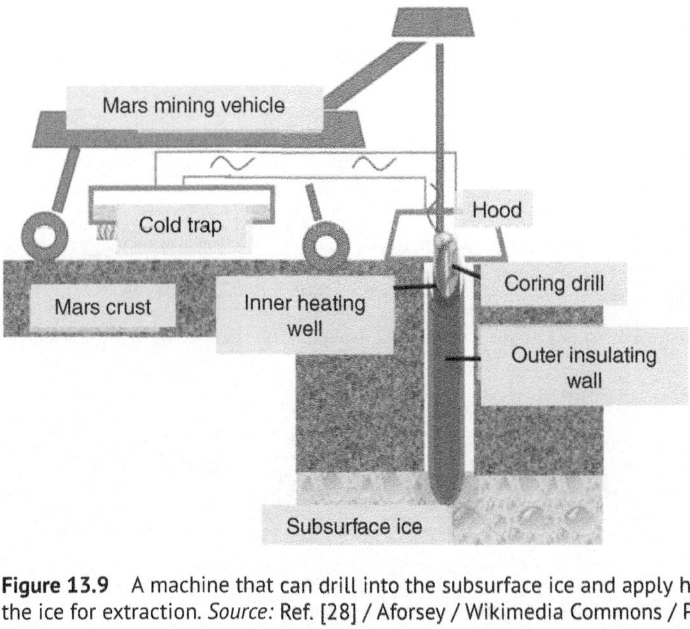

Figure 13.9 A machine that can drill into the subsurface ice and apply heat to warm up the ice for extraction. *Source:* Ref. [28] / Aforsey / Wikimedia Commons / Public Domain.

heating the soil utilizing preheated nitrogen and carbon dioxide to promote the release of water vapor. Water vapor is converted into liquid water in a condenser [29]. The caveat is that this process consumes a lot of energy.

13.3.4 Oxygen

Oxygen can be extracted from the carbon dioxide in the thin atmosphere of Mars. The Mars Oxygen In-Situ Resource Utilization Experiment (MOXIE) device (see Figure 13.10) carried by the Perseverance rover extracts carbon dioxide from the atmosphere and is then converted into oxygen via electrolysis. MOXIE successfully demonstrated that the process worked. However, the current MOXIE unit can only produce enough oxygen for one astronaut to breathe for about 10 minutes. Larger-scale oxygen production equipment will be needed in the future [31].

A research team from the University of Science and Technology of China, using the robotic chemist platform "Xiao Lai," has successfully prepared a new type of oxygen-producing catalyst from Martian meteorites. The robot employs chemical decomposition to extract metal hydroxides from Martian meteorites. The hydroxides are then utilized as anode catalysts in the electrolysis of simulated Martian brines to generate oxygen [32].

Figure 13.10 Technicians working on the MOXIE apparatus and integrating with the rover Perseverance. *Source:* Ref. [30] / NASA / Public domain.

13.3.5 Agriculture

The basic elements necessary for plant growth are oxygen, water, carbon dioxide, nitrogen, phosphorus, potassium, and organic nutrients. The Martian soil lacks water, nitrogen, and organic matter. Nitrogen and carbon dioxide have the tendency to be broken apart by solar radiation and escape from the planet's thin atmosphere from the early days. The missing elements will need to be introduced technologically to grow plants.

13.3.5.1 Building a Greenhouse

The average surface temperature and the atmospheric density of Mars are not conducive to growing Earth's plants in the open. A pressurized, heated, and humidified enclosed environment is required. Artificial lighting systems such as LEDs in the greenhouse can compensate for insufficient natural light due to the distance of Mars from the Sun [30].

13.3.5.2 Soil Improvement

Martian regolith contains high concentrations of perchlorates, which are hazardous to human health. Perchlorates can be mitigated by washing the soil or using certain bacteria to convert perchlorates into less aggressive chlorides. The physical and chemical properties of the soil need to be adjusted to ensure that it has the correct pH value, porosity, and the ability to retain adequate moisture. Adding organic matter and microorganisms to the regolith, such as compost, treated wastewater, and earthworms, can help create fertile soil [30].

13.3.5.3 Soilless Cultivation

Soilless cultivation techniques, including hydroponics and aeroponics, are described in Chapter 12 for lunar missions. These techniques can be deployed to Mars.

13.3.6 Shelters

ISRU practice requires permanent settlement on Mars to be constructed with local materials such as bricks, concrete, metals, and plastics [33]. Unlike the case on the Moon, early settlers could opt to live in caves on Mars. Caves offer protection from harsh surface conditions such as extreme temperatures, radiation, sandstorms, and micrometeoroids. Satellite data support the evidence of caves on Mars.

13.3.6.1 Bricks

Traditional brick-making techniques on Earth are equally viable on Mars. Recent studies have shown that binders and kilns are unnecessary because the Martian regolith contains a significant amount of iron oxide, which can act as a binder. It is possible that bricks can be produced to have a strength greater than that of reinforced concrete [33–35].

13.3.6.2 Concrete

Alternatives to limestone and water are preferred for cement production on Mars due to scarcity of these materials [33]. One solution is to use sulfur, which is abundant on Mars, as a substitute. Sulfur concrete can be produced by heating sulfur to 240 °C to liquefy it and then mixing it with the Martian soil. This type of concrete not only has high compressive strength but can also be recycled through heating [33]. Additionally, human serum albumin can replace cement as a binder, creating a new material known as "astrocrete." Extracting urea from astronauts' tears, sweat, and urine and incorporating it into astrocrete can further enhance its compressive strength. According to calculations, six astronauts can produce over 250-kg space concrete annually on the Martian surface.

13.3.6.3 Metals

Metals such as silicon, iron, magnesium, aluminum, titanium, and chromium can be extracted from the metal oxides in the Martian regolith through reduction reactions. Taking the extraction of iron as an example, three main technologies can be employed: carbothermal reduction, hydrogen reduction, and molten regolith electrolysis [33].

The carbothermal reduction technique utilizes carbon dioxide from the Martian atmosphere as a carbon source and harnesses the focused solar energy as a heat source to convert carbon dioxide into carbon monoxide. Carbon monoxide then acts as a reducing agent to react with the iron oxides in the regolith to produce metallic iron.

Hydrogen reduction technology employs hydrogen gas as a reducing agent to convert the iron oxides in the regolith into metallic iron. Molten regolith electrolysis involves heating the regolith above its melting point to transform it into a liquid state, followed by an electrolytic process to extract metals.

13.3.6.4 Plastic

Plastic, especially polyethylene, has become a necessity for space exploration due to its wide range of applications such as radiation protection, 3D printing materials, everyday utensils, and building materials. Carbon monoxide in the Martian atmosphere and hydrogen from ice layers are key raw materials for the production of polyethylene. Through chemical reactions, these gases can be polymerized into ethylene, which can then be polymerized into polyethylene through technologies such as the Ziegler–Natta process [33].

13.3.7 Clothing

The fabrication of clothing for Mars can be divided into two main categories: one for use inside the bases and manned spacecraft and the other for use in the external environment.

When designing clothing for use inside a base and manned spacecraft, comfort and warmth are key considerations. Local materials should be considered to supply clothing materials for the long term. Researchers from the Xinjiang Institute of Ecology and Geography of the Chinese Academy of Sciences and their collaborators heated the simulated Martian soil to a molten state and stretched it into thin fibers through a special drawing process. These fibers have shown good softness and comfort and can be further processed into fabric for making clothing suitable to be worn inside the base [36].

For clothing used outside of Mars bases and manned spacecraft, the design is more complex and must account for the extreme environments and life-support needs of astronauts [36]. These garments typically consist of multiple layers, including an inner comfort layer, an insulating layer, a liquid cooling garment, a gas-tight restraint layer, a thermal insulation layer, and an outer protective layer [36]. Spacesuits developed by NASA under the Artemis program for extravehicular activities and human missions are suitable for Mars.

13.3.8 Minerals

When settling on Mars, the main minerals that can be utilized include olivine, chlorides, sulfates, and kaolinite and talc, among others. These minerals can provide construction materials, oxygen, water resources, and energy for a Mars base [35]. For instance, olivine can produce oxygen through low-temperature alteration. Chlorides can be made into table salt and disinfectant. Sulfates can be used to make cement and plastics. Kaolinite is an important component in the production of ceramics and refractory materials.

To fully and efficiently utilize the ISRU of the Martian soil and mineral resources, various plans have been proposed. Mainstream methods include the silicate-sulfuric acid process and the Mars aqueous processing system. Both methods essentially involve producing sulfuric acid and then using it to extract useful substances from minerals [36–38].

13.4 Three Recent Missions

As of 2023, there have been 18 satellite missions to Mars, and 7 of them are still active. These satellites orbit Mars and study different aspects of the Martian surface. There have been 21 landing missions launched since the early 1970s. Among them, 10 are classified as successful landings, some still operational, and one is considered a partial success.

In 2021, three spacecraft arrived in the Mars orbit at about the same time. Two were rover missions, and one was a satellite. Prior to the two rover missions in 2021, rovers of earlier missions looked for key resources such as water and studied the Martian climate. For example, in November 2018, the US InSight lander safely landed on Elysium Planitia on Mars. InSight was equipped with a seismometer, which became the first instrument to detect quakes on Mars and provided insights into the planet's internal structure [39]. It also discovered liquid water deep underground in Mars's crust and included a probe designed to burrow deep into the ground to study the heat flow in the soil. InSight operated for 4 years until 2022. The mission also included two CubeSats, marking them the first CubeSats launched to another planet to relay communications to Earth from InSight during its landing.

This section will describe the three missions in 2021.

13.4.1 Perseverance Mission

The Perseverance mission is officially known as Mars 2020. Nicknamed Percy, Perseverance is a rover about the size of a sedan weighing 1,025 kg (dry mass). The NASA Jet Propulsion Lab mission was launched in July 2020. Perseverance landed on Mars in February 2021. The mission's main objective was to look for microbial life and signs of past life.

The landing zone was Jezero Crater. The area is believed to have been "friendly" to life in the past. Satellite imagery had identified the delta of a large lake in the past. The crater is 45 km wide and is located just north of the equator.

Perseverance was equipped with a drill designed to collect samples. The rover also carried a small helicopter named Ingenuity (Figure 13.11). The helicopter was packed in the underside of Perseverance during landing. It was released upon landing safely. Both Perseverance and Ingenuity were equipped with cameras and communications equipment.

Ingenuity made history by being the first powered flying machine on Mars. It weighed 1.8 kg with rotor blades spanning 1.2 m. Its design range was 300 m from the rover with a flight altitude up to 5 m. The helicopter could fly in a thin atmosphere. One of the Perseverance's assignments was to go around selected locations and drill for samples. The samples were placed in containers to be picked up by another rover in a future mission to be transported back to Earth.

13.4.1.1 Images

NASA posts photos taken by various cameras from Perseverance on Mars on its websites, as seen in Figures 13.12–13.14.

Figure 13.11 Ingenuity Mars helicopter. *Source:* Ref. [40] / NASA / Public domain.

Figure 13.12 Landing of the Ingenuity Mars Helicopter after one of its flights in 2021 [41, 42]. *Source:* Ref. [41] / NASA / Public domain.

Figure 13.13 Image was acquired in 2024 by the Perseverance rover using its onboard navigation camera on the left. *Source:* Ref. [43] / NASA / Public domain.

Figure 13.14 This image was acquired by NASA's Mars Perseverance rover in 2024. *Source:* Ref. [43] / NASA / Public domain.

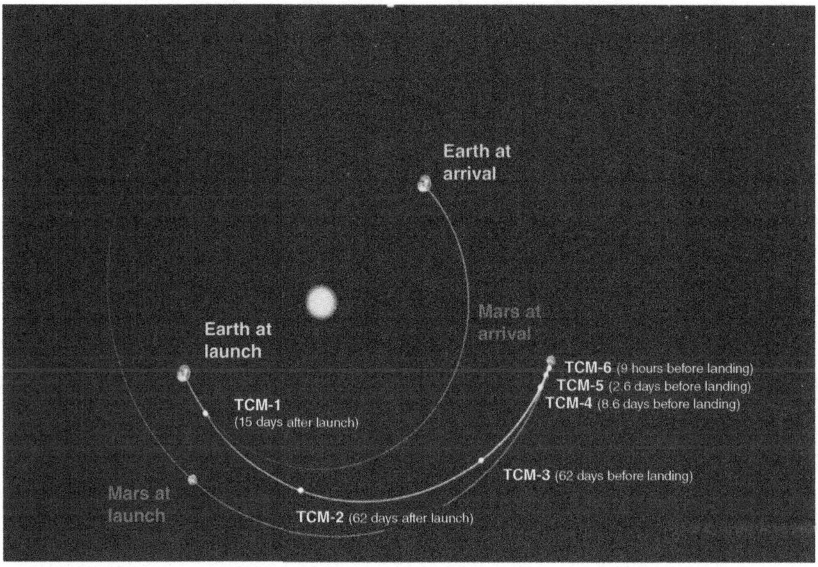

Figure 13.15 The flight path of the Perseverance spacecraft from Earth to Mars, including several trajectory-correcting maneuvers. *Source:* Adapted from Ref. [44].

13.4.1.2 Flight Path of the Perseverance Mission

Figure 13.15 shows the flight path taken by the Perseverance mission. At the time of launch, Mars was slightly "ahead" of Earth in the anticlockwise direction in this diagram. Mars and Earth are said to be "in opposition," which will be explained below. The spacecraft "caught up" with Mars 203 days and 470 million km later after making a total of 6 trajectory correction maneuvers (TCMs) along the way to change its orbital trajectory to match Mars.

13.4.1.3 Landing

The Perseverance spacecraft was contained inside a protected aeroshell. The shell entered the top of Mars's atmosphere at an altitude of about 125 km with a speed of 5.4 km/s. This was when Entry, Descent, and Landing (EDL) began. The thin atmosphere provided the aerobraking to slow down the spacecraft, thereby beginning the "Seven Minutes of Terror." The heat shield of the shell reached as high as 1,300 °C about 80 seconds after atmospheric entry, while the temperature inside the aeroshell remained at room temperature. The EDL process is illustrated in Figure 13.16.

Along the way, the thruster at the back of the shell fired to guide the descent automatically. The heat shield was dropped about 6 minutes after entry when the shell was about 11 km above ground at a speed of 0.58 km/s. At that time, a parachute was deployed, slowing the shell to about 0.09 km/s (320 km/h). The

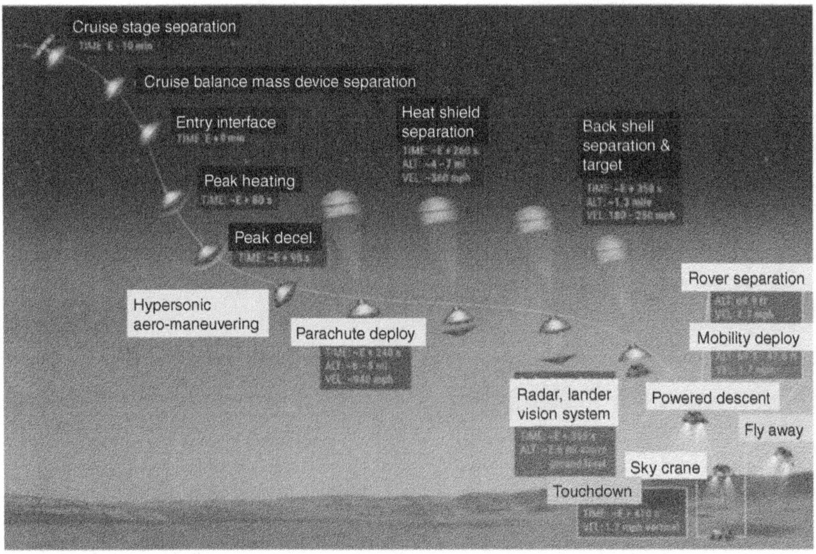

Figure 13.16 The atmospheric reentry of Perseverance from 125 km altitude. The aeroshell containing Perseverance decelerates due to the thin atmosphere of Mars. The aeroshell sheds the heat shield and deploys its parachutes. The rover is then lowered to the ground [45].

Perseverance rover was then lowered to the ground via cables from the aeroshell, as shown in Figure 13.17. This approach was known as the "sky crane" and proved to work with the Curiosity rover mission earlier.

Upon landing without damage, Perseverance sent a message to Earth. Earth officially learned of Percy's safe landing 11 minutes later.

13.4.1.4 Mission Status

Figure 13.18 shows the locations of the Perseverance rover and the Ingenuity helicopter as of October 2024, at which time the rover had been active on Mars for 1,287 sols (1,322 Earth days) since its landing. As of September 2024, the rover had traveled 28.9 km from its landing spot.

Ingenuity made its first flight to a height of 3 m on April 19, 2021, becoming the first machine to achieve powered flight in the atmosphere of another planet. Over its mission, Ingenuity flew a total distance of 17.78 km, reaching a maximum altitude of about 24 m in Figure 13.18. It completed 72 flights before its mission concluded in January 2024. Ingenuity sustained damage to the blade after a hard landing.

Perseverance's mission duration was at least 1 Mars year (1.8 Earth years). The rover's power system was similar to that of the Curiosity rover and was expected to operate for more than 14 Earth years.

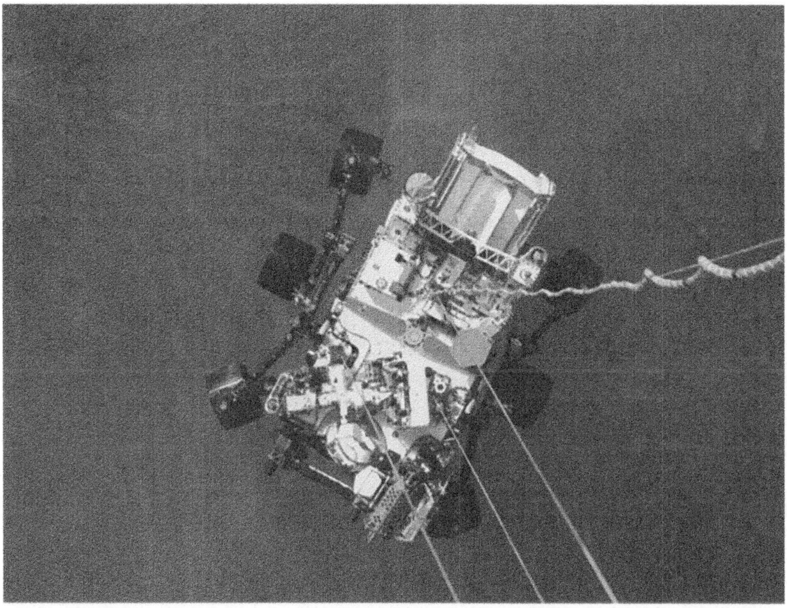

Figure 13.17 The Perseverance rover is shown dangling below the sky crane platform to the surface of Mars. *Source:* Ref. [46] / NASA / Public domain.

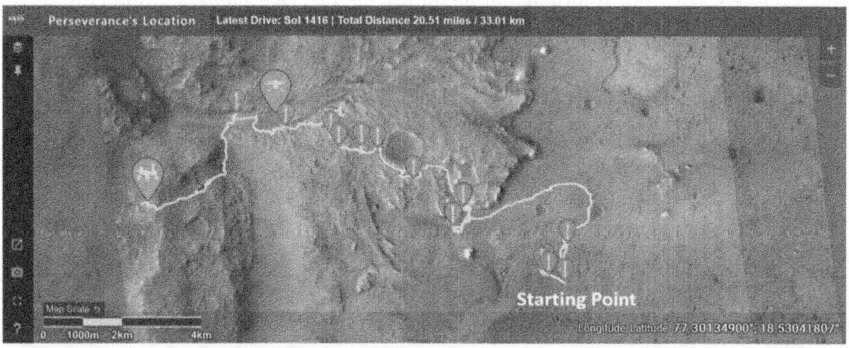

Figure 13.18 This image shows the landing site for NASA's Perseverance rover, shown as "Starting Point." The final locations of the rover and the helicopter are shown in [47]. *Source:* Ref. [47] / NASA / Public domain.

13.4.2 Tianwen-1 Mission

Tianwen means "Questions to Heaven." The Tianwen-1 mission was the first mission to Mars by China. It was launched in July 2020 and entered into Martian orbit in February 2021. The payload was an orbiter and a lander. The lander carried a

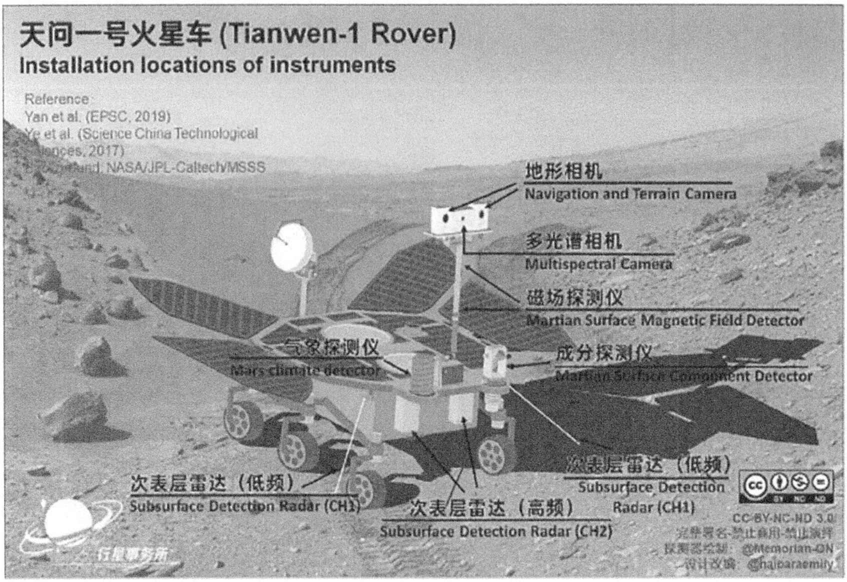

Figure 13.19 China's Zhurong Rover of the Tianwen-1 mission. *Source:* Ref. [48] / NASA / Public domain.

rover named Zhurong, which weighed about 240 kg. The payload landed on Mars in May 2021. The mission's stated goals included scientific studies of the Martian soil, environment, water ice, topography, and looking for signs of life. The Tianwen-1 rover is shown in Figure 13.19.

13.4.2.1 Flight Path

The mission's flight path is shown in Figure 13.20. The mission took 7 months and 5 TCMs to reach Mars, allowing the orbiter to move from a transfer orbit to orbit insertion. After a braking thruster was fired to slow the orbiter for a Mars orbital capture, the orbiter entered an initial orbit of 400 × 180,000 km (periapsis × apoapsis) with an inclination of 10°. The orbiter maneuvered to change its orbital inclination, resulting in the orbiter in a reconnaissance orbit 265 × 60,000 km at an inclination of 86.9°. The orbiter then began a deorbiting and landing trajectory. The lander and its rover separated from the orbiter at the periapsis and proceeded to land. The orbiter then changed its orbit to enter a science orbit to perform scientific studies.

Once in a science orbit, the spacecraft would execute a series of maneuvers to study the upper atmosphere. The orbit also precesses so full coverage of Mars's surface is achieved.

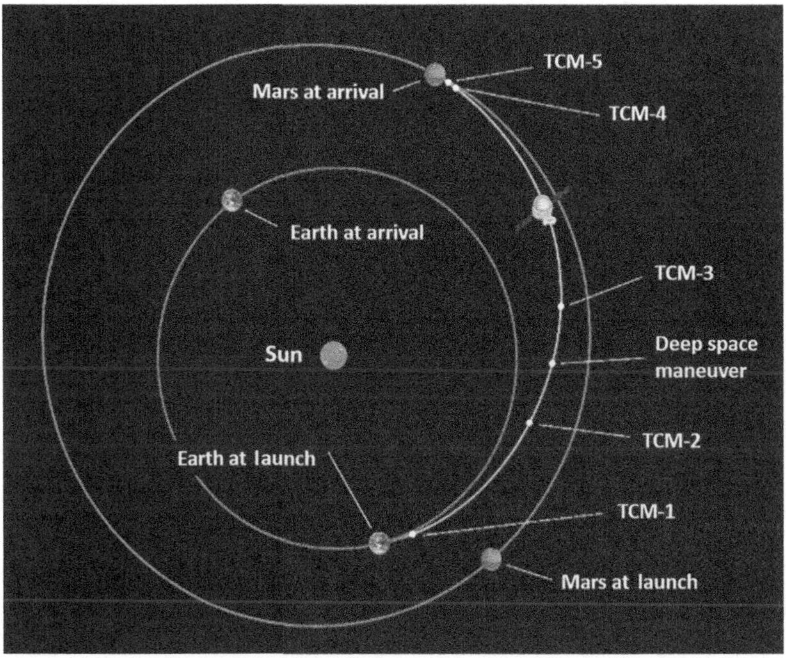

Figure 13.20 The orbiter's transfer orbit and TCM. A schema of the transfer orbit between Earth and Mars of China's Tianwen-1 probe and the TCM along the way. *Source:* Ref. [49] / Kaynouky / Wikimedia Commons / CC BY 4.0.

13.4.2.2 Landing

The Entry, Descent, and Landing (EDL) process is illustrated in Figure 13.21. The lander was packed in a protective shell and entered the Martian orbit at about 125 km altitude and steadily descended while being protected from temperature rise. The parachute was deployed at an altitude of 10 km followed by the jettison of the heat shield. The lander proceeded to a soft landing with onboard thrusters. The Zhurong rover then stepped off the lander.

Figures 13.22–13.24 show images taken by Zhurong on Mars.

13.4.2.3 Status

The orbiter in the Tianwen-1 mission was planned to operate for 2 Earth years, while the rover Zhurong was designed for a life expectancy of 93 Earth days. Zhurong was active for 358 days before activating its hibernation mode in May 2022 due to a reduction in the solar power available to the spacecraft during the Martian winter. Zhurong did not "wake up" as scheduled in December 2022. The chief designer of China's Mars exploration program believed that the solar

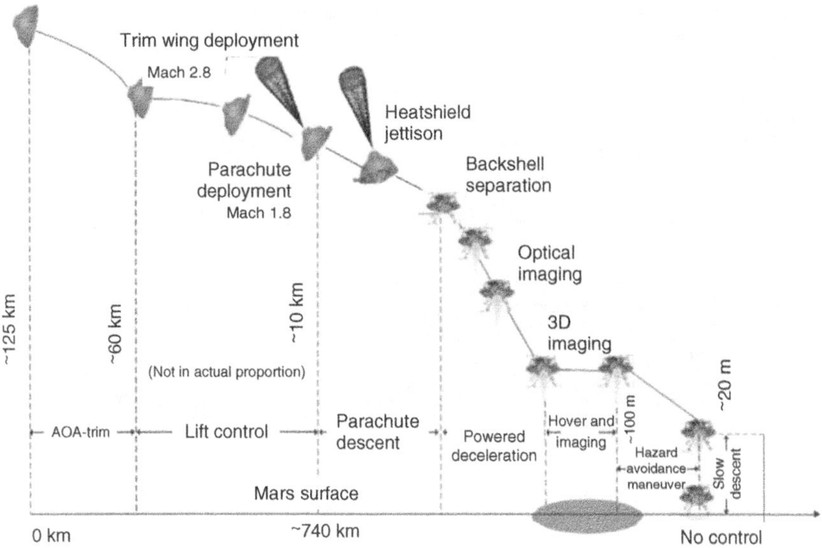

Figure 13.21 Sequence of Entry, Descent, and Landing (EDL) of the Tianwen-1 lander and Zhurong rover. *Source:* Ref. [50] / American Association for the Advancement of Science - AAAS / CC BY 4.0.

Figure 13.22 The Zhurong rover looks back at its lander while providing a view of the landing site. *Source:* Ref. [51] / China National Space Administration.

Figure 13.23 The Zhurong rover looks back as it drives onto the Martian surface to investigate Mars' ancient geology. *Source:* Ref. [52] / China National Space Administration.

Figure 13.24 An artist's depiction of what the Tianwen-1 rover will look like on Mars's surface. *Source:* Ref. [53] / China National Space Administration.

panels of the Zhurong spacecraft could be covered in dust and could not generate enough electricity.

13.4.3 Emirates Mars Mission

The Emirates Mars Mission is also known as the EMM-Hope mission. Funded by the government of the United Arab Emirates, the Hope mission is the first interplanetary probe by the Arab world [54].

The mission was one of scientific data collection and to last for 1 Martian year. The stated goals were to study climate dynamics and global weather distribution by characterizing Mars's lower atmosphere and how such weather changes would affect the planet's ability to retain hydrogen and oxygen. The spacecraft was equipped with a high-resolution camera, an ultraviolet spectrometer, and an infrared spectrometer. The spectrometer was used to measure oxygen and carbon monoxide in the thermosphere (100–230 km) as well as hydrogen and oxygen in the upper atmosphere (50–60 km). The infrared spectrometer would measure the surface and atmospheric temperatures, as well as the global distribution of water ice, water vapor, and dust in the lower atmosphere.

The spacecraft measured 2.37 m × 2.90 m and weighed 550 kg (dry mass) and 1,350 kg with fuel. Launched in July 2020, the Hope spacecraft coasted for 200 days and a distance of 493.5 million km and arrived at Mars orbit in February 2021. Its speed reached 11.9 km/s.

13.4.3.1 Flight Path

The spacecraft executed an extended burn to slow down enough to be captured by Mars's gravity to enter the first orbit. The initial capture orbit was a 1,000 × 49,380 km ellipse (period 40 hours). Then, the spacecraft maneuvered to raise its orbit to the science orbit, which was 22,000 × 44,000 km (period 55 hours) with 25° inclination as shown in Figure 13.25.

13.4.3.2 Status

The probe was to spend two Earth years studying the daily and seasonal weather cycles on Mars. This 2-year study was completed in January 2023. The UAE government announced in January 2023 that the Hope probe will make a slight orbital change to include the outer moon Deimos for observations [56]. Deimos can be observed without giving up the continual study of the Martian surface.

13.4.4 Follow-Up Mars Missions

NASA and ESA have planned a set of joint Mars Sample Return (MSR) missions to retrieve samples collected by the Perseverance rover at different locations. These logistics missions are planned for the early 2030s and are illustrated in Figure 13.26. In this figure, the Perseverance mission is referred to as the Mars 2020 mission.

NASA is responsible for the MSR mission, which is to send a rover called the Sample Retrieval Lander (SRL) to retrieve the samples that have been left at specific locations. This mission will reportedly also carry 2 Ingenuity-class helicopters to Mars. The fetch rover will transfer these samples to the Mars Ascent Vehicle (MAV). The MAV will then launch into orbit from Mars and later rendezvous with

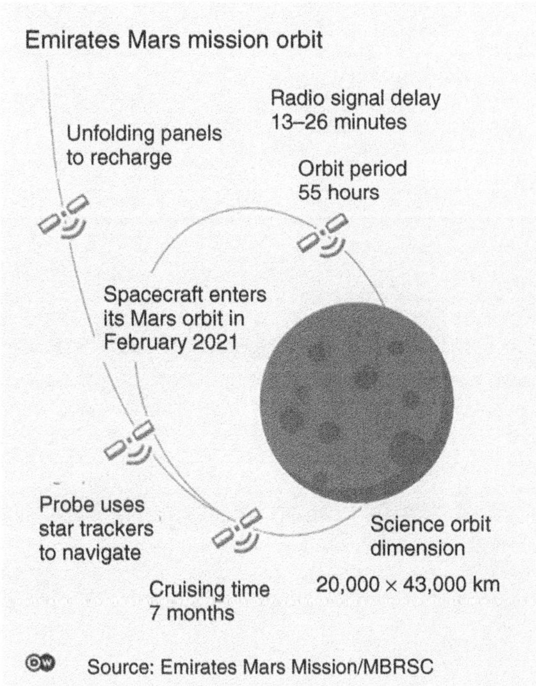

Figure 13.25 The Mars journey of the UAE's Hope mission. *Source:* Ref. [55] / Deutsche Welle.

the Earth Return Orbiter (ERO), for which ESA is responsible. ERO is also known as the Sample Return Orbiter (SRO) mission. Upon receiving the MSR samples from the MAV, the ERO will change its orbit and return to Earth. Similar to the previous practice with asteroid samples, the ERO will release the package of the samples for Earth reentry. The ERO orbiter will not reenter Earth.

13.5 Future Human Missions to Mars

Governments have set their sights on Mars for reasons of both scientific curiosity and technological development. Although it is too early to have convincing economic motivations, Mars is the next frontier after the Moon.

There have been several rover and satellite missions to Mars. Martian missions are so far one-way journeys using ion drives for propulsion. These missions often place minimum energy consumption as a priority over flight time.

NASA's goal is to send humans to Mars and return safely in the latter part of the 2030s. As of this writing, this future program is only in the planning stage.

There are obvious differences between going to Mars and going to the Moon, as summarized in Table 13.1.

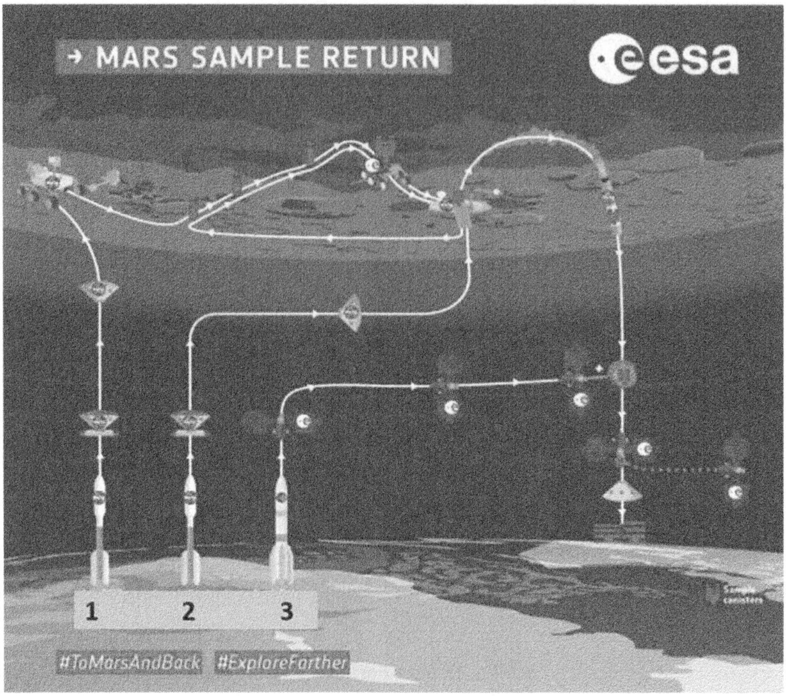

Figure 13.26 Schematic of the overall Mars Sample Return campaign strategy. Mission 1 is the Mars 2020 (Perseverance) mission. Mission 2 is the Sample Retriever Lander (SRL) mission with a rover to retrieve the samples collected by the Perseverance rover. The samples will depart Mars on a Mars Ascent Vehicle (MAV). Mission 3 is the Earth Return Orbiter (ERO), which will rendezvous with the MAV to transfer the collected samples and return them to Earth. *Source:* Adapted from Refs. [57, 58].

Table 13.1 The difference between going to Mars and going to the moon.

	Going to the Moon	Going to Mars
Launch window	No	Yes
Flight time	Short	Long
Stay time	No constraint	Constrained
Return time	No constraint	Constrained
One-way distance (km)	348,000	Average 225 million km
Transportation vehicle available	Yes	To be developed
Radiation during flight	Yes	Yes and higher
Gravitational pull	1/6 g	1/3 g
Aerobraking	No	Yes

Among the differences, there are two most challenging points: (a) launch window related and (b) transportation system related.

a) Launch window: Mars's orbit is much farther from the Sun than Earth's, and the launch window refers to the most suitable time to leave Earth for Mars and vice versa.
b) Transportation systems: Several issues need to be addressed to transport humans and cargoes in both directions

Each point will be elaborated below. As far as the stay time on the Moon is concerned, the choice of orbit for the Gateway space station determines the length of stay, that is, 6.5 days or higher multiples of 6.5 days. Unlike Mars, the return journey from the Moon to Earth is not dependent on the relative positions of Earth and the Moon.

13.5.1 Energy ΔV

All missions amount to orbital transfer and require ΔV. It is common practice to express the ΔV from low earth orbit (LEO) as a starting point. All missions send the spacecraft to LEO first. The amount of ΔV from Earth's surface to LEO is about 9.5 km/s, taking atmospheric drag into account. The traditional practice is for the upper stage to deliver the payload to the LEO. The payload will utilize its onboard propulsion system, which is typically an ion drive, to travel to its destination. From Chapter 5, the Starship upper stage is the vehicle that takes passengers and cargoes to the destinations by chemical combustion. On-orbit refueling is generally assumed for journeys to high orbits and deep space.

The energy, i.e. ΔV needed to go from LEO to Mars, depends on the chosen orbit. Orbital transfer has many options. Recall from Chapter 9 about missions to near-Earth asteroids, flight paths and ΔVs are tradeoffs between energy-efficiency and time-efficiency transfers.

Mars follows a larger elliptical orbit around the Sun than Earth and has varying orbital speeds. The relative distance between Earth and Mars is constantly changing. The Earth-to-Mars orbit is also influenced by Mars gravity as a spacecraft approaches the planet, so the flight path becomes more complex. This is one of the reasons why non-crewed missions in Section 13.4 need multiple trajectory correction maneuvers. Each correction maneuver consumes different amounts of ΔV.

In a 2022 white paper termed "Mars Transportation," [59], NASA offers the orbit known as the Lunar Distance High Earth Orbit (LDHEO) used in the Artemis program as a starting point. Energy is added to expand the orbit to reach Mars. Figure 13.27 illustrates this concept. The paper starts with $\Delta V = 9.4$ km/s from the Earth's surface to LEO where the ISS orbits. From that altitude, a boost of $\Delta V = 3.2$ km/s will send the spacecraft to high Earth orbit.

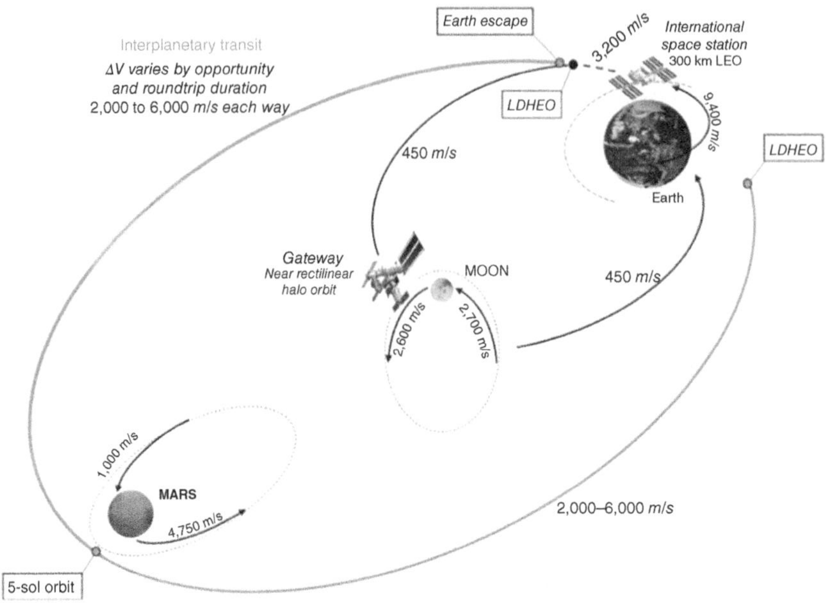

Figure 13.27 Earth–Moon–Mars ΔV (energy) map. In this analysis, the Lunar Distance High Earth Orbit (LDHEO) is used as the starting point for Earth departure staging for Mars missions, and a 5-sol orbit is used for Mars arrival staging and departure. *Source:* Ref. [59] / NASA / Public Domain.

For illustration, a boost of $\Delta V = 450$ m/s added at the right place as shown in Figure 13.27 will take the spacecraft to the lunar orbit. In this case, this orbit will allow the spacecraft to rendezvous with the Lunar Gateway. From the Gateway, a deceleration of $\Delta V = 2.7$ km/s would be required for the spacecraft to land on the Moon. To depart from the Moon's surface, a ΔV of 2.6 km/s would be needed to return to the Gateway.

Instead of going to the Moon, a larger ΔV applied in Earth's high orbit will result in a much larger elliptical orbit that will bring the spacecraft toward Mars. This amount of ΔV ranges from 2 to 6 km/s, depending on the launch window, which will be described in the next section. If there were no Mars's gravitational influence, the spacecraft would be just under Earth's gravity and reach the apogee of this large orbit and have a very slow speed. Trajectory corrections and the emergence of Mars's gravity as the spacecraft approaches, resulting in orbit capture forming a local elliptical orbit, as shown. This Martian orbit has a period of 5 solar days.

From this 5-sol orbit, a deceleration ΔV of 1.0 km/s will bring the spacecraft into the Martian atmosphere where the residual speed will be dissipated by the planet's atmosphere. On the return trip, the spacecraft can lift off with $\Delta V = 4.7$ km/s to leave Mars to return to the 5-sol orbit due to Mars's 1/3-g gravity. To return to Earth, another ΔV ranging from 2 to 6 km/s will be needed, depending on the launch window at the time of departing Mars.

13.5.2 Launch Window

Recall that the distance between Earth and Mars is ~60 million km (closest) and 400 million km (farthest). Moving at different orbital speeds, Earth and Mars are closest to each other every 26 months (780 Earth days). The two planets are said to be in opposition. This periodic pattern opens up a launch window for mission planners to select a suitable time to launch from Earth to Mars. The least amount of fuel (energy) would be consumed.

This closest distance itself varies due to the elliptical nature of Mars's orbit but exhibits another periodic pattern. Hence, there is a "lowest" value for the closest distances between Earth and Mars, which happens every 15–17 years.

An energy-efficient path is the default choice in the forward journey in non-crewed missions. But crewed missions have other considerations as well. Crewed missions need to be concerned about factors such as the crew's exposure to radiation in space, fuel, and food consumption for the round trip until ISRU is in place on Mars. The tradeoff for mission planners is fuel consumption versus crew survival.

Another consideration is that by the time the spacecraft lands on Mars months later, the relative position between Mars and Earth has changed. The crew will stay on Mars for a period to carry out mission assignments. The initial length of stay on the Martian surface depends on factors such as the supply of fuel and food as well as crew safety.

NASA's practice has been to ensure that human missions have sufficient on-board capability to return to Earth. Such capability includes fuel, food, and water. It is not clear if this practice will be adjusted when it comes to Mars. For example, if the Orion MPCV in lunar missions is used, its crew capacity is 4 astronauts and 100-kg cargo together with fuel.

The entire Martian mission, therefore, is the sum of three durations: flight time from Earth to Mars, stay time on Mars, and flight time from Mars to Earth. As will be seen in the animation below, the total mission time can add up to 3 Earth years. Hence, unlike the Artemis program where astronauts will only stay on the Moon for 6.5 days due to the choice of lunar orbit, crewed missions to Mars will be an exercise of tradeoffs.

13.5.2.1 Conjunction Versus Opposition

There are two classes of orbits for launch from LEO going to Mars: the conjunction and opposition classes. They are related to the relative position between Mars and

Earth. The conjunction class is when the two planets are far apart. The opposition class is when the planets are close to each other. The travel time is relatively short for the opposition class orbits and long for the conjunction class.

13.5.3 Animation from Earth to Mars

NASA shows an animation termed "Chasing Mars" on a website about sending humans to Mars [60]. The animation describes a simulated round-trip mission to Mars in 2039 lasting 850 Earth days (2 years and 4 months), including a stay on Mars for 66 days. Several snapshots taken from the animation are shown in Figure 13.28(a)–(f). The animation does not show the ΔV consumption for the forward and return journeys.

(a)
(b)

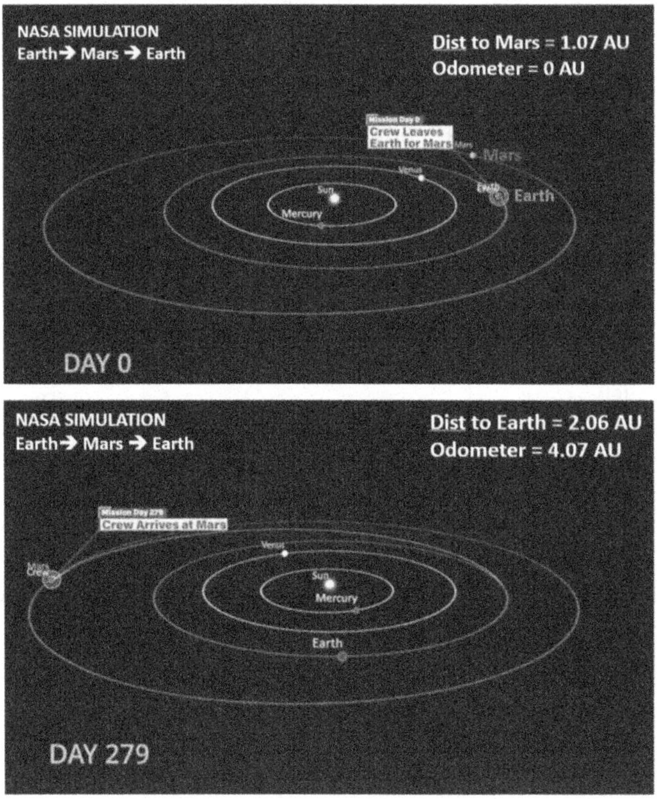

Figure 13.28 Snapshots of the 850-Day Earth–Mars–Earth trajectory for the 2039 mission opportunity. *Source:* Ref. [60] / NASA / Public Domain.

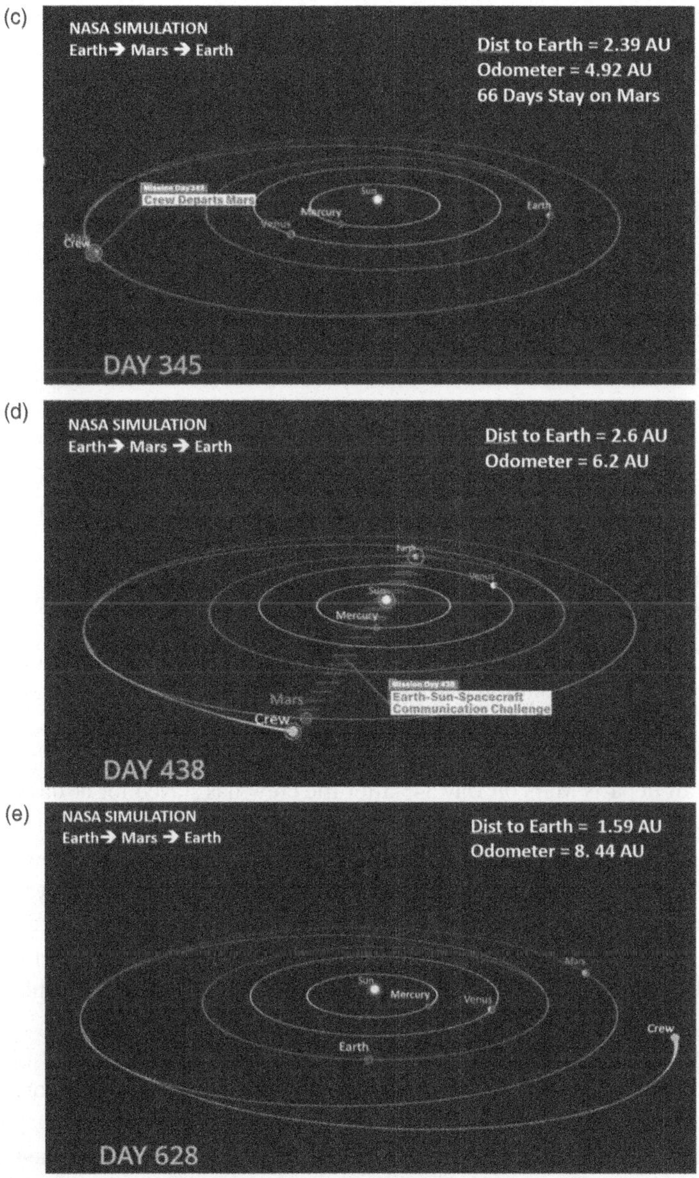

Figure 13.28 *(Continued)*

(f)

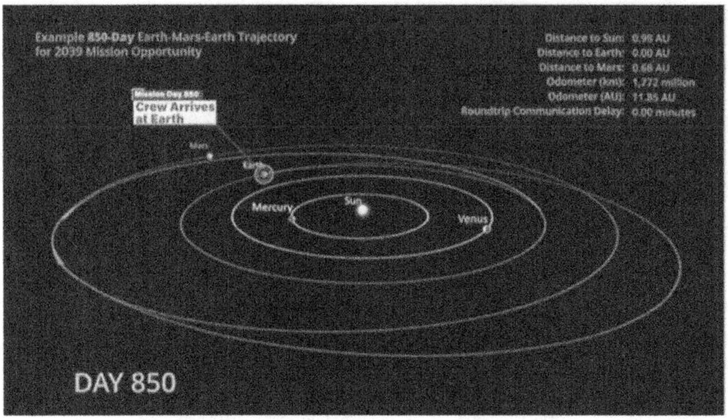

Figure 13.28 (*Continued*)

The two planets go around the Sun in a counterclockwise direction in these figures. NASA states that 2039 is chosen as an "energy opportunity" for the purpose of evaluation, viz. a local minimum launch window. For ease of appreciating the distance, they are expressed in AUs in Table 13.2.

Several observations can be made from the figures and Table 13.2.

a) Mars is 1.07 AU from Earth when the spacecraft (opposition class) is launched.
b) The spacecraft arrives on Mars on Day 279 (9.3 months), that is, 279/850 = 32.8% of the total mission time. It has traveled 609 million km (4.06 AU). Spacecraft/Mars is now 2.06 AU from Earth
c) The spacecraft departs Mars on Day 345 after the crew has stayed for 66 days, or 7.76% of the total mission time. The spacecraft is 2.39 AU from Earth at the time of departing Mars, and the two planets are almost the farthest from each other (conjunction class). Figure 13.28(d) shows that the two-way communication between Earth and the spacecraft takes almost 40 minutes. The spacecraft's odometer reading shows 4.92 AU, meaning the odometer reading includes the movement of Mars/spacecraft during the crew's stay, that is, (4.92 – 4.06=) 0.86 AU.
d) The spacecraft's return orbit takes an outside swing before cutting back to Mars's orbit toward Earth. This return orbit covers more distance and therefore requires more time.
e) The spacecraft returns to Earth on Day 850. It takes (850 – 345=) 505 days (16.8 months), or 59.4% of the total mission time. The return journey travels 1,036 million km (6.91 AU).
f) The odometer reading for the total mission is 1,772 million km (11.85 AU), which can be checked by adding the odometer readings

Table 13.2 Snapshots of the 850-day Earth–Mars round-trip trajectory simulation.

Action	Day	Figure 13.28	Spacecraft position	Earth position	Mars position	Spacecraft distance to the Sun (AU)	Spacecraft distance from Earth (AU)	Spacecraft distance from Mars (AU)	Odometer reading (AU)
Launch	0	a	3 o'clock	3 o'clock	2 o'clock	1.0	0.0	1.07	0
Arrives Mars	279	b	9 o'clock	5–6 o'clocks	9 o'clock	1.64	2.06	0.01	4.07
Departs Mars	345	c	8–9 o'clocks	3 o'clock	8–9 o'clocks	1.59	2.39	0.0	4.92
No Comm	438	d	7 o'clock	12–1 o'clocks	7 o'clock	1.61	2.60	0.16	6.2
Returning to Earth	628	e	3 o'clock	6 o'clock	2 o'clock	1.68	1.59	1.17	8.44
Arrives Earth	850	f	10–11 o'clocks	10–11 o'clocks	10–11 o'clocks	0.98	0.0	0.68	11.85

g) In Figure 13.28(d), the line of communication between the spacecraft and Earth is blocked by the Sun
h) Earth has gone through about 2 1/4 orbital revolutions for the whole mission, while Mars has gone through less than 1 ¼ revolutions

Point a) shows that the orbit from Earth to Mars takes a relatively short time, while Point c) shows that the orbit from Mars to Earth takes a long time. The return path follows a bigger outside loop to catch up with Earth.

13.5.4 Transportation Systems

Rover and satellite missions to Mars have been one-way journeys using ion drives for propulsion. The joint Mars Sample Return missions by NASA and ESA in Section 13.4.4, on the other hand, are multi-stepped missions to reach Mars to collect samples and return them. What propulsion systems are suitable for space vehicles to transport humans to Mars and back? Propulsion systems determine a vehicle's ability to deliver ΔV and thrust. Do we rely on ion drives or use chemical combustion engines? Should the propulsion capability be set for orbits for minimal ΔV or minimal flight time?

We can appreciate the problem further by comparing the Earth weight of the Perseverance rover (1,025 kg) with the dry weight of the Orion MPCV (9,670 kg) without any passenger and cargo, assuming the Orion is the vehicle going to Mars. The spacecraft needs propulsion to deliver thrust in order to land on Mars and later take off with a similar payload weight, which current ion drives cannot deliver. Therefore, propulsion systems with sufficient thrust and ΔV performance will be needed for a round trip between Earth and Mars.

To be safe, the initial phase of human missions to Mars must carry enough fuel and other supplies for a round trip. It comes down to the mission duration versus the spacecraft propellant efficiency.

13.5.4.1 NASA's 4 Propulsion Systems

NASA's broader perspective is to investigate a variety of propulsion systems to determine which type will work best. NASA identifies four propulsion systems: All chemical propulsion (ALL-CHEM), nuclear electric propulsion (NEP), nuclear thermal propulsion (NTP), and solar electric propulsion (SEP). The four systems are evaluated for long (1,250 total days) and intermediate (870 days) round-trip missions. Each case study uses a separable habitat for a crew of four equipped to last for 1,200 days. The key issues are energy (fuel) and transit time.

The four systems differ by their primary energy source. Chemical combustion is the source of energy for an ALL-CHEM system. Nuclear fission power is the source of energy for the NEP and NTP. Solar power is the energy source for the SEP. Of the four systems, ALL-CHEM delivers the highest thrust by discharging large amounts of combusted byproducts. The other three propulsion systems have low thrust, among which NTP has the most thrust.

We are familiar with ALL-CHEM systems. Martian missions need enough thrust to land on and take off from Mars. Hence, spacecraft must carry large amounts of liquid propellants for the whole mission at the expense of payload capacity. ALL-CHEM should offer a shorter flight time to Mars. ALL-CHEM propulsion has not been tried on non-crewed Martian missions.

In nuclear thermal propulsion, nuclear fission generates a large amount of heat in the reactor core. The temperature in the core can reach over 2,000 °C. A liquid (propellant) is pumped through the reactor core and turns into a rapidly expanding hot gas. The hot gas is then expelled out of the nozzle of the spacecraft, producing thrust. The reactor core acts like a combustion chamber, although no chemical combustion occurs.

Examples of liquid propellants for nuclear propulsion are liquid hydrogen (lightest), ammonia, and water. Hydrogen, being the lightest, gains the most speed, so the exhaust speed is the highest. But a larger quantity of hydrogen by mass would be needed to produce thrust, so a large storage tank is needed, which adds weight to the dry mass of the spacecraft. Water is also used as the propellant, which can be held in a small tank due to its much higher density than hydrogen.

In practice, the NTP as a thruster does not produce enough thrust to overcome the gravity well of Earth and other celestial bodies. It cannot be the primary thruster for launch and should only be activated after the spacecraft is in orbit.

In the case of NEP, the thermal energy from a nuclear reactor is converted into electricity and high voltage. Electric power is then used to ionize the propellant. The ionized propellant is then accelerated by an electromagnetic field. After being brought to electrical neutral at the exit, the propellant exits the spacecraft, producing thrust. Inert gases such as krypton and xenon are examples of propellants. NEP provides less thrust than NTP. It is used in long-duration missions where efficiency is more important than power.

As of this writing, plasma electric engines are in the news. Instead of accelerating ions in ion thrusters, plasma is being accelerated by electromagnetic means. The initial estimate of the exhaust speed of a plasma electric engine is claimed to be as high as 100 km/s versus 20–50 km/s from regular ion thrusters. Plasma electric engines are more efficient. But the rate of mass exhaust is still low, so they cannot deliver the kind of thrust like chemical engines. In principle, a plasma electric engine can shorten the travel time from Earth's orbit to Mars compared with regular ion drives.

Solar electric propulsion works in a similar fashion as NEP. Electric power comes from the solar panels. The remaining functional blocks are similar to those of NEP. The solar power intensity decreases as the spacecraft gets further away from the Sun. Therefore, the supply of electric power becomes progressively weaker. SEP provides less thrust than NEP.

In comparing the four propulsion systems, the weight of the system represents "dry" weight, i.e. excluding the weight of the propellants. SEP has the lowest thrust-to-weight ratio (TWR), but its advantage is simplicity. A chemical rocket engine has the highest TWR and thrusts, but the amounts of chemical propellants and the support system such as fuel tanks take up room and increase the dry mass in the spacecraft.

The nuclear fission reaction requires a nuclear reactor with sufficient shielding on board the spacecraft. A reactor does not need massive quantities of nuclear fuel. However, for extended missions, nuclear fuel needs replenishment, thereby generating nuclear waste during the trip. Managing nuclear waste is an important component of the mission.

The propellants in NTP, NEP, and SEP are the liquids to be ionized and then discharged from the spacecraft. For deep space missions, sizable supplies of propellants are still needed if these propulsion systems are to provide the thrust from high Earth orbit to Mars.

It is likely that a hybrid propulsion system will be incorporated in deep-space human missions. The hybrid system combines an ALL-CHEM system with one of the NTP, NEP, and SEP systems. The chemical combustion system provides the high thrust for climbing up and down Mars's gravity well. The other ion drive will then be turned on for the long journeys between Earth and Mars, depending on the flight time requirement.

13.5.4.2 Cryogenic Fluid Management

NASA's investigation of the four propulsion systems led to the recommendation that cryogenic fluid management (CFM) be a critical common factor for all cryogenic propellants. CFM is a collection of technologies covering the storage and transfer of cryogenic fluids (also known as cryogens). ALL-CHEM systems, for example, use propellants like hydrogen, oxygen, methane, and kerosene, all of which are in supercooled liquid state.

Boil-Off Heat exchange is a generic term describing the flow of thermal energy between two bodies at different temperatures. There is always a heat exchange between a cryogen and its surroundings, which includes the storage tank and the connecting pipelines during fluid transfer. All the equipment in a spacecraft are different heat sources operating at different temperatures. The cryogens in the fuel tanks gain heat from the surroundings and turn into gas. This evaporation process is called cryogenic boil-off. Insulation and other factors, such as the tank's liquid level, all determine the extent of the cryogenic boil-off.

The boiled-off cryogen increases the vapor pressure inside the tank. The higher vapor pressure is sometimes eased by venting it off into space—a process we sometimes see when a rocket is fueled on the ground before launch. A thicker and heavier tank is required to handle the pressure increase, thereby increasing the spacecraft's dry mass. Boil-off also leads to a steady loss of the liquid propellants available for propulsion. Such mass loss means requiring earlier than expected or frequent refueling.

Boil-off is typically addressed by good insulation and the choice of container materials, both of which are passive measures. These passive measures only reduce boil-off but do not prevent it. This practice is acceptable for short missions such as LEO or going around the Moon. Boil-off is a concern for the Artemis program, where spacecraft such as the Lunar Gateway and Orion will stay in space for a long time. Fuel depots to be installed on the lunar surface will have similar challenges of liquid boil-off.

The above challenges will only be magnified by many times for human missions to Mars, where total flight durations last for 3 or more years. Although NASA recognizes the need for refueling on Mars missions, how it will be carried out is not yet clear. This challenge is a drawback in the use of ALL-CHEM propulsion to Mars.

Active CFM is currently a technology development to manage such a potential hurdle. NASA has development programs for Zero-Boiloff (ZBO) technology. Active ZBO combines active and passive systems to reduce or eliminate the boil-off of cryogenic fluids. The propellant tank's gas pressure is actively controlled to keep boil-off to a minimum. These technologies are currently being tested on the ISS.

In short, NASA considers cryogenic fluid management a critical problem to be better understood and integrated in deep-space human missions to minimize boil-off as part of the effort to preserve fuels. Mars transportation systems must carry more propellants, which should be kept in the cryogenic states for extended periods of time. Boil-offs, no matter how minute, still translate to substantial propellant loss over the entire mission. Such loss must be compensated by carrying extra propellants or implementing CFM [61–65].

13.5.4.3 SpaceX Mars Vision

SpaceX was founded with going to Mars in mind. The Starship system is reportedly developed for that purpose. The Starship has the advantage of a large payload capacity for crew and cargo.

The current Starship runs on chemical propulsion with some reaction control systems (RCS) thrusters for maneuvers and is therefore an ALL-CHEM system. The current configuration of the Starship upper stage has three vacuum-level and

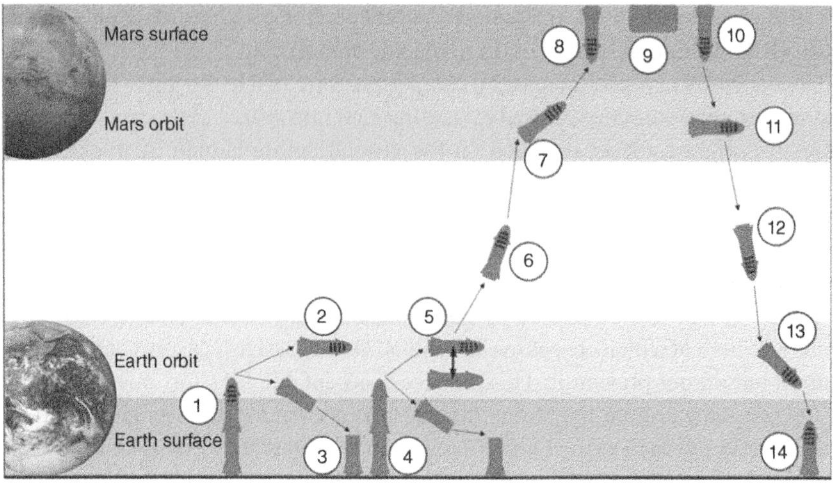

Figure 13.29 Starship round-trip flight concept to Mars. The numbered steps mean: 1) launch of the crew Starship, 2) entering Earth orbit, 3) booster returning to the ground, 4) launch of fuel tankers, 5) refueling the crew Starship, 6) Earth to Mars orbit, 7) orbiting Mars, 8) landing on Mars, 9) local refueling, 10) ascending to Mars orbit, 11) orbiting Mars, 12) Mars to Earth, 13) entering Earth Orbit, and 14) landing on Earth. *Source:* Adapted from Ref. [66].

three sea-level engines. More engines can be added to spacecraft in the future. The Starship is a high-capacity vehicle capable of carrying 100+ people or 100+ tons of cargo to LEO. On-orbit refueling must be in place for the Starship to go beyond LEO.

Figure 13.29 shows the flight concept of a Starship's round-trip mission to Mars. The numbered steps are as follows: 1) launch of the crew Starship, 2) entering Earth orbit, 3) booster returning to the ground, 4) launch of tankers, 5) refueling the crew Starship, 6) Earth to Mars, 7) orbiting Mars, 8) landing on Mars, 9) local refueling, 10) ascending to Mars orbit, 11) orbiting Mars, 12) Mars to Earth, 13) entering Earth orbit, and 14) landing on Earth.

For deep-space missions, SpaceX has the human landing system (HLS) for the Moon as well as a deep-space variant. The HLS has no flaps and heat shields, as the Moon has no atmosphere. Flaps and heat shields will be necessary to land on Mars to take advantage of the thin atmosphere to dissipate its orbital speed.

The flow in Figure 13.29 consists of the following vital functions: 1) on-orbit refueling in LEO, 2) in situ resource utilization on Mars, producing methane and oxygen to return to Earth, and 3) aerobraking while landing on Mars and later Earth.

Locating the fuel depots between Earth and Mars and refueling these depots are open questions to be addressed. It is also likely that one or more robotic missions will be sent to Mars to set up ISRU and basic facilities prior to human arrival [66–68].

To recap, two conditions are required for Mars missions on a Starship to work. They are on-orbit refueling and in situ resource utilization, producing liquid methane and liquid oxygen on Mars.

13.6 Transportation and Communications on Mars

13.6.1 Moving Around Mars

After landing and setting up a base on Mars, pioneers will explore the surrounding areas by mobile vehicles. This activity resembles those on the Moon. There are two types of crewed rovers for Mars: unpressurized and pressurized. Crews in the unpressurized rover need to wear spacesuits but not in a pressurized vehicle. A pressurized rover is for short trips from the Mars base, or it may be equipped as a mobile base or laboratory.

Between 2008 and 2015, NASA began developing the space exploration vehicle (SEV). The SEV consists of a pressurized cabin for crew members to work without wearing spacesuits. An unpressurized cabin is for crew members wearing space suits.

Figure 13.30 (a) and (b) show the concept of the SEV. Figure 13.30(c) shows the vehicle's functional characteristics. The SEV is the main mode of local transportation on Mars. The vehicle weighs 3,000 kg with a payload capacity of 1,000 kg. Its length is 4.45 m, height 3.3 m, and a wheelbase of 3.94 m, as well as having pivoting wheels that enable the vehicle to move 360° in all directions. Its speed is 10 km/h, and a range of over 200 km in all directions.

The pressurized cabin of the SEV allows astronauts to stay inside for up to 14 days. In the event of solar events such as solar storms, medical, and other emergencies, the heavily shielded cabin also doubles as a radiation shelter for up to 72 hours.

Astronauts drive in the vehicle cabin at the fore (Figure 13.30(b)) without putting on their spacesuits. The cabin has a suit-port that allows the crew to get into (ingress) and out of (egress) their spacesuits in less than 10 minutes and then quickly get out of the vehicle. The time saving allows for multiple short walks at different locations. A chariot-style driving station is at the aft where suited astronauts work outside the vehicle and get back on the rover and drive in the open in Figure 13.30(c).

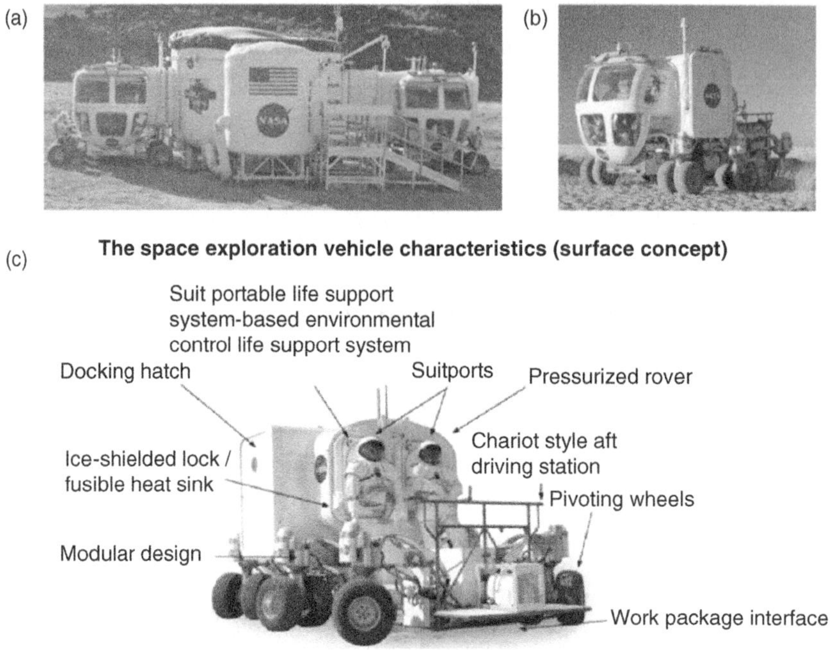

Figure 13.30 NASA Space Exploration Vehicle (SEV). Astronauts can drive the vehicle inside a pressurized cabin (without a spacesuit) at the fore of the vehicle or from the aft in the open. *Source:* Ref. [69] / NASA / Public domain.

The modular design allows different tools such as bulldozer blades and cranes to be attached in accordance with the tasks at hand.

13.6.2 Communications

The time delay for electromagnetic waves between Earth and Mars ranges from 4 minutes (nearest) to 24 minutes (farthest) plus the occasion that the Sun is between the two planets. Such a delay makes near-real-time communication impossible. If radio frequency is used, the beam broadens over the long distance and the signal intensity is significantly weakened by the time it reaches the receiving side. Laser is a better alternative as the beam does not broaden as much. A laser can pack large amounts of information such as images and videos.

Section 12.10 covers the deep space network (DSN). The DSN is currently used to communicate with the satellites and activities of the Artemis program. It also supports Mars missions such as Perseverance. Chapter 12 covers the development of the LunaNet and disruption-tolerant network (DTN), which aims to be the deep-space internet. The DSN will need to migrate to the LunaNet in the future.

Currently, there are five probes (satellites) orbiting Mars. These probes form a network known as the Mars Relay Network. These probes are the Mars Reconnaissance orbiter (MRO), trace gas orbiter (TGO), Mars atmospheric and volatile evolution (MAVEN), Mars Odyssey, and Mars Express. The relay network links rovers roaming on the Martian surface to relay information to Earth via the network's more powerful antennas. Communicating with Earth individually would consume too much power and require a large antenna in each rover. Thus, the mode of communication between Mars and Earth is similar to that being built around the Moon.

13.7 Best Regions to Land

Where should human landers land on Mars?

In 2015, NASA hosted a conference called "Landing Site/Exploration Zone Workshop for Human Missions to the Surface of Mars" to solicit inputs about where to land initial human missions. Forty-five potential exploration zones (EZs) were identified [70]. An EZ is a circle 200 km in diameter, as shown in Figure 13.31. These EZs must meet constraints similar to those of unmanned Mars probes/rovers that have successfully landed [71, 72].

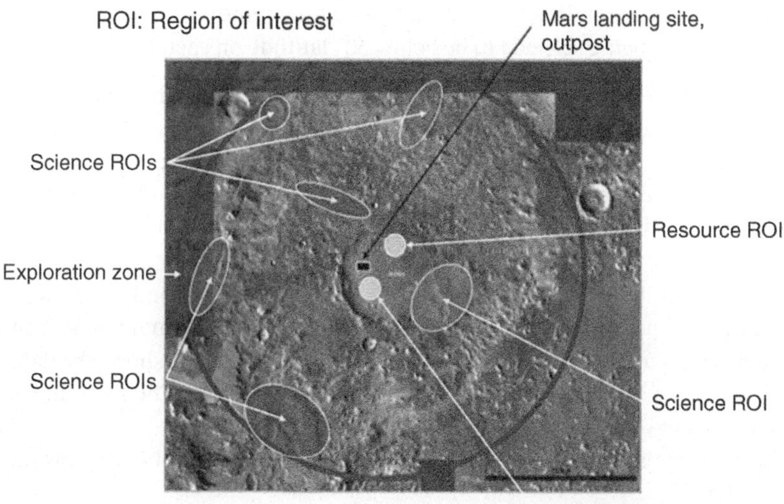

Figure 13.31 In a 2015 workshop, 45 locations on Mars were proposed as candidate locations for the landings of future human missions. Those locations are between 50°N and 50°S latitude around Mars. Regions of interest (ROIs) are within reach of the landing sites. *Source:* Ref. [76] / NASA / Public domain.

Table 13.3 Selected engineering constraints on the landing site of the Mars probes.

Engineering parameter	Target value
Latitude	50°N to 50°S
Elevation	≤2 km
Landing circle radius	Target ≤100 m radius
Landing "blast zone"	~1 km diameter
Terrain relief	To be determined
Slopes	Landing ≤15°; Crew rover <30°
Surface winds	<15 m/s (steady); <30 m/s (gusts)

Source: Adapted from Ref. [73].

Some of the technical constraints for the Mars landing site provided by NASA are shown in Table 13.3 [74].

Simply put, the constraints aim to ensure:

a) Safe landing and smooth takeoff of cargo and crew, as well as facilitating subsequent missions
b) Spacecraft having sufficient time and space to decelerate after entering the Martian atmosphere
c) The probe receives ample solar power and avoids extreme weather elements
d) The landing site being selected to be below 50° latitude on each hemisphere

Mission planners must consider these constraints in order for the landing site to have sufficient resources to ensure survival. Drinkable water is the primary resource from which rocket fuels, and oxygen and hydrogen are derived [75]. The scientific value of the landing site must also be considered. These values include searching for signs of life of the past and present and studying the atmospheric and geological sciences.

Two broad possibilities are considered: Equatorial regions and near-polar regions. The near-polar regions are areas with latitudes near but not lower than 50° latitude owing to the availability of water ice [76]. Figure 13.32 shows the data of the Subsurface Water Ice Mapping (SWIM) project by NASA. The map shows different depths of water ice across the Martian globe.

The north pole is well endowed with subsurface water ice but has several drawbacks [76]:

a) Poor sunlight intensity
b) Extremely low average temperature at −150 °C, long winters, and long polar nights due to Mars's tilt angle
c) Atmosphere too thin for parachuting and aerobraking to work well

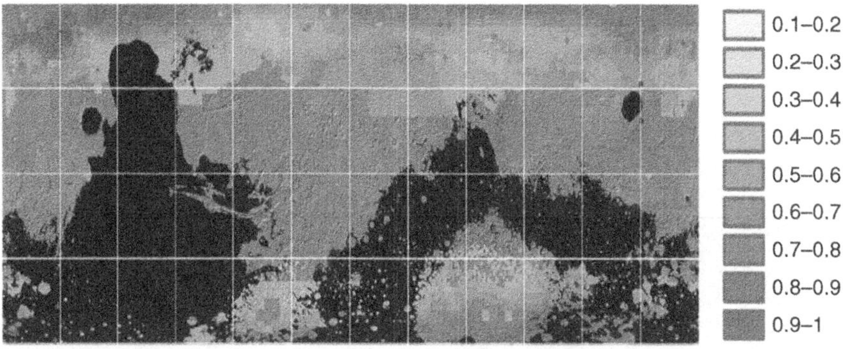

Figure 13.32 SWIM stands for the Subsurface Water Ice Mapping project. The different shades of gray represent the presence of ice of depth from 1 to 5 m [77].

d) Higher fuel consumption to depart from the poles

Low-latitude regions better meet initial human mission objectives because [70]:

a) Better temperature environment from −73 to 20 °C
b) High sunlight intensity (compared to near polar latitudes)
c) The equatorial and mid-latitude regions have more areas to explore for scientific interests and ISRU within the EZ's diameter
d) Launching a spacecraft on the return journey can leverage Mars's rotation and save some propellant

The workshop favored four locations shown in Figure 13.33 for further consideration. The four locations are in low latitudes near the equator.

In summary, it appears that landing near the equator is the best choice for early human landing missions. A chosen EZ will see the buildup of infrastructure around an outpost, ISRU facilities, landing/launch pad for spacecraft, and energy (nuclear and solar) [10, 70 72, 74, 76].

The choice of landing on the lunar south pole is mostly determined by the Moon's long day–night cycle and access to water. In that regard, Mars is more "human friendly."

13.8 Conclusion

To Cislunar enthusiasts, the Moon is not the end but the beginning for human civilization to become spacefaring. Mars will be the next frontier after the Moon. Mars is presently explored by non-crewed probes orbiting the planet and rovers roaming around the surface. Humans have set their sights on landing on Mars. By today's estimate, the first human landing on Mars should happen in the 2030s.

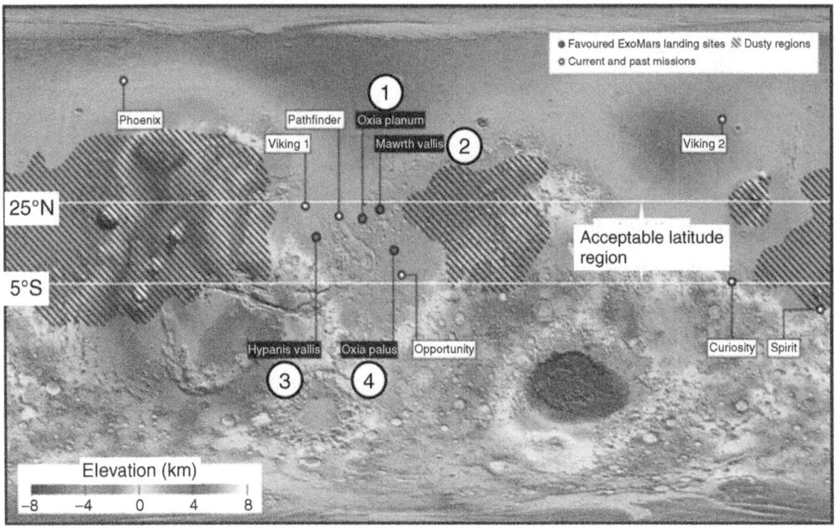

Figure 13.33 From the 2015 workshop, four favored locations were selected. For legibility, these locations are numbered locations, as shown in the figure. They are: (1) Oxia Planum, (2) Mawrth Vallis, (3) Hypanis Vallis, and (4) Oxia Palus. Other labels are past and present probes' landings [78].

Mars is many times farther from Earth than the Moon. Living on Mars is totally different from living on the Moon. The Martian environment is harsh in different ways. Therefore, some technologies developed for settling on the Moon will be useful, while other technologies may not work on Mars.

All missions to Mars must include the launch window in the planning process due to the vast variation in the distance between the two planets at any given time. There is a tradeoff between fuel consumption (energy) and the duration of travel. The NASA animation covered in Section 13.5.3 shows a total round-trip time of 850 days, including a stay on Mars for 66 days. Hence, the choice of the propulsion system is a priority topic of investigation as far as NASA is concerned.

Initial human missions will likely be planned with enough fuel, food, and water to last the entire mission without any assumed relief by local resource utilization. This requirement is mandated in Artemis missions but is not an easy undertaking when applied to Mars. Managing cryogenic propellants is many times more critical for Mars missions than for the lunar ones. Not having enough propellants to or return from Mars spells disaster that there is little Earth can do.

There is a strong possibility that more non-crewed Martian missions will occur before the first human mission. These missions could be robots establishing some form of ISRU facility. Like the Lunar Gateway, having an orbiting space station such as the Mars Base Camp in Chapter 4 is the same concept.

On a positive note, technologies such as harvesting water from ice and underground water, as well as utilizing Martian regolith to make building materials, can be built on the experience on the Moon.

The distance of Mars from the Sun limits the intensity and therefore the usefulness of solar power as the primary energy source. Nuclear fission power will be the best choice to provide electric power continuously and reliably. Other alternatives such as wind and geothermal power will have to demonstrate themselves as viable complementary sources.

Reading List

Aldrin B, Pruden J. *Mission to Mars: My Vision for Space Exploration*. Blackstone Audio Inc.

Bizony P. *NASA Missions to Mars: A Visual History of Our Quest to Explore the Red Planet*. Motorbooks; 2022. ISBN: 10: 0760373140.

Carpenter G. *The Legacy of Mission Mars* (Kindle Edition). Carpenter Publications; 2024.

Crumpler LS. *Missions to Mars: A New Era of Rover and Spacecraft Discovery on the Red Planet* (Kindle Edition). Harper; 2021.

Pyle R, Green J. *Mars: The Missions That Have Transformed Our Understanding of the Red Planet*. Andre Deutsch; 2019. ISBN: 10: 0231194056.

Reynolds D. *The Perseverance Mars Rover (Mission Control)*. PowerKids Press; 2024. ISBN: 10: 1499449917.

References

1 Tillman JE. Mars. Live from Earth and Mars [updated 1997 Sep 23; cited 2025 Feb 28] [online]. Available from: https://www-k12.atmos.washington.edu/k12/resources/mars_data-information/mars_overview.html.

2 NASA. Seasons in the Martian year as the red planet orbits the Sun. NASA [updated 2016 Jul 1; cited 2025 Feb 28] [online]. Available from: https://science.nasa.gov/resource/seasons-in-the-martian-year-as-the-red-planet-orbits-the-sun/.

3 Mars volcanoes. Marspedia [cited 2025 Feb 28] [online]. Available from: https://marspedia.org/Mars_volcanoes.

4 Mars facts. NASA Science [cited 2025 Feb 28] [online]. Available from: https://science.nasa.gov/mars/facts/.

5 Goddard space flight center image gallery. NASA [cited 2025 Feb 28] [online]. Available from: https://science.gsfc.nasa.gov/attic/mola/images.html.

6 Astrogeology Science Center. USGS [cited 2025 May 1] [online]. Available from: https://www.usgs.gov/media/images/olympus-mons-taken-viking-orbit.
7 Yang L, Xing W, Zhenghao L, Yongliao Z. Geological evolution and habitable environment of Mars: Progress and prospects. Reviews of Geophysics and Planetary Physics. 2021;*52*(4):416–436.
8 Volcanoes. Mars Education at Arizona State University [updated 2008 Jun 6; cited 2025 Feb 28] [online]. Available from: https://marsed.asu.edu/mep/volcanoes.
9 SpaceRef. NASA's mars rovers probing water history at two sites. SpaceNews [update 2004 Oct 8; cited 2025 Feb 28] [online]. Available from: https://spacenews.com/nasas-mars-rovers-probing-water-history-at-two-sites/.
10 Jet Propulsion Laboratory. NASA's treasure map for water ice on mars. Jet Propulsion Laboratory [update 2019 Dec 10; cited 2025 Feb 28] [online]. Available from: https://www.jpl.nasa.gov/news/nasas-treasure-map-for-water-ice-on-mars/.
11 Carr MH, Belton MJS, Malin MC. Character of the surface. Encyclopaedia Britannica [updated 2025 Feb 25; cited 2025 Feb 28] [online]. Available from: https://www.britannica.com/place/Mars-planet/Character-of-the-surface.
12 Scheller EL, Bosak T, McCubbin FM, Williford K, Siljeström S, Jakubek RS, et al. Inorganic interpretation of luminescent materials encountered by the Perseverance rover on Mars. Science. Advances. 2024;*10*(39):eadm8241.
13 Choi CQ. Water on Mars: The story so far. Astrobiology at NASA [updated 2016 Oct 14; cited 2025 Feb 28] [online]. Available from: https://astrobiology.nasa.gov/news/water-on-mars-the-story-so-far/.
14 Greicius A. NASA's treasure map for water ice on Mars. NASA [updated 2019 Dec 10; cited 2025 Feb 28] [online]. Available from: https://www.nasa.gov/solar-system/nasas-treasure-map-for-water-ice-on-mars/.
15 Meyer TR, McKay CP. The resources of Mars for human settlement. Journal of the British Interplanetary Society. 1989;*42*:147–160. Available from: https://pubmed.ncbi.nlm.nih.gov/11539074/.
16 Bressan D. Comparing Earth and Mars, new study shows how mineral diversity is controlled by plate tectonics and emergence of life. Forbes [updated 2023 Sep 11; cited 2025 Feb 28] [online]. Available from: https://www.forbes.com/sites/davidbressan/2023/09/12/comparing-earth-and-mars-new-study-shows-how-mineral-diversity-is-controlled-by-plate-tectonics-and-emergence-of-life/.
17 Williams M. How bad is the radiation on Mars? Phys.org [cited 2025 Feb 28] [online]. Available from: https://phys.org/news/2016-11-bad-mars.html.
18 Chancellor JC, Scott GB, Sutton JP. Space radiation: The number one risk to astronaut health beyond low earth orbit. Life. 2014;*4*(3):491–510.

19 Balme M, Greeley R. Dust devils on Earth and Mars. Reviews of Geophysics. 2006;*44*(3):1–22.

20 NASA. Mars before and after dust storm. NASA [updated 2018 Jul 19; cited 2025 Mar 1] [online]. Available from: https://science.nasa.gov/resource/mars-before-and-after-dust-storm/.

21 The fact and fiction of martian dust storms. NASA [cited 2025 Mar 1] [online]. Available from: https://www.nasa.gov/solar-system/the-fact-and-fiction-of-martian-dust-storms/.

22 Strain D. Mars' infamous dust storms can engulf the entire planet. University of Colorado Boulder [updated 2023 Oct 29; cited 2025 Mar 1] [online]. Available from: https://www.colorado.edu/today/2024/12/09/mars-infamous-dust-storms-can-engulf-entire-planet-new-study-examines-how.

23 Kuthunur S. Mars is bombarded by more meteorites than previously thought. Astronomy [cited 2025 Mar 1] [online]. Available from: https://www.astronomy.com/science/mars-was-bombarded-by-more-meteorites-than-previously-thought/.

24 Phobos and deimos. Wikimedia Commons [updated 2020 Jun 10; cited 2025 Mar 1] [online]. Available from: https://commons.wikimedia.org/wiki/File:Phobos_and_Deimos.jpg.

25 Dorminey B. Why geothermal energy will be key to mars colonization. Forbes [updated 2016 Sep 30; cited 2025 Mar 1] [online]. Available from: https://www.forbes.com/sites/brucedorminey/2016/09/30/why-geothermal-energy-will-be-key-to-mars-colonization/.

26 Mars. Oregon State University [cited 2025 Mar 1] [online]. Available from: https://volcano.oregonstate.edu/mars.

27 Fuel cell power system options for Mars rovers. European Space Agency [cited 2025 Mar 1] [online]. Available from: https://www.esa.int/Science_Exploration/Human_and_Robotic_Exploration/Exploration/Fuel_cell_power_system_options_for_Mars_rovers.

28 Mining the Martian surface for in situ resources. MIT Technology Roadmapping [updated 2020 Jun 10; cited 2025 Mar 1] [online]. Available from: https://roadmaps.mit.edu/index.php/Mining_the_Martian_Surface_for_in_Situ_Resources.

29 Mars Rodwell experiment final report. NTRS - NASA Technical Reports Server [cited 2025 Mar 1] [online]. Available from: https://ntrs.nasa.gov/citations/20205011353.

30 This is one mars rover with MOXIE. NASA [updated 2019 Apr 5; cited 2025 Mar 1] [online]. Available from: https://science.nasa.gov/resource/this-is-one-mars-rover-with-moxie/.

31 Sia J. How to make fuel, oxygen, and water on Mars: ISRU part 1. Mars Society of Canada [updated 2020 Aug 24; cited 2025 Mar 1] [online]. Available from: https://www.marssociety.ca/2020/08/24/how-to-make-fuel-oxygen-and-water-on-mars-isru-part-1/.

32 Chinese "machine chemist" successfully develops mars oxygen production catalyst. Rope Count [updated 2023 Nov 14; cited 2025 Apr 25] [online]. Available from: https://www.ropecount.com/news/6553a2e572410407e51b51ee.

33 Sia J. Construction materials on Mars: ISRU part 2. Mars Society [cited 2025 Apr 25] [online]. Available from: https://www.marssociety.ca/2020/09/02/construction-materials-on-mars-isru-part-2/.

34 Sia J. Construction materials on Mars: ISRU part 2. Mars Society of Canada [updated 2020 Sep 2; cited 2025 Mar 1] [online]. Available from: https://www.marssociety.ca/2020/09/02/construction-materials-on-mars-isru-part-2/.

35 Katz B. Scientists make sturdy bricks from Mars-like soils. Smithsonian Magazine [updated 2017 May 15; cited 2025 Mar 1] [online]. Available from: https://www.smithsonianmag.com/smart-news/scientists-make-sturdy-bricks-mars-soils-180963077/.

36 Cheng S, She W, Gao Y, Zhou Y, Wen S, Zhou C. Research and prospect of in-situ construction materials on mars. Chinese Journal of Space Science. 2024;*44*(3):511–524.

37 In-situ resource utilization (ISRU). NASA [cited 2025 Mar 1] [online]. Available from: https://www.nasa.gov/mission/in-situ-resource-utilization-isru/.

38 Anderson SL, Sansom EK, Shober PM, Hartig BA, Devillepoix HA, Towner MC. The proposed silicate-sulfuric acid process: Mineral processing for In Situ Resource Utilization (ISRU). Acta Astronautica. 2021;*188*:57–63.

39 Astrogeology Science Center.New InSight Lander data suggest deeply buried water on Mars. U.S. Geological Survey [updated 2021 Jul 27; cited 2025 Mar 1] [online]. Available from: https://www.usgs.gov/centers/astrogeology-science-center/news/new-insight-lander-data-suggest-deeply-buried-water-mars.

40 Ingenuity Mars helicopter. NASA [updated 2020 Jul 30; cited 2025 Mar 1] [online]. Available from: https://science.nasa.gov/mission/mars-2020-perseverance/ingenuity-mars-helicopter/.

41 Ingenuity Mars helicopter. NASA [cited 2025 Mar 1] [online]. Available from: https://www.nasa.gov/gallery/ingenuity-mars-helicopter/.

42 PIA24649: Mastcam-Z GIF show Ingenuity's fifth landing. Jet Propulsion Laboratory [cited 2025 Mar 1] [online]. Available from: https://photojournal.jpl.nasa.gov/catalog/PIA24649.

43 Mars perseverance raw image of the week. NASA [cited 2025 Mar 1] [online]. Available from: https://mars.nasa.gov/mars2020/multimedia/raw-images/.

44 Earth–Mars transfer trajectory. Marspedia [cited 2025 Mar 1] [online]. Available from: https://marspedia.org/Earth-Mars_Transfer_Trajectory.

45 Mars 2020 perseverance launch press kit. Jet Propulsion Laboratory [cited 2025 Mar 1] [online]. Available from: https://www.jpl.nasa.gov/news/press_kits/mars_2020/launch/mission/.

46 High-resolution still image of perseverance's landing. NASA [updated 2021 Feb 19; cited 2025 May 1] [online]. Available from: https://science.nasa.gov/resource/high-resolution-still-image-of-perseverances-landing-2/.

47 Where is perseverance? NASA [updated 2021 Feb 17; cited 2025 Mar 1] [online]. Available from: https://science.nasa.gov/mission/mars-2020-perseverance/location-map/.

48 Tianwen-1 ("Questions to Heaven"-1). eoPortal [updated 2020 Aug 3; cited 2025 Mar 1] [online]. Available from: https://www.eoportal.org/satellite-missions/tianwen-1#spacecraft.

49 Tianwen-1 transfer orbit and TCM. Wikipedia [updated 2020 Jul 21; cited 2025 Mar 1] [online]. Available from: https://en.m.wikipedia.org/wiki/File:Tianwen-1_transfer_orbit_and_TCM.png.

50 Huang X, Li M, Wang X, Hu J, Zhao Y, Guo M, et al. The Tianwen-1 guidance, navigation, and control for Mars entry, descent, and landing. Space: Science & Technology. *2021*: 9846185, 2021.

51 Strickland A. China's rover makes surprising water discovery at Mars landing site. CNN World [updated 2022 May 11; cited 2025 Mar 1] [online]. Available from: https://edition.cnn.com/2022/05/11/world/mars-water-china-zhurong-rover-scn/index.html.

52 Radio France Internationale. China's Mars rover Zhurong begins roaming the Red Planet. Radio France Internationale [updated 2021 May 22; cited 2025 Apr 25] [online]. Available from: https://www.rfi.fr/en/china-s-mars-rover-starts-roaming-the-red-planet.

53 Ackerman E. China becomes third nation to successfully land rover on Mars. IEEE Spectrum [updated 2021 May 1; cited 2025 Mar 1] [online]. Available from: https://spectrum.ieee.org/china-becomes-third-nation-to-successfully-land-rover-on-mars.

54 UAE Space Agency [cited 2025 Mar 1] [online]. Available from: https://www.emiratesmarsmission.ae/.

55 Abbany Z. The facts on hope: The UAE's Mars mission 'Al-Amal'. DW [updated 2021 Sep 2; cited 2025 Mar 1] [online]. Available from: https://www.dw.com/en/the-facts-on-hope-the-uaes-mars-mission-al-amal/a-54067401.

56 Hope probe to move to a new mars orbit and observe deimos. UAE Space Agency [updated 2023 Feb 9; cited 2025 Mar 1] [online]. Available from: https://space.gov.ae/en/media-center/news/9/2/2023/hope-probe-to-move-to-a-new-mars-orbit-and-observe-deimos.

57 Baker M. Driving Mars exploration: How the perseverance rover will pave a path into the future. National Air and Space Smithsonian [updated 2021

Feb 17; cited 2025 Mar 1] [online]. Available from: https://airandspace.si.edu/stories/editorial/driving-mars-exploration-perseverance.

58 Mars sample return overview infographic. European Space Agency [updated 2019 May 29; cited 2025 Mar 1] [online]. Available from: https://www.esa.int/ESA_Multimedia/Videos/2019/05/Mars_Sample_Return_overview_infographic.

59 Mars transportation white paper. NASA [cited 2025 Mar 1] [online]. Available from: https://www.nasa.gov/wp-content/uploads/2023/10/acr22-wp-mars-transportation.pdf.

60 Mars. NASA [cited 2025 Feb 27] [online]. Available from: https: //www.nasa.gov/humans-in-space/humans-to-mars/#preparing.

61 Zero-boil-off tank experiments to enable long-duration space exploration. NASA [updated 2024 Mar 12; cited 2025 Feb 27] [online]. Available from: https://science.nasa.gov/science-research/science-enabling-technology/zero-boil-off-tank-experiments-to-enable-long-duration-space-exploration/.

62 Robotic Refueling Mission 3 (RRM3). NASA [cited 2025 Feb 27] [online]. Available from: https://www.nasa.gov/nexis/robotic-refueling-mission-3/.

63 Zbot. Glenn Research Center [cited 2025 Feb 27] [online]. Available from: https://www1.grc.nasa.gov/space/iss-research/msg/zbot/.

64 Cryogenic fluid management. NASA [cited 2025 Feb 27] [online]. Available from: https://www.nasa.gov/wp-content/uploads/2024/01/cryogenicfluidmanagement-factsheet.pdf.

65 Hansen HC, Johnson WL, Meyer M, Werkheiser A, Stephens JR. Cryogenic fluid management technologies enabling for the Artemis program and beyond. In: ASCEND 2020; 2020. p. 4000.

66 Making life multiplanetary. SpaceX [updated 2017 Sep 28; cited 2025 Feb 27] [online]. Available from: https://www.spacex.com/media/making_life_multiplanetary_transcript_2017.pdf.

67 SpaceX beginning to tackle some of the big challenges for a Mars journey. Ars Technica [updated 2019 Jun 3; cited 2025 Feb 27] [online]. Available from: https://arstechnica.com/science/2019/06/spacex-working-on-details-of-how-to-get-people-to-mars-and-safely-back/.

68 NASA materials indicate SpaceX targets Mars landing for starship by 2030. Top Popular [updated 2024 Jun 25; cited 2025 Feb 27] [online]. Available from: https://toppopular.com/nasa-materials-indicate-spacex-targets-mars-landing-for-starship-by-2030/.

69 NASA testing next-gen lunar rover in Arizona. New Atlas [updated 2008 Oct 27; cited 2025 May 1] [online]. Available from: https://newatlas.com/nasas-spr-lunar-rover-tested-in-arizona/10271/.

70 If humans go to mars, where's the best place to land? Popular Science [updated 2015 Oct 28; cited 2025 Mar 1]. Available from: https://www.popsci.com/if-humans-go-to-mars-wheres-best-place-to-land/.

71 Berggren M, Zubrin R, Wilson C, Rose H, Carrera S. Mars aqueous processing system. In: Mars: Prospective Energy and Material Resources (ed. Badescu V). Springer; 2009. p. 563–586.

72 Dantong G, Pingyuan C. Overview and prospect of planetary landing site selection. Journal of Deep Space Exploration. 2016;*3*(3):197–203.

73 Clarke J, Willson D, Smith H, Hobbs S, Jones E. Southern meridiani planum-A candidate landing site for the first crewed mission to Mars. Acta Astronautica. 2017;*133*:195–220.

74 Where should future astronauts land on Mars? Follow the Water. NASA [updated 2021 Feb 8; cited 2025 Mar 1] [online]. Available from: https://www.jpl.nasa.gov/news/where-should-future-astronauts-land-on-mars-follow-the-water/.

75 SWIM map shows subsurface water ice on Mars. Planetary Society [updated 2023 Oct 26; cited 2025 Mar 1] [online]. Available from: https://science.nasa.gov/resource/swim-map-shows-subsurface-water-ice-on-mars/.

76 Beaty D, Hays L, Davis R, Bussey B, Abbud-Madrid A, Boucher D, et al. The possible strategic significance of mid-latitude ice deposits to a potential future human mission to Mars. In: Sixth International Conference on Mars Polar Science and Exploration (Volume 1926); 2016. p. 6059.

77 Petersen CC. Where are the best places to land humans on Mars? Universe Today [updated 2022 Dec 15; cited 2025 Mar 1] [online]. Available from: https://www.universetoday.com/articles/where-are-the-best-places-to-land-humans-on-mars/.

78 Gibney E. ExoMars scientists narrow down landing sites. Nature. 2014;*508*(7494):19.

14

Space Debris

14.1 Introduction

> US issues first-ever fine for space junk after TV company leaves satellite in the wrong place.
>
> Euronews Green, October 03, 2023 [1]

> Cantwell, Hickenlooper Bill to Clean Up Space Junk Passes Senate Unanimously.
>
> U.S. Senate Committee on Commerce, Science, and Transportation, November 1, 2023 [2]

Space debris is now a known problem in the public eye owing to the media coverage of occasional incidents. A grave picture of the Earth surrounded by a thick cloud of junk has been painted.

The long-term harm of space debris is existential. Domestic regulations and international guidelines are steadily being put in place to impose an end-of-life solution for launched hardware. Leading space agencies around the world have formed the Inter-Agency Space Debris Coordination Committee to address orbital debris issues.

For a long time, space debris has been viewed as inevitable in space activities. But now a tipping point is almost upon us. It remains unclear who will clean up the already polluted space environment. More specifically, how much will it cost and who will pay for cleaning up the large quantities of existing man-made space debris? The best we can do in the near term is to reduce the production of new debris.

New Space: From Low Earth Orbit to the Moon and Beyond, First Edition.
Edited by Stephen W. Cheung and Chih-Yung Wen.

The underlying problem with space debris is its ability to self-multiply due to breakups and collisions. Debris hinders the free use of space, crew safety, and impacts daily living and our economy on Earth. In July 2023, the news website Space.com reported that the mega satellite constellation operator Starlink had to perform 25,000 collision-avoidance maneuvers in the previous 6 months. Near misses range from a few meters to several kilometers [3]. The upcoming mega satellite constellations will add to the debris stockpile if preventive measures are not put in place.

The existential problem of thousands of satellites coexisting with millions of debris already in low earth orbit (LEO) can no longer be ignored. Occurrences of near collisions are increasing. The more satellites we launch into space, the more space debris we create and become more vulnerable.

In a 2021 report, NASA considers different remediation approaches to reduce the growing risks posed by orbital debris. As LEO is commercialized rapidly, governments need to take active roles in regulating space debris production as well as supporting efforts to test viable solutions. Then, the private sector can scale up workable services. Space debris management is a nascent but growing sector of the space industry.

In this chapter, we are mainly concerned about debris in LEO and their remedial efforts due to their large populations and potential hazards.

14.1.1 Amount of Space Debris and Their Distribution

Estimates of the total amount of space debris and their sizes vary. Debris is often categorized by its size. Table 14.1 lists estimates by NASA, the European Space Agency (ESA), and Spaceworks, a technology company that has been spun off from the Georgia Institute of Technology.

According to the United Nations in conjunction with the European Space Agency, approximately 95% of orbital debris is concentrated in LEO.

Table 14.1 Estimated amount of space debris by size based on information from NASA, ESA, and Spaceworks.

Category	Size	NASA's estimate (2022)	ESA's estimate (2023)	Spaceworks (2023)
Large	Over 10 cm	>25,000	36,500	34,000
Small	1–10 cm	500,000	1,000,000	900,000
Tiny	0.1–1 cm	100+ million	130+ million	128 million
Total Tonnage		9,000 tons	11,500 tons	Not available

Spaceworks is a spinoff from the Georgia Institute of Technology.
Source: Adapted from Refs. [4, 5].

The population of debris varies with altitude. The largest aggregation of debris is found near 700–1,000 km.

14.2 Origin of the Space Debris

Most space debris orbiting Earth is man made. Recall that payloads are carried by rocket upper stages to LEO where they are deployed. The upper stages and their spent parts are left behind in the orbit and become debris. Derelict satellites also join the debris population. These defunct satellites and spent parts often fall apart, and some explode due to wear and tear, extended exposure to solar radiation and heat cycles, and uncontrolled ignition of the residual fuels.

Batteries are of particular concern, and their liability in space has been recognized. Batteries, especially the end-of-life ones and those in disintegrated satellites, are susceptible to breakups and explosions due to ionizing radiation and heat exposure. Such explosions often lead to many fragments in the 1–10 cm range. Batteries are the source of most fragments.

NASA in 2007 estimates that 50% of the orbital debris is directly linked to fragmentation and 12.5% are results of explosions. Although batteries only make up a minor percentage of most spacecraft by volume, they account for nearly 8% of all known debris currently in orbit [6–8].

14.2.1 Debris Source

Collisions in space have disastrous outcomes. As of January 2024, over 8,000 active satellites are in various Earth orbits. Their distributions are geostationary orbit (GEO) (12%), medium Earth orbit (MEO) (3%), and LEO (84%). Large satellites (weighing several tons) are in the GEO, and small satellites (about 1,000 kg or less) are in the LEO. About 72% of active satellites are small. As of December 2024, the total number of nanosatellites sent to orbit at the beginning of the New Space era is approaching 7,000 [9]. These nanosatellites usually weigh less than 10 kg and use off-the-shelf components on missions that often last just a few months in orbit. But the derelict hardware will stay in orbit for months and years. Derelict nanosatellites are more susceptible to breakups because they are built of off-the-shelf parts. In 2009, a derelict satellite collided with an active Iridium satellite and reportedly created over 2,000 pieces of debris.

All these active and inactive spacecraft contribute to potential collisions with one another. Collisions cause fragments to fly in different directions. Therefore, collisions are random, and their relative speeds of collision could be higher than the orbital speed.

Collision with a 10-cm orbiting object could cause catastrophic fragmentation of a satellite. A 1-cm object could disable a satellite and penetrate a space station's shields. A 1-mm object could destroy subsystems onboard a spacecraft.

Table 14.2 A spacecraft's decay time at different LEO attitudes.

Altitude (km)	Expected decay time
Above 1,000	Centuries
800	Decades
600	Years
400	Months
Below 200	Hours

14.2.2 Risky Orbits

LEO has the highest concentration of manmade debris and therefore the highest risk of collisions and near misses. LEO is full of satellites, particularly those of certain inclinations for communications. These orbits present high risks of incidents owing to heavy satellite traffic.

14.2.3 Orbital Decay Time

How long will a piece of orbiting debris remain in orbit? The decay time is the time it takes for an orbiting object to deorbit back to Earth due to residual atmospheric drag. The decay time is altitude dependent. Table 14.2 provides the coarse decay time versus the LEO altitude. A long decay time leads to debris accumulation.

14.3 Space Debris Management

In 2022, the Orbital Debris Interagency Group, under a subcommittee of the US National Science and Technology Council, published *The National Orbital Debris Implementation Plan*. The plan names three broad and encompassing methods to reduce the risks posed by orbital debris: 1) limit the creation of new debris, 2) remediate existing debris, and 3) track and characterize debris better [10].

14.3.1 Debris Dynamics

An object's orbital speed in LEO is about 7–8 km/s (Mach 24–26), depending on the altitude. The ballistic cross-section of the object is a key parameter. A large ballistic cross-section means that the object is more exposed to collisions. A space station with large solar panels, for example, has a large ballistic cross section.

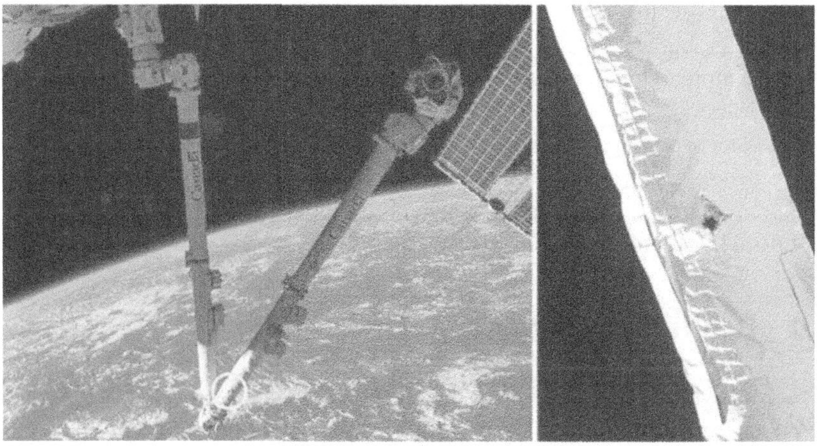

Figure 14.1 Debris hitting part of the robotic arm of the ISS in 2021. *Source:* Ref. [11] / NASA / Public domain.

How much damage can an object moving at Mach 25 incur? Compared to Mach 1, an object at Mach 25 possesses 25 times more momentum and (25 × 25=) 625 times more kinetic energy.

In 2021, the International Space Station reported that its robotic arm had been hit by a small piece of space debris, causing a hole in a part of the robotic arm, as shown in Figure 14.1. The size of the hole is not identified, nor is the material makeup of the debris. Fortunately, the robotic arm remained functional.

NASA has stated that "A number of space shuttle windows were replaced because of damage caused by material that was analyzed and shown to be paint flecks.... In fact, millimeter-size orbital debris represents the highest mission-ending risk to most robotic spacecraft operating in low Earth orbit."

In summary, debris is harmful to space assets. Debris of different sizes and masses poses different levels of threats. Tiny debris of less than 1 cm is likely the cause of collisions due to their sheer number, but it is usually stopped by proper shielding. Large debris is fewer in number; it tends to crash into other large objects and produces multiple fragments that add to the debris population. Debris size of 1–10 cm in number in the high hundred thousand is the biggest contributor to the exponential debris growth.

14.3.2 Kessler Syndrome

A cascading effect of debris collision was proposed by NASA scientist Donald Kessler in 1978. Space debris carries enormous momentum and kinetic energy due to its hypersonic speed. A collision between two pieces of debris produces fragments that scatter in many directions. These fragments not only add to the debris cluster but could also continue to collide with other objects, thereby generating more fragments. A domino effect ensues, similar to the chain reaction in nuclear fission.

The Kessler syndrome occurs when the amount of orbital debris reaches a point of no return with debris accumulating in a runaway fashion. When the rate of debris production is faster than the rate of removal, the Earth would eventually be surrounded by a permanent shroud of space junk.

14.4 Debris Remediation

The Kessler syndrome can be reversed if the rate of debris removal is faster than that of accumulation. Remediation is the field of debris removal and collision avoidance.

14.4.1 Reduce the Creation of New Debris

The first step is legislation. In September 2022, the Federal Communications Commission approved a new rule to reduce the growing debris in LEO. The rule requires derelict satellites in LEO to deorbit no later than 5 years after the end of their missions. New LEO satellites must have an end-of-life solution in place before launch with either a built-in solution or a contract as a service. Previous longstanding guidelines for deorbiting satellites were up to 25 years after the mission ended. The new rule has taken effect on new LEO launches in September 2024.

The FCC rule aims to slow down the rate of debris accumulation in LEO by actively removing new debris. This rule change is timely considering the unprecedented number of LEO satellites being launched in this decade and the next [12].

The ESA has also adopted the "Zero Debris approach" that provides the framework for achieving a net-neutral contribution to space debris. Strict requirements are to be imposed on ESA missions to reduce the amount of additional debris these missions leave in orbit.

Several of these ESA requirements are listed here for reference: 1) guarantee successful disposal, 2) avoid in-orbit collisions, 3) avoid internal breakups, and 4) prevent intentional release of space debris [13].

Remediation is the magic word. Incineration by the atmosphere via reentry is common for debris removal in LEO. Large debris is relatively easy to track, but the millions of tiny debris are not. Hence, it is crucial to prioritize remedial actions on large and small pieces that could pose an imminent danger of disintegration or are in orbits where many spacecraft frequent. About 20,000 large debris fit this priority as do tens of thousands of small debris.

The options of remediation are 1) physically removing an orbiting object, 2) changing its trajectory, and 3) moving it to harmless places, which are labeled as satellite graveyards.

14.4.2 Spacecraft Graveyards

Sending debris to satellite graveyards is a longstanding practice. One graveyard is a zone 300 km further above GEO for the derelict spacecraft. It would take too much fuel to deorbit a spacecraft from GEO to the atmosphere 36,000 km away. An end-of-life GEO spacecraft can move itself to the graveyard orbit utilizing the fuel set aside for this purpose.

Another graveyard is a designated area on Earth for numerous derelict spacecraft deorbiting from LEO. Point Nemo is in the middle of the South Pacific, around 2,700 km from the nearest dry land. The area has no human activity. This area is reserved for very large spacecraft that do not incinerate completely by atmospheric reentry. Where possible, these spacecraft are timed to deorbit so that the partially incinerated structures end up in this area. The pieces then sink to the bottom of the ocean.

14.4.3 Remediating Existing Debris

Relying on the natural orbital decay would take too long for most debris. Some large objects may erode to a state that produces a trail of smaller debris well before decaying into the atmosphere. Problematic debris must be removed deliberately. The 2022 US National Science and Technology Council report [10] in Section 14.3 states that the most effective remediation methods entail removing small debris (1–10 cm) and nudging large debris up to the size of satellites to avoid collisions.

NASA and other space agencies have stated that 5–10 of the most massive debris must be removed every year to prevent debris accumulation from getting out of hand [14]. The NASA report offers three recommendations: 1) removing small debris and nudging large debris to avoid collisions, 2) controlled and uncontrolled reentry via reusable remediation services, and 3) recycling space debris. The NASA report will be elaborated further in Section 14.8.

This section focuses on remedial technologies that aim to remove LEO debris [15, 16].

14.4.4 Removing Debris by Direct Contact

Direct contact means physically capturing the target debris. A host spacecraft, also known as the chaser, is equipped with a mechanical device to "grab and capture" the target. One important consideration is that the target may be tumbling.

A chaser capturing a tumbling target will inherit the angular momentum, making the chaser "uncontrollable." The second consideration is identifying the best place to grab the target to avoid an inadvertent breakups.

14.4.4.1 Docking and Mission Extension

Satellites can add service life to extend their missions by external means. In principle, mission extension works for satellites in any orbit, but only GEO satellites have taken up this service so far.

GEO satellites typically weigh several tons. Such satellites have price tags worth several hundred million US dollars. The typical design life of a GEO satellite is 15 years. Satellite operators are financially motivated to extend their revenue streams for several more years. Mission extension delays a working satellite from becoming derelict.

A mission extension vehicle (MEV) approaches and docks with the target satellite in orbit. Services typically include refueling, recharging, and replacing batteries and certain key components. Startups and established companies have entered this arena. For example, two GEO communications satellites, Intelsat 9-01 and Intelsat 10-2, were serviced by mission extension vehicles made by Northrop Grumman called MEV-1 and MEV-2 in 2020 and 2021, respectively. The two satellites, otherwise healthy, were running low on fuel. The service reportedly extended the satellite's life by 5 years. Grumman's MEV-1 and MEV-2 are shown in Figure 14.2 [17, 18].

Figure 14.2 Mission extension vehicle-1 and -2 by Northrop Grumman. *Source:* Ref. [19] / NASA / Public domain.

A GEO satellite normally has a fuel reserve to move itself to the graveyard orbit. An MEV can perform the same task. Alternatively, the MEV can attach a propulsive device to the target satellite and move it accordingly. This method can be considered if the target satellite is no longer serviceable.

Using an MEV to deorbit an LEO satellite would be overkill unless the cost–benefits are justified. A mini version of the MEV is worth considering. Older LEO satellites are usually designed for 10 years, and some may be worth the life-extending effort. One option to deorbit large LEO debris with propulsion capability is to attach a small rocket engine to a satellite before launch. An Italian company, D-Orbit, for example, offers their D3 device using solid fuel [16].

14.4.4.2 Robotic Arm

Using a robotic arm to snag a target is an obvious option. Clearspace of Switzerland and Astroscale of the United Kingdom are among the leaders.

a) Clearspace: In 2020, the ESA contracted Clearspace to remove space debris from orbit [20]. The first active debris removal mission is expected to be launched in 2025. The vehicle ClearSpace-1 will rendezvous with a spent rocket part previously left there by a European launcher and use a robot to capture the target, as depicted in Figure 14.3. Clearspace-1 then takes the captured debris to a lower orbit for reentry.
b) Astroscale UK: Astroscale UK offers the "Cleaning Outer Space Mission through Innovative Capture" (COSMIC) technology. The mission will be launched by 2026. See Figure 14.4. The company specializes in proximity maneuvers by a chaser to a target. Rendezvous and proximity operation (RPO) refers to orbital maneuvers (rendezvous) in which one spacecraft intercepts and approaches another at a close distance (proximity). This technology is critical in close observation and docking and has been used in asteroid sample collection missions.

The Astroscale-UK mission aims to use a robotic arm to remove two defunct British satellites currently orbiting Earth. Astroscale Japan is also developing active debris removal using a chaser to obtain images of a rocket body. The chaser then transmits the data to the mission control to determine the best way for its robotic arm to capture the target. This "observe and inspect" technology is particularly important for large debris of irregular shapes.

It is not an easy task for a robotic arm to grab a tumbling target. This is particularly true with the old satellites. Brute force may cause the target to break up. Technologies such as the RPO come in so that the host comes close and observes the target for mission control to find a suitable course to do the capture.

Figure 14.3 Concept of a Clearspace satellite grabbing a dead satellite. *Source:* Ref. [20] / Clearspace.

Figure 14.4 Astroscale satellite (with solar panel) extends its robotic arm to grab a target satellite. *Source:* Astroscale.

Hence, targets that are tumbling and/or possess irregular shapes explain the complexity of direct capture.

14.4.4.3 Tethered Net

The European Commission and several commercial parties funded the University of Surrey in the UK to perform a demo test using a common platform known as the RemoveDEBRIS satellite. The experiment cost roughly 15 million euros.

The experiment consists of two separate tests: a tethered net and a tethered harpoon on mock targets. Figure 14.5 depicts the sequences of the two tests. The RemoveDEBRIS satellite is the host carrying two CubeSats DSat-1 and DSat-2

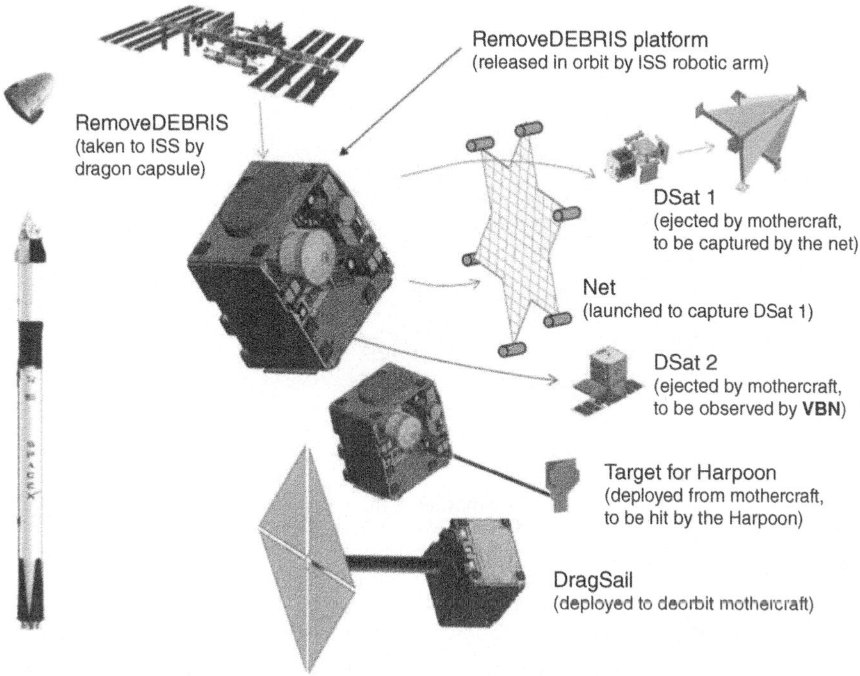

Figure 14.5 RemoveDEBRIS is a satellite platform from which the physical capture of debris is tested in space. The two CubeSats DSat-1 and DSat-2 are released from the host (mothercraft) once in orbit. One CubeSat tests the concept of using a net. Another CubeSat tests the use of a harpoon. *Source:* Ref. [22].

with a tethered net and a tethered harpoon. The 100-kg host satellite was deployed from the ISS in June 2018.

Test 1: Tethered Net DSat-1was the target. In September 2018, the 2U mock target was released from the host. The mock target deployed a boom structure with a sail made of aluminum foil, making the mock target look like an inflated balloon. The balloon increased DSat-1's ballistic cross-section. When the target was more than 7 m away, the host satellite shot its net and captured the target. The net had masses at its edges to facilitate the capture, as shown in Figure 14.6. The captured target was then reeled in by the host. The net experiment was considered a successful demonstration of the concept.

Test 2: Harpoon After a mission decision, DSat-2 was no longer a target but a recorder. The target would come from the host. The target was a 10×10 cm aluminum panel. The target would be held by a carbon fiber boom extended 1.5 m from the host, as illustrated in Figure 14.5.

Figure 14.6 The net is packed in the mother craft. It has masses at the corners before deployment. The net is sprung from the mother craft to capture DSat-1, which is the mock debris. *Source:* Ref. [23] / WTWH Media, LLC.

Test 2 entailed shooting a harpoon from the mother spacecraft at the target panel with the intention to pierce through and retain the target without creating fragments. One end of the harpoon was connected to the host via a tether. The sharp end of the pen-sized harpoon was made of titanium, as shown in Figure 14.7. The harpoon was shot out from the host and hit the aluminum panel at a speed of 20 m/s. The harpoon secured itself in place upon penetration of the target. The harpoon successfully snagged the target and reeled it back to the host. Cameras on the host satellite documented the test.

The harpoon experiment also proved to be successful in capturing orbital junk in space for the first time as a debris-removal demonstration [24, 25].

14.4.4.4 Magnets

Magnetic force is a convenient tool to pick up another ferromagnetic object and therefore would apply to certain debris.

The ELSA (End-of-Life Service by Astroscale) program intends to provide commercial services to satellite operators for spacecraft retrieval. ELSA-d was a demo mission. ELSA-d comprised the chaser (Servicer) and the target (Client) satellites in Figure 14.8. The servicer was a mini-satellite weighing 180 kg and the client about 20 kg. The satellite assembly was launched in March 2021 to an orbit of about 550 km in altitude. The servicer and the client were held together by a magnetic target separation mechanism during launch. The client was equipped with a ferromagnetic docking plate.

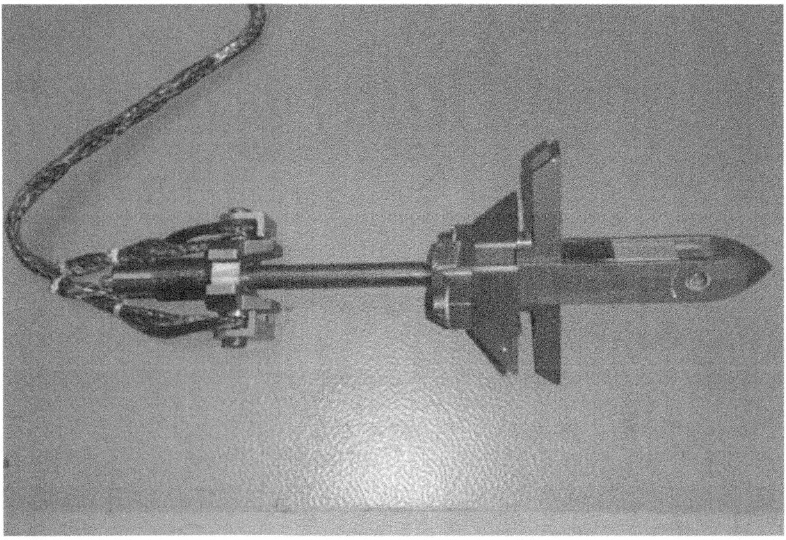

Figure 14.7 The harpoon is packed in the mother craft. It is designed to skewer and reel in the mock target. *Source:* Ref. [25] / Airbus / the RemoveDEBRIS consortium.

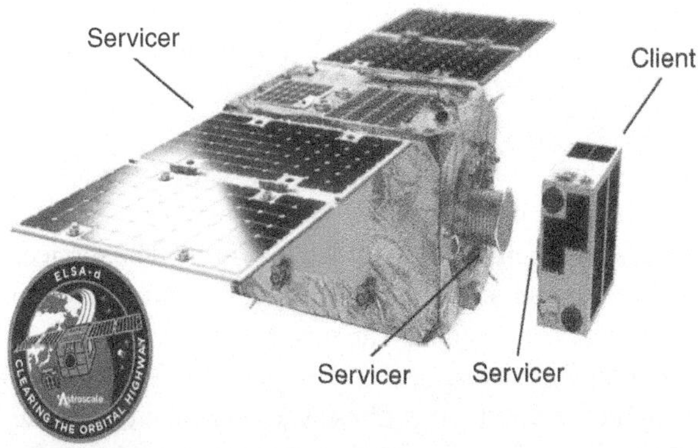

Figure 14.8 This Astroscale test has the servicer release and then recaptures a small client satellite with magnetic properties [26].

The demo flight aimed to test two scenarios regarding the client. In Scenario 1, the client was "semi-cooperative," which meant the target was non-tumbling. Scenario 2 was active debris removal (ADR) with the client being "fully uncooperative," that is, tumbling.

The two satellites separated from each other in orbit. There were seven phases of the ELSA-d mission, as illustrated in Figure 14.9. The test started in August 2021.

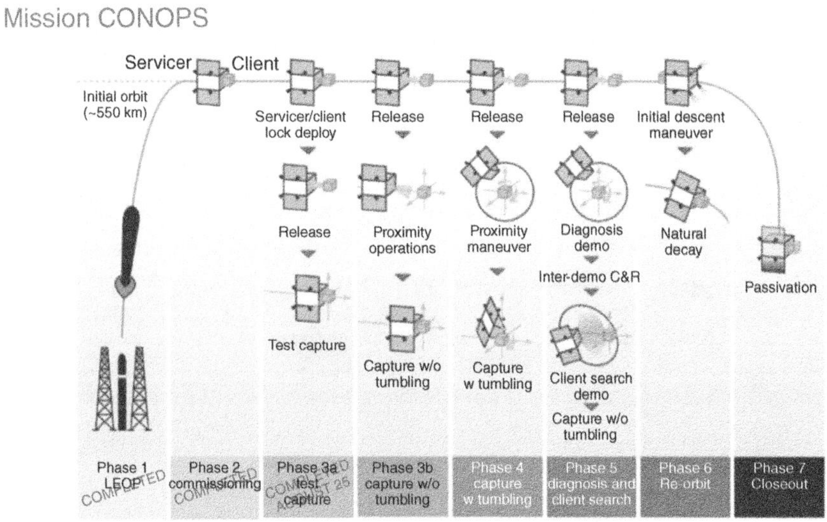

Figure 14.9 ELSA-d is a demo mission. The mission has seven phases. The first test is for the servicer (host) to capture a non-tumbling target. The second test is to observe and capture a tumbling target. *Source:* Ref. [27] / Astroscale.

Phases 1–2: Launch and commissioning—servicer and client were launched together.

Phase 3: Capture without tumbling—servicer and client separate in space. Some rendezvous maneuvers were tested. The servicer moved in to capture the client magnetically using the docking plate on the client for guidance.

Phase 4: Capture with tumbling—the client is commanded to tumble. The servicer maneuvered to observe the client and sent the data to the mission control to compute the trajectory for the servicer to follow. The servicer oriented itself and moved in to capture the client magnetically.

Phase 5: Relative navigation demonstration—this phase was a critical test of the target-and-search capability. The servicer separated and moved sufficiently far away from the client via a thruster and simulated the "search." The servicer followed a corkscrew motion, simulating an initial approach to an uncooperative client.

Phases 6–7: Deorbit and passivation—the final phase had the servicer perform a maneuver to bring the satellite assembly to a lower altitude. Both the chaser and the target proceeded to an uncontrolled deorbit burning up on reentry.

The ELSA-d mission has attracted much attention due to its complexity. The demo mission is considered a success overall [28, 29].

14.4.4.5 Electromagnetic Tethers

The physics behind the electromagnetic tethers is the Lorentz force. This force is the result of a current-carrying wire moving through a magnetic field. The three parameters are the current (or velocity of an electric charge), the magnetic field, and the induced Lorentz force. The three vector quantities are mutually perpendicular to one another following the right-hand rule in physics. Conventional current, that is, the flow of positive charges, is assumed.

For simplicity, we assume a satellite follows a circular orbit above the Equator, so the velocity vector of the satellite is perpendicular to the Earth's magnetic field from north to south. A current-carrying tether is deployed vertically toward the Earth. The tether cuts across the Earth's magnetic field perpendicularly. A Lorentz force results on the tether perpendicular to the magnetic field and the current.

The key components of the tether setup include a conductive tether, two electron emitters, and two electron collectors, as explained in the TEPCE mission below. The environment needed for the tether to work is the presence of a magnetic field and free electric charges in space. These charged particles are available in the ionosphere, which is from 80 to 1,000 km altitude.

The direction of the resulting Lorentz force on the tether is parallel to the satellite's velocity. The direction of the current in the tether determines the direction of the Lorentz force, that is, either the same as the satellite velocity or against. Therefore, the current's direction can determine the Lorentz force to either accelerate (boost) or decelerate (de-boost) the satellite-tether assembly. The de-boost mode serves to slow the assembly to drop to lower altitudes like a drag on the satellite.

Note that the Lorentz force only acts on the current-carrying segments of the tether. The longer the tether, the larger the cumulated force.

TEPCE: Tethered Electrodynamic Propulsion CubeSat Experiment Several universities have proposed electrodynamic tether experiments taking advantage of nanosatellites. TEPCE is a 3U CubeSat built by the US Naval Research Laboratory to demonstrate electrodynamic propulsion in space. TEPCE was launched in June 2019 as a technology demonstration mission. The operating altitude of the TEPCE was an elliptical orbit 300×860 km, which was in the ionosphere. The spacecraft was deployed in November 2019 and re-entered the atmosphere in February 2020.

The payload consisted of two identical CubeSats nicknamed Ren and Stimpy. Sandwiched between Ren and Stimpy was a 1U box. The 1U box housed a spool of bare conductive tether 1 km long. The tether spool was initially wound up, as shown in Figure 14.10.

The conductive tether connects Ren and Stimpy electrically. Figure 14.11 illustrates the two CubeSats in the deployed state after they were separated in space. Ren and Stimpy each contained identical systems. Each CubeSat had a tungsten filament that would emit thermal electrons into free space when powered by a

Figure 14.10 Photo of NRL's TEPCE 3U CubeSat with the tether spool in the middle. *Source:* Ref. [30] / Jerome Pearson.

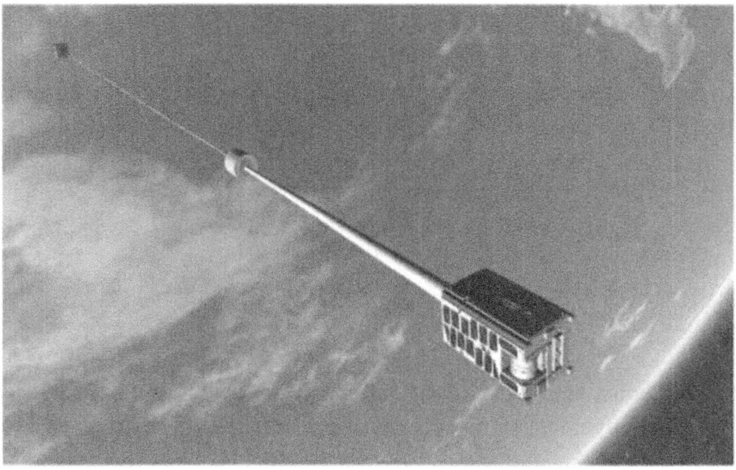

Figure 14.11 Configuration of the TEPCE mission after tether deployment. The conductive tether connects the two CubeSats. *Source:* Ref. [31] / U.S. Naval Research Laboratory.

battery and a protrusible 5-m metal collector tape. The collector tape, when protruded, collected free electrons from the ionosphere, while the tungsten filament emitted them. The collection and emission of the charges formed a circuit.

Figure 14.12 describes the two configurations for Ren and Stimpy. The Earth's magnetic field B goes from right to left, and the satellite assembly travels perpendicularly "into the paper" as depicted by the V_{orbit} vector. We will focus on Scenario (b), which is the de-boost mode. In this scenario, Ren's collector tape protrudes, thereby collecting free electrons from the ionosphere. Stimpy's tungsten filament

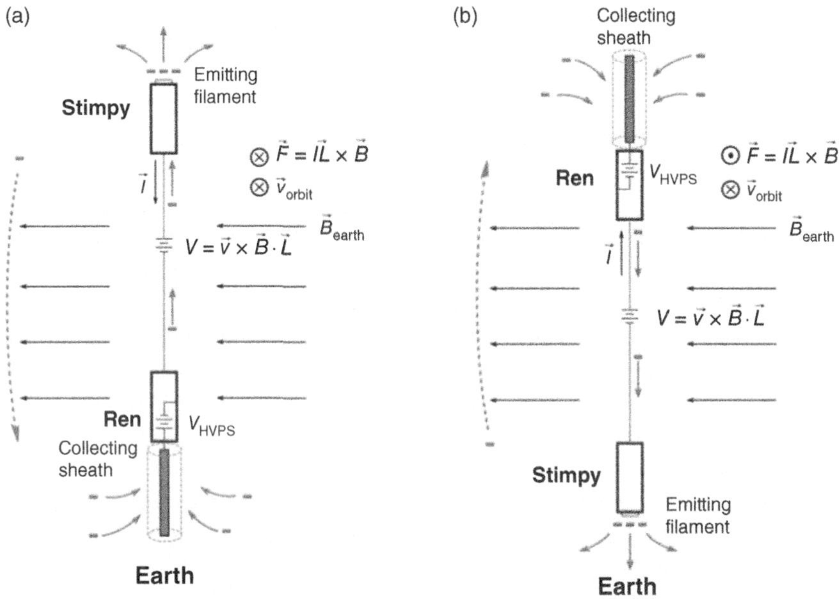

Figure 14.12 Tethered electrodynamic propulsion has two modes: (a) boost and (b) de-boost. The direction of the current flow between the two CubeSats, Ren and Stimpy, determines the mode at work. *Source:* Ref. [31] / Defense Technical Information Center/Public domain.

is emitting. Electrons flow from Ren to Stimpy via the tether and are promptly emitted back into the ionosphere.

The electron flow from Ren to Stimpy is equivalent to the conventional current "I" from Stimpy to Ren along the tether, as illustrated in Scenario (b) of Figure 14.12. The Lorentz force F on the tether is in the opposite direction to the satellite's velocity, that is, coming out of the paper and de-boosting.

The boost mode is when the electron collector and filament arrangement are reversed, as shown in Scenario (a).

14.4.4.6 Dog Tags

It is clear from the different concepts of debris removal by direct contact that they work best if a designated fixture is placed on a target to facilitate future removal. This fixture serves as the bullseye for capture.

One idea for such a designated component is a "dog tag." Altius Space Machines of Colorado offers a grappling fixture to be mounted on future satellites. The DogTag design has a universal interface for small satellites. The fixture design enables various grappling techniques to enable future decommissioning by direct contact methods. It is compatible with magnetic attraction, adhesives, robotic arm, and harpooning capture methods. In 2021, the first batch of DogTags was launched on OneWeb satellites. Over 300 DogTags have already been launched into space.

14.4.4.7 Drag Devices for Deorbiting

Drag devices are packages that subject the target to more atmospheric drag to accelerate the deorbit process. Many LEO satellites, especially the thousands of nanosatellites, are usually too small to carry any active deorbit capability. In compliance with the recent 5-year deorbit requirement, future satellites can install drag devices to increase their ballistic cross sections. Drag sails are lightweight and small in volume before deployment. The sail works at altitudes 800 km and below.

The drag-sail concept has been flight-tested in several missions. One successful case is the CanX-7 mission by the Space Flight Laboratory (SFL) of the University of Toronto, Canada. CanX-7, which stands for Canadian Advanced Nanospace eXperiment-7, was launched in September 2016. One of the tests was for the 3U CubeSat to demonstrate the deorbiting capability of a drag sail. The mission's orbit was initially between 800 and 900 km.

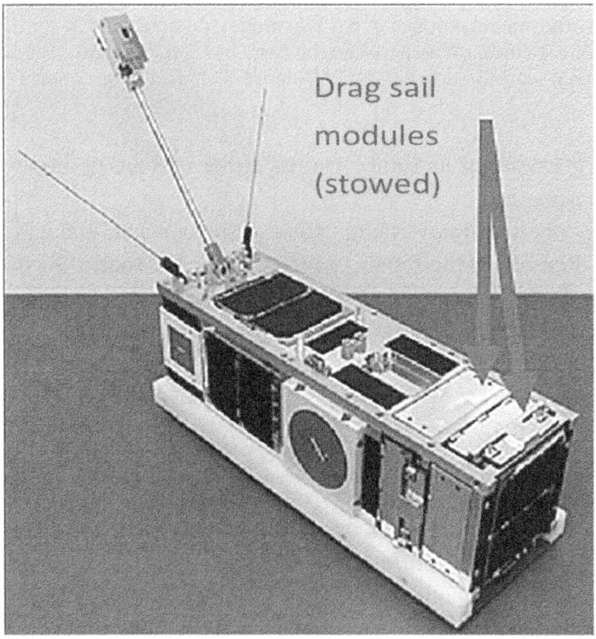

Figure 14.13 The exterior of the CanX-7 satellite by the University of Toronto. The drag sail modules are packed on one side of the bus. *Source:* Ref. [32] / University of Toronto Institute for Aerospace Studies Space Flight.

Figure 14.14 CanX-7 with four drag sails deployed in the laboratory. The size perspective of the sails can be made by noting the CubeSat in the center. *Source:* Ref. [32] / University of Toronto Institute for Aerospace Studies Space Flight Laboratory.

Figure 14.13 shows where the drag sail modules were packed in the CanX-7 satellite. Each sail was made of triangular thin-film materials and was packed and deployed individually. When deployed, each sail was about 1 m^2 in area, as shown in Figure 14.14.

Four drag sails were deployed in May 2017 at an altitude of 800 km. A little over a month after deployment, the CanX-7 satellite's altitude dropped to around 685 km. The Space Flight Laboratory announced that CanX-7 successfully deorbited in mid-June 2022.

14.5 Recycling

Space junk is traditionally considered "trash" to be disposed of. With most space debris being derelict man-made articles, recycling is a sensible alternative, at least before they deteriorate further to become unusable. Some of these articles could still bear economic value.

Therefore, the issue is comparing the cost of recycling against that of remediation such as incineration in the atmosphere. There are two options for recycling: recycling on Earth and recycling in space. Recycling on Earth means that a target piece is physically captured and then returned to Earth in a protective enclosure.

This invariably involves a host spacecraft equipped with the propulsion capability and payload capacity to carry large and valuable space parts. The next question is whether the host spacecraft has a high payload capacity to capture several targets. The host spacecraft must consume fuel to move from orbit to orbit to go after the targets.

For recycling in space, researchers at the University of Southampton connect the proposition to the circular economy. "A fully circular economy for space—removing, reusing, or recycling it—is a viable possibility to safeguard the future for satellites and space exploration," according to new research from the University of Southampton [33].

Estimates by the Southampton team suggest that several thousand tons of recyclable materials are worth multiple billions of dollars. Space recyclables include spent objects such as rocket upper stages and mission articles as well as active and derelict satellites. Space stations could become candidates for future reuse. Identifying which spent part is recyclable would be important and require both tracking and assessing the targets.

The crux of a circular economy is to recycle, reuse, refurbish, repair, and repurpose the materials on hand for as long and as many times as possible. The idea is therefore about managing all the large debris by putting recyclable materials to good use.

The ESA advances a picture of a sustainable space economy via the concepts of "On-Orbit Manufacture, Assembly and Recycling" (OMAR) and Management End of Life-Space Debris Reduction. Briefly stated, the OMAR concept entails to 1) recycling the assets already in space, 2) reducing the use of raw materials from Earth for space missions, and 3) developing the necessary technologies to advance to a space economy.

For these ideas to work, one must first take an expanded view of the space economy. The paradigm shift is to view space debris not as a source of harm or useless materials but as part of a bigger space economy in the future. This change in thinking is the same as treating asteroids as sources of materials and not just threats to civilization.

Putting the practice of circularity to work in space at this juncture is sensible. Why turn spent space parts that previously cost a lot of money to make into ash? Instead, we need to develop a system to manage the end-of-life space parts [34].

Recycling space debris would need the following space infrastructure to become a reality:

(a) Vehicles to collect and transport tons of man-made debris back to Earth or to orbiting depots in space
(b) Factory-scale recycling facilities in space
(c) Factories in space to manufacture new products for use on Earth and in space
(d) Ways to identify recyclables from nonrecyclables before retrieving them

Such infrastructure may sound esoteric at this time, but it is necessary if we aim to build a sustainable space economy. These ideas are not that different from the transporting, mining, and processing of asteroids in space in Chapter 10.

The concept of retrieving large articles by the SpaceX Starship has been advanced. The Starship Cargo variant has a large cargo bay and is equipped to return to Earth. For a large ship like Starship to move from one orbit to another (both altitude and inclination), a large amount of ΔV is needed, that is, high fuel consumption.

14.6 Removing Debris with Lasers

Section 14.3 mentions that the 2022 US government report considers that the most effective remediation methods entail the removal of small debris (1–10 cm) and nudging large debris (>10 cm) off their trajectories to avoid collisions. The nudging method is a short-term fix because the debris remains in space and continues to be a future threat.

The consensus is on removing small 1–10 cm debris as a high priority because they are responsible for causing large debris to break up and are the primary contributor to Kessler syndrome.

This section covers the use of lasers as a remediation tool. Lasers have two effects on an object: radiation pressure and ablation. Both effects produce thrust of different magnitudes on the illuminated target. If the mass of the target is small, the thrust could result in substantial influence on its motion.

14.6.1 Radiation Pressure

Radiation pressure, also known as photon pressure, is a known effect due to momentum exchange by the incident photons on an object. Pressure is the force per unit area. The more intense the beam, i.e. the larger the amount of photons, the higher the radiation pressure. The radiation pressure is generally very weak and is inversely related to the wavelength of the radiation.

The radiation pressure generally does not produce enough pressure on small debris and will not be elaborated further.

14.6.2 Ablation

Ablation is the preferred method in laser orbital debris removal (LODR) for clearing both large and small debris. Ablation occurs when a narrow train of high-power laser pulses hits a tiny point on the target surface, resulting in the formation of a micrometer-sized hot spot. The high power carried by this narrow beam causes local evaporation that instantly vaporizes the material from the surface, forming a tiny cavity. The beam's pulse width is usually on the order of microseconds, so each strike's duration is short enough that most of its power goes to vaporize the hot spot's surface, and only a very small portion is absorbed by the remainder of the target. The follow-on pulses further compress the vapors in the cavity and turn them into plasma. The plume is ejected from the cavity with a velocity in the range of 100 km/s in vacuum, forming a thrust.

The thrust is in the same direction as the incident laser beam, as shown in Figure 14.15. The thrust must be in a direction to slow down or nudge the target. Ablation is very effective in removing small debris by slowing down their orbital speeds toward re-entry.

14.6.3 Vaporization and Ablation

It is useful to have an idea of the energy needed to completely vaporize an object. The answer obviously depends on the material and the size. In 2011, NASA scientists used lasers to completely vaporize aluminum blocks in the laboratory. The vaporization energy for aluminum was found to be 9,200 kilojoules per kilogram or 9.2 kJ/g.

Although the total vaporization of the space debris would be a good remediation solution, it would take too much laser energy to do so. The economic value of this effort is unattractive. The alternative is to nudge the debris as less energy is needed. Using ablation, the NASA team further investigated the laser power to bring a 2.7 kg (2,700 g) aluminum block orbiting at 10 km/s to a full stop versus slowing it down by 30% to 7 km/s. The 30% speed reduction should be enough to drop the orbiting block to lower altitudes and hasten the reentry into the atmosphere.

NASA's results show that the laser energy to bring the Al block from 10 km/s to a full stop needs 3.3 kJ/gm, and 1.1 kJ/gm for 30% speed reduction; both of which are less than the vaporization energy of the block. The lower energy consumption, compared to the total vaporization, is to remove kinetic energy from the object but not enough to vaporize it. Yet, the energy consumption to slow down an object by several km/s is substantial regardless.

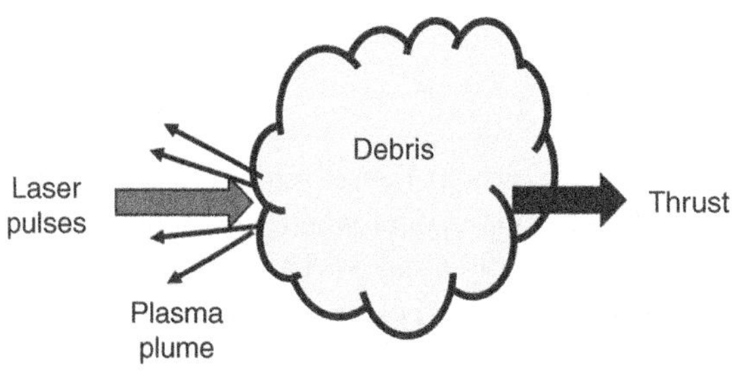

Figure 14.15 Ablation is the formation of the local plasma of a piece of material by a train of high-power laser pulses. The plasma plume comes off the material like a jet of plasma, resulting in a thrust in the direction of the laser beam.

One can surmise that slowing down a large amount of debris in the kilogram range requires a few million joules of laser energy. Nudging requires less energy. Note that a tremendous amount of laser energy would still be needed to nudge numerous debris that weigh kilograms.

To conclude, ablation by lasers is deemed a favorable method to slow down small debris substantially and nudge large debris. The question remains whether high-power lasers are ground based or space based. And if space-based, what kind of spacecraft to put the laser in. The cost and benefit analysis of the laser energy per piece of debris needs to be done. Nevertheless, the idea of using lasers to deorbit small debris has technological merit.

14.7 Space Situational Awareness

The outer space is full of natural and man-made objects as well as dangerous energies. Cislunar thinking allows us to think well beyond Earth. Chapter 10 shows the enormous economic value of asteroids and near-Earth objects. Asteroids are villains until we recognize their economic values and their role in our expanding space economy.

Fluxes of radiation and charged particles emitted by the Sun and the cosmos are also present in space. Although the Earth is protected by its magnetic field and atmosphere, orbiting spacecraft are vulnerable to these dangerous fluxes.

An expanded and integrated view has been formulated considering the many sources of potential harm present in the volume of space enclosing our planet. This integrated framework is known as Space Situational Awareness (SSA).

Space debris is just a part of the bigger picture of keeping track of what is happening out there in space around Earth. When national security is included in the equation, SSA is a serious business.

The Australian Space Academy describes SSA [35] as

> In the broadest sense, Space Situational Awareness may be defined as a knowledge of the energy and particle fluxes in near-Earth space, natural and artificial objects passing through or orbiting within this space, including the past, present and future state of these components. The realm of near-Earth space may be left rather vague at this stage. It is definitely within cis-lunar space but extends to an Earth-radius of at least 100,000 km to include nearly all man-made objects currently in orbit.

Figure 14.16 shows the categories of natural and man-made objects in the space enclosing Earth under the purview of SSA. This diagram is redrawn based on the Australian Space Academy article.

The six boxes in the top row describe the sources of physical objects and radiation enclosed in a spherical dome encompassing the Earth. This imaginary dome

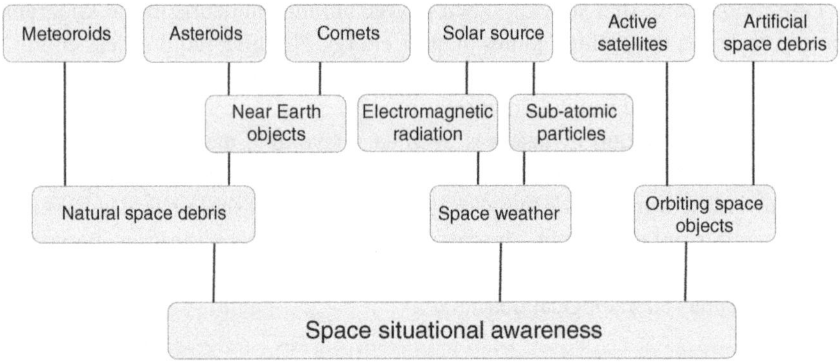

Figure 14.16 Space situational awareness main building blocks. *Source:* Adapted from Ref. [35].

extends 100,000 km from the Earth's surface, which is about 1/4 of the distance between the Earth and the Moon. The six sources are classified into three categories: orbiting man-made objects, natural objects, and space weather.

There are three phases of SSA:

Phase 1: Awareness comes from observation, that is, information based on data collected from different tracking technologies and assets.
Phase 2: Information processing allows decision makers up and down the chain of command to evaluate the threat level and decide to take actions commensurate with the threat level.
Phase 3: Acting on the recommendations.

Some nations and international alliances have placed SSA under bigger umbrellas. For example, SSA has evolved to become the Space Safety Program under the purview of an ESA initiative to "monitor hazards from space, determine their risk, make this data available to the appropriate authorities and where possible, mitigate the threat."

14.8 Cost–Benefit Analysis of Debris Removal

The 2023 NASA report [15] offers a cost–benefit model targeting satellite operators. The model compares the financial risk of doing nothing in the near term versus paying for remedial efforts from now on.

At first glance, maintaining the status quo of doing nothing appears to cost satellite operators less despite the various costs of warnings, evasive maneuvers, and damage or destruction from collisions with debris. But the exposure increases substantially over time and eventually hampers the whole space industry as debris continues to accumulate. Operators of mega satellite constellations in LEO and defense-oriented spacecraft are the most vulnerable.

The first part is to minimize new debris production. Upfront remediations are to implement deorbiting measures in new spacecraft design or contract a servicer for removal after the end-of-mission. This would at least reduce the introduction of new debris.

The second part is to reduce the quantity of the existing debris. The cost of remediating existing debris already in orbit needs to be shared by governments and commercial operators whose assets are exposed to the risks of collision.

The NASA report also only considers the cost of bringing large debris back to Earth for recycling on Earth and does not address recycling in space or returning small debris.

Remediation must be targeted, considering the tremendously high debris counts. NASA considers two remediation scenarios in the analysis:

(a) Remediate 50 pieces of statistically most-concerning large debris over 10 cm in size every year
(b) Remediate 100,000 pieces of debris 1–10 cm in size every year

For each scenario, NASA selected three remediation approaches for comparison since their costs differ. These technological approaches fall under three categories:

(1) Removal: Removing debris by controlled and uncontrolled reentry
(2) Nudging: Moving debris with lasers or rockets for collision avoidance or for trajectory change
(3) Recycling: Recycling debris on Earth to recover materials for other uses

It should be noted that sending a rocket to nudge a very large piece of debris off its course is possible on the condition that no fragmentation results.

14.8.1 Large Debris

Large debris varies significantly in size and weight. Very high costs should be expected to remediate numerous debris.

Removing or nudging numerous debris is a major undertaking. If reentry is considered, it should be a "controlled" one for numerous debris to avoid casualties and damage on the ground.

Figure 14.17 shows the cumulative cost in US$'M over 25 years from remediating the top 50 large man-made debris every year. The five methods employed are: 1) uncontrolled reentry: one-time vehicle to deorbit one large debris for uncontrolled reentry, 2) controlled reentry: remove one large debris for controlled reentry or recycling (on Earth), 3) rockets to nudge each large debris, 4) lasers to nudge a large debris off an imminent collision course (but not necessarily

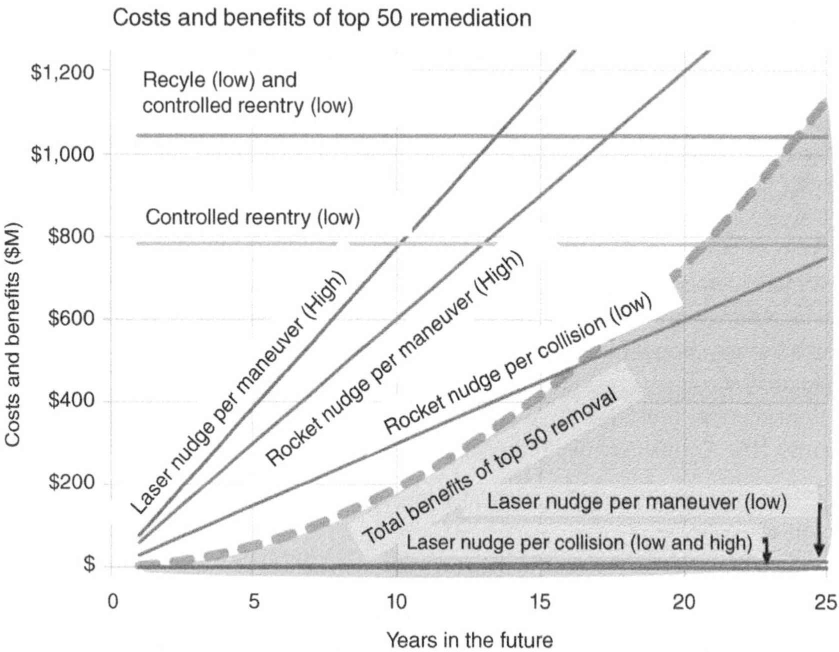

Figure 14.17 NASA's cost and benefit analysis of remediating the top 50 large debris per year. *Source:* Ref. [36] / Federation of American Scientists (FAS).

to deorbit), and 5) lasers to nudge a large debris to steadily deorbit even when no collision danger is present.

The high and low estimates are also shown in Figure 14.18.

The analysis for remediating 50 large debris can be summarized as follows:

- Controlled and uncontrolled reentry are the most costly
- The cost of launching small rockets to nudge large objects is well known. The cumulative cost increases predictably between a low and high envelope
- The cost of nudging with laser ablation varies widely between very low and high

The heavy dotted line is the cumulative financial benefits to satellite operators if the 50 pieces of most-concerning large debris are removed per year. The dotted line intersects each method at different time durations, which is the point when the benefit begins to outweigh the cost of the method. Laser ablation may pay off much earlier than others.

14.8.2 Small Debris

For the 100,000 pieces of small debris per year, NASA's position is that laser ablation is the most suitable method. It makes neither economic nor practical sense to use rockets to knock off small debris that is under 10 cm. The "sweeper" line in

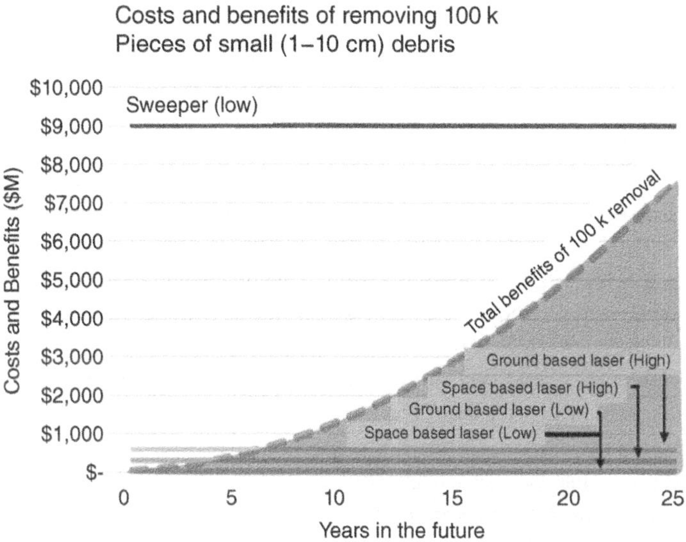

Figure 14.18 NASA's cost and benefit analysis of removing 100,000 pieces of small debris per year. *Source:* Ref. [36] / Federation of American Scientists (FAS).

Figure 14.18 is to have one or more vehicles go around sweeping up small debris for recycling. This method, predictably, is costly for sending host spacecraft to sweep up the small debris in different orbits.

Figure 14.18 also shows the cost–benefits from using lasers to move small pieces. The crossover point happens very early—in a matter of several years. The figure shows that the high side of using a ground-based lasers is more costly than a space-based ones. The reason could be related to ground-based lasers being fixed in location and each can only service a small part of the sky.

14.9 Conclusion

Most space debris comes from derelict man-made objects orbiting in LEO. Debris has become an existential threat to space activities because of its sheer quantity. If not mitigated, debris will accumulate rapidly, as predicted by the Kessler effect.

Debris of different sizes incurs different damage to the active spacecraft. Collisions with large debris over 10 cm produce disastrous outcomes. Small debris 1–10 cm in size, numbering around 1 million, could incur significant damage to active spacecraft and are the biggest contributor to the Kessler effect. Tiny debris under 1 cm in number over 100 million is harmful, but adequate shielding can limit their damage.

The commercialization of space is making LEO congested. Tens of thousands of small satellites (under 1,000 kg) are soon to coexist with debris and significantly increase the statistical risk of collisions. Governments have recognized the debris problem, and regulations are being steadily enacted for satellites to be launched in the near future. That is the good news.

Cost and benefit analysis by NASA gives the satellite industry a model to decide if investing in remediation now is financially better off than maintaining the status quo.

Ideas to remove debris or change their trajectories to avoid collisions are actively pursued, and some prototypes have been flight tested. These endeavors are in the exploratory phase. The 2023 NASA report suggests that the removal of a significant amount of existing debris would cost multiple billions of dollars—amounts few governments would be willing to pay for. The Kessler effect makes it impossible to pinpoint the blame on the party responsible for a particular debris.

For the near term, consensus supports reducing new debris, nudging large debris off course, and removing the small pieces. The solution may come down to a paradigm shift similar to the dawn of the New Space era. Policy changes to encourage the private sector to participate have stimulated the growth of the New Space economy. The rule change from 25 to 5 years for future LEO satellites to deorbit should stimulate innovations and the growth of a new business vertical in the areas of remediation.

Commercialization has not only resulted in a serious reduction in the launch cost using reusable rockets, for example, but it can also help promote "design for end of life" practices. The same applies to recycling and on-orbit services as a part of the space economy. Even though the NASA report does not consider recycling favorable in the near term, recycling does play a critical role in the future space economy. Even if there is an ample supply of materials in the Cislunar space, material usage and waste management should still be managed as early as possible instead of letting the problem get out of hand.

Reading List

Aglietti GS, Taylor B, Fellowes S, Ainley S, Tye D, Cox C, et al. RemoveDEBRIS: An in-orbit demonstration of technologies for the removal of space debris. The Aeronautical Journal. 2020;*124*(1271):1–23.

Bendisch J. *Space Debris 2000*. Amer Astronautical Society; 2001.

Biesbroek R. *Active Debris Removal in Space: How to Clean the Earth's Environment from Space Debris*. CreateSpace; 2015.

Moltz JC. *Crowded Orbits: Conflict and Cooperation in Space*. Columbia University Press; 2014.

National Academies of Sciences, Engineering, and Medicine. *Defending Planet Earth: Near-Earth-Object Surveys and Hazard Mitigation Strategies*. Washington, DC: The National Academies Press; 2010.

National Academies of Sciences, Engineering, and Medicine. *NASA Space Technology Roadmaps and Priorities Revisited*. Washington, DC: The National Academies Press; 2016.

National Academies of Sciences, Engineering, and Medicine. *Near-Earth Object Surveys and Hazard Mitigation Strategies: Interim Report*. Washington, DC: The National Academies Press; 2009.

Welser W IV, Baiocchi D. *Confronting Space Debris: Strategies and Warnings from Comparable Examples Including Deepwater Horizon*. RAND Corporation; 2011.

References

1 US issues first-ever fine for space junk after TV company leaves satellite in the wrong place. Euro News [updated 2023 Oct 3; cited 2025 Mar 1] [online]. Available from: https://www.euronews.com/green/2023/10/03/us-issues-first-ever-fine-for-space-junk-after-tv-company-leaves-satellite-in-the-wrong-pl.

2 Cantwell, Hickenlooper bill to clean up space junk passes senate unanimously. U.S. Senate [updated 2023 Nov 1; cited 2025 Mar 1] [online]. Available from: https://www.newscientist.com/article/2325889-uk-wants-to-send-a-spacecraft-to-grab-two-dead-satellites-from-space/.

3 Pultarova T. SpaceX Starlink satellites had to make 25,000 collision-avoidance maneuvers in just 6 months — and it will only get worse. Space.com [updated 2023 Jul 7; cited 2025 Mar 1] [online]. Available from: https://www.space.com/starlink-satellite-conjunction-increase-threatens-space-sustainability.

4 Astromaterials research and exploration science. NASA Orbital Debris Program Office [cited 2025 Mar 1] [online]. Available from: https://orbitaldebris.jsc.nasa.gov/faq/.

5 Space debris by the numbers. European Space Agency [cited 2025 Mar 1] [online]. Available from: https://www.esa.int/Space_Safety/Space_Debris/Space_debris_by_the_numbers.

6 Berger B. Battery likely caused explosion aboard military weather satellite. SpaceNews [updated 2015 Mar 3; cited 2025 Mar 1] [online]. Available from: https://spacenews.com/battery-likely-the-culprit-in-military-weather-satellite-explosion/.

7 Stay safe with battery testing for space. NASA Spinoff [updated 2022 Jan 24; cited 2025 Mar 1] [online]. Available from: https://spinoff.nasa.gov/Stay-Safe-with-Battery-Testing-for-Space.

8 Hoffpauir D. Li-ion batteries: Managing risk to secure reward. Space-News [updated 2016 Aug 8; cited 2025 Mar 1] [online]. Available from: https://www.nasa.gov/centers-and-facilities/nesc/li-ion-batteries-managing-risk-to-secure-reward/.
9 Nanosats Database [updated 2024 Dec 31; cited 2025 Mar 1] [online]. Available from: https://www.nanosats.eu/.
10 National orbital debris implementation plan. White House [updated 2022 Jul 7; cited 2025 Apr 28] [online]. Available from: https://bidenwhitehouse.archives.gov/wp-content/uploads/2022/07/07-2022-NATIONAL-ORBITAL-DEBRIS-IMPLEMENTATION-PLAN.pdf.
11 Strickland A. Space junk hit the International Space Station, damaging a robotic arm. CNN World [updated 2021 Jun 1; cited 2025 Mar 1] [online]. Available from: https://www.cnn.com/2021/06/01/world/iss-orbital-debris-robotic-arm-scn/index.html.
12 FCC adopts new '5-year rule' for deorbiting satellites. Federal Communications Commission [updated 2022 Sep 30; cited 2025 Mar 1] [online]. Available from: https://www.fcc.gov/document/fcc-adopts-new-5-year-rule-deorbiting-satellites-0.
13 Mitigating space debris generation. European Space Agency [cited 2025 Mar 1] [online]. Available from: https://www.esa.int/Space_Safety/Space_Debris/Mitigating_space_debris_generation.
14 Pardini C, Anselmo L. Using the space debris flux to assess the criticality of the environment in low Earth orbit. Acta Astronautica. 2022;*198*:756–760.
15 Colvin TJ, Karcz J, Wusk G. *Cost and Benefit Analysis of Orbital Debris Remediation*. Nasa; 2023.
16 State-of-the-art of small spacecraft technology. NASA [updated 2025 Feb 26; cited 2025 Mar 1] [online]. Available from: https://www.nasa.gov/smallsat-institute/sst-soa/deorbit-systems/.
17 Mission extension Vehicle 1 and Intelsat 901. Intelsat [cited 2025 Mar 1] [online]. Available from: https://www.intelsat.com/global-network/satellite-network/launches/intelsat-901/.
18 Mission Extension Vehicles succeed as Northrop Grumman works on future servicing/debris clean-up craft. NSF [updated 2021 May 7; cited 2025 Mar 1] [online]. Available from: https://www.nasaspaceflight.com/2021/05/mev-success-ng-future-servicing/.
19 MEV-1 & 2 (Mission Extension Vehicle-1 and -2). eoPortal [updated 2020 Feb 6; cited 2025 Mar 1] [online]. Available from: https://www.eoportal.org/satellite-missions/mev-1.

20 O'Callaghan J. UK wants to send a spacecraft to grab two dead satellites from space. NewScientist [updated 2022 Jun 23; cited 2025 Mar 1] [online]. Available from: https://www.newscientist.com/article/2325889-uk-wants-to-send-a-spacecraft-to-grab-two-dead-satellites-from-space/.

21 ADRAS-J. Astroscale [cited 2025 Mar 1] [online]. Available from: https://astroscale.com/missions/adras-j/.

22 Aglietti GS, Taylor B, Fellowes S, Ainley S, Tye D, Cox C, et al. RemoveDEBRIS: An in-orbit demonstration of technologies for the removal of space debris. The Aeronautical Journal. 2020;*124*(1271):1–23.

23 Greenwood M. Satellite uses net to clean up space junk. Engineering.com [updated 2018 Sep 27; cited 2025 Mar 1] [online]. Available from: https://www.engineering.com/satellite-uses-net-to-clean-up-space-junk/.

24 Amos J. Space harpoon skewers 'orbital debris'. BBC [updated 2019 Feb 15; cited 2025 Mar 1] [online]. Available from: https://www.bbc.com/news/science-environment-47252304.

25 Pultarova T. Watch a satellite fire a harpoon in space in wild debris-catching test (video) news. Space.com [updated 2019 Feb 18; cited 2025 Mar 1] [online]. Available from: https://www.space.com/space-junk-harpoon-removedebris-satellite-video.html.

26 Szondy D. ELSA-d spacecraft captures "space debris" in orbit for the first time. New Atlas [updated 2021 Aug 31; cited 2025 Mar 1] [online]. Available from: https://newatlas.com/space/astroscale-elsa-d-spacecraft-space-debris-capture-demonstration/.

27 Wokes S. ELSA-d: Mission design and performance to date. New Atlas [cited 2025 Mar 1] [online]. Available from: https://indico.esa.int/event/321/contributions/6383/attachments/4369/6591/ESA%2520CleanSpace%25202021%2520-%2520Astroscale%2520ELSA-d.pdf.

28 Marc J, Scott K. 'Satellite catcher' will use magnets to clean up space junk. CNN Business [updated 2017 Jun 20; cited 2025 Mar 1] [online]. Available from: https://www.cnn.com/2017/06/13/tech/space-sweepers-astroscale-japan/index.html.

29 Astroscale's ELSA-d mission successfully completes complex rendezvous operation. Astroscale [updated 2022 May 4; cited 2025 Mar 1] [online]. Available from: https://astroscale.com/astroscales-elsa-d-mission-successfully-completes-complex-rendezvous-operation/.

30 Pearson J, Carroll J, Levin E. An update on EDDE, the electrodynamic delivery express. In: Proceedings of the 5th International Conference on Tethers in Space; 2016. p. 2–3.

31 Coffey S, Crippa C, Dutchover G, Brunner MD, Sibert Z, Kindl S, et al. Tepce: A tethered electrodynamic propulsion CubeSat experiment. US Naval Research

Laboratory; 2022 [updated 2022 Feb 28; cited 2025 Apr 29]. Available from: https://apps.dtic.mil/sti/citations/AD1161973.
32 CanX-7 (Canadian Advanced Nanospace eXperiment-7). eoPortal [updated 2013 Sep 17; cited 2025 Mar 1] [online]. Available from: https://www.eoportal.org/satellite-missions/canx-7.
33 Circular economy to tackle space junk. University of Southampton [updated 2023 Feb 7; cited 2025 Mar 1] [online]. Available from: https://www.southampton.ac.uk/news/2023/02/space-junk.page.
34 Esa is looking into futuristic in orbit services recycling satellites. European Space Agency [updated 2019 Sep 9; cited 2025 Mar 1] [online]. Available from: https://blogs.esa.int/cleanspace/2019/09/09/esa-is-looking-into-futuristic-in-orbit-services-recycling-satellites/.
35 An overview of space situational awareness. Space Academy [cited 2025 Mar 1] [online]. Available from: https://www.spaceacademy.net.au/intell/ssa.htm.
36 Iyer V. How do you clean up 170 million pieces of space junk? FAS [updated 2023 May 23; cited 2025 Mar 1] [online]. Available from: https://fas.org/publication/how-do-you-clean-up-170-million-pieces-of-space-junk/.

15

Looking Ahead

15.1 LEO and Beyond

The space age is now an integral part of the 21st century's economic and technological frontiers. Going to space is now driven by both the government and the private sectors. The low Earth orbit is the front door to space. Our current rocket technology is constructed so that the core stage does the initial heavy lifting to climb up Earth's gravity well. The payload begins its journey to its destination from low earth orbit (LEO). Therefore, LEO can be regarded as the base of all space adventures.

Lowering the cost of access to space fuels commercialization, which then drives the New Space era. Demand for space launches and spacecraft, such as satellites for commercial applications, has increased tremendously. Fully reusable rockets will continue to push commercialization forward.

Affordable access to space leads to emergence of innovative applications. Constellations of LEO satellites with intersatellite laser links effectively replicate the terrestrial fiber optics network in space. The Earth can now be served by an expanded, seamless ICT network. This is an example of "space for Earth."

New Space also means humanity's ambition to expand our presence to other celestial bodies, in what is known as the Cislunar economy. Several nations have definitive programs to send humans to the Moon in the next 10 years. The goal is to build up a sustained presence, including the extraction of vital materials that could benefit Earth. The Moon is deemed the precursor to Mars.

Asteroids also have an abundance of materials that can benefit Earth and the buildup of Cislunar infrastructure. Near-Earth asteroids are within reach of Earth. The challenge is to extract target materials industrially from these rocks in space. Near-Earth asteroids are precursors to the Asteroid Belt.

Industrialization in the 20th century has taught us a lesson regarding the damage of man-made waste to the environment at large and the steady depletion of resources. We are now on red alert, as space debris in LEO is soon approaching a

New Space: From Low Earth Orbit to the Moon and Beyond, First Edition.
Edited by Stephen W. Cheung and Chih-Yung Wen.

tipping point. Debris threatens access to space, but it is also a real problem looking for innovative solutions.

By the middle of the 2030s or earlier, we can expect to see these space technologies become nominal:

(a) Fully reusable, high-capacity rocket
(b) On-orbit refueling
(c) Communications satellite constellations as part of a seamless ICT infrastructure
(d) Integration of Earth observation satellites with other platforms, such as navigation, enabling new applications
(e) New source(s) of energy from space and the Moon
(f) New materials and technologies to facilitate the proliferation of commercial space stations in LEO. More civilians will travel to commercial space stations
(g) Mining of off-world celestial bodies from the Moon to Mars to asteroids
(h) Debris remediation becoming a new business vertical

We can also expect new fields emerging in the next several decades:

(a) Asteroid mining (starting with near-Earth objects)
(b) Space solar power
(c) Helium-3 fusion research as technology verification
(d) R&D and manufacturing in space
(e) Environmental practice in space
(f) Space tourism to natural and man-made destinations
(g) Space logistics

15.2 Below LEO for Further Commercialization

LEO is not only where spacecraft orbit Earth but also the starting point to go beyond to reach high orbits and beyond to Cislunar space. How about below LEO?

We introduce two topics for general interest. They are suborbital flights (for humans and cargo) and low altitude economy (LAE).

15.2.1 Suborbital Flights

Suborbital flights do not enter orbit, at least not full orbit. The payload will presumably ride in a specialized propulsive launch system like a rocket or a space plane equipped with one or more rocket engines to reach the edge of space. Currently, suborbital flights focus on the experience of reaching the Kármán line and a few minutes of weightlessness. Such flights are basically up-and-down, and a propulsive system needs to reach about 1.7 km/s vertically at around 60 km altitude. Inertia brings the vehicle to reach the Kármán line.

Suborbital travel, on the other hand, aims to reach destinations anywhere around the globe in a short time. It can be regarded as primarily productivity driven, that is, going from Point A to Point B as fast as possible. But it also bears recreational aspects for travelers interested in looking at Earth from the edge of space.

Flight times of 1 hour or less have been publicized. In that case, the propulsive stage needs to travel "horizontally" along Earth's circumference. A comparison is revealing. The former Concorde supersonic plane, with a top speed of 0.6 km/s (Mach 2), can fly from London to New York (5,570 km) in 3 hours at an altitude of 18 km, without accounting for the wind.

Recalling from Chapter 5 and Figure 5.37, the SpaceX's Starship IFT 5 shows that the spacecraft traveled about 25,000 km in 1 hour 5 minutes with an apogee at 210 km. The Earth's circumference is about 40,000 km, so half the circumference is 20,000 km. The Starship spent 42 minutes at altitudes above 100 km, with speeds up to 7.4 km/s. The deorbit process was 15 minutes of speed reduction of 4.5 km/s down to 3 km/s. Hence, the Starship flight is a good reference profile.

One can roughly estimate that to go halfway around a static Earth without worrying about fuel consumption. Supersonic/hypersonic winged flights at Mach 5 (~1.7 km/s) would take over 3 hours. A simple calculation of covering a nominal of 20,000 km in 60 minutes works out to be 5.6 km/s. Taking out the launch and deorbit time, the Starship's nominal speed near 7 km/s is reasonable, especially because it covers a trajectory of longer distance. We need the 7 km/s speed to cover the semi-circle around Earth and then the reentry process to dissipate that speed.

The remaining questions are payload capacity, reusability, and market demand. Payload capacity means passenger and cargo capacity, e.g. carrying 100 passengers. Reusability is whether the spacecraft is single use or just drops off the cargo by parachute, or it can perform propulsive landing as in the case of passenger travel. Full reusability of rocket stages will drive down costs. Suborbital travel usually targets business travelers, where efficiency overrides ticket prices. High-end vacationers may find the experience fascinating.

15.2.2 Low Altitude Economy

Low altitude economy (LAE) emerged less than two decades ago. Officially, LAE refers to economic activities at altitudes up to 1,000 m. In some regions, the upper boundary increases to 3,000 m. LAE employs both manned and unmanned flying vehicles.

Many countries are bullish on LAE. It is an emerging industry and is farther along relative to suborbital flights because the market economics is clearer. LAE is projected to be a disrupting force to transform the workflows in major industries such as logistics, agriculture, surveillance, rescue and emergency services.

The disruption usually manifests as improvements in efficiency and productivity by utilizing the vertical dimension instead of being limited to two-dimensional ground transport. LAE is often regarded as an enabler of "smart cities."

One reason that many countries are ambitiously pushing LAE is that only conventional aerodynamic means, such as those used in helicopters, are needed to generate thrust at sea-level altitudes. The flying vehicles are relatively low cost, and a range of size and distance is in the market. Unlike the space industry, LAE entrepreneurs require far less capital to develop their products and services.

The terms "unmanned aerial vehicles" (UAV) and "drones" are often used interchangeably. Aerial devices are usually flown with operators controlling them remotely. Advanced aerial vehicles can fly autonomously without human intervention.

LAE vehicles operate as eVTOL, which stands for "electric vertical take-off and landing." The advantage of vertical take-off is that the vehicle requires no runway to take off and land, and its speed is relatively low—comparable to automobiles. Like helicopters, lifts are generated by rotor blades and not wings.

LAE currently covers the fields of logistics of passengers and materials, surveying, and rescue services, as illustrated by the LAE applications in Figure 15.1. The weight and range vary depending on the size of the vehicle, which is typically from 100 m to 1 km. Winged UAVs can fly missions for thousands of kilometers without refueling.

Drones are further classified by user groups: consumer and commercial. Consumer drones are for recreational use (affordable for ordinary consumers) and prosumer use (consumer and producer). Commercial drones are for business applications. The range of different types is 1) recreational and toy drones, 20–100 m; 2) consumer drones, 2–10 km; 3) prosumer up to 16 km; and 4) commercial 10–16+ km. Table 15.1 shows different size drones, their payload capacity, and their use. Table 15.2 compares consumer drones and commercial drones.

Table 15.1 Drone size and payload capacity.

Drone size	Payload capacity	Description
Mini	Up to 100 g	Recreation use
Small	100 g to 1 kg	Recreational, light commercial tasks
Medium	1–5 kg	Professional (photography, surveying, agriculture)
Large	5 kg to over 30 kg	Industrial inspections, cargo transport
Heavy lift	30–220+ kg	Heavy lifting used in construction, agriculture, and emergency

Source: Adapted from Ref. [2].

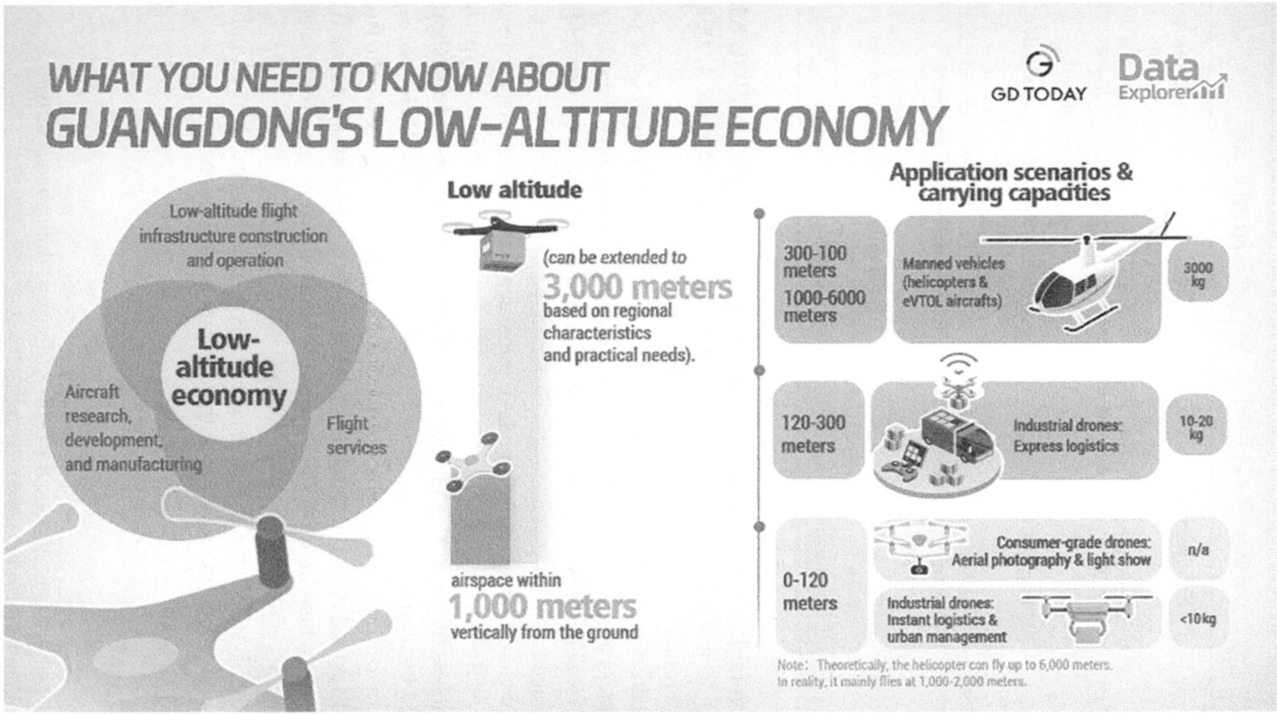

Figure 15.1 Illustration of applications cf low altitude economy in China. *Source:* Ref. [1].

Table 15.2 Comparison of consumer drones and commercial drones.

	Consumer drones	Commercial drones
Purpose	Entertainment	Business, such as surveying, construction, or inspections
Features	Affordable, portable, easy to use	Advanced features such as high-resolution cameras and longer flight times
Flight range	1.6–10 km	Can fly beyond 16+ km under optimal conditions
Flight restrictions	Limited to 122 m and often cannot fly in restricted airspace	May be able to fly beyond 122 m when close to a structure
Piloting (air taxi)	Fly under safety guidelines from a community-based organization	Fly under the rules found in Part 107 of the Federal Aviation Regulations

Some communities are opening their airspace to air taxis. These eVTOL vehicles usually do not require pilots and have a lifting capacity of up to 1,000 kg. Flying taxis aim for long battery life and range from 150 to 250 km.

LAE uses drones and flying cars to primarily serve local communities. We note that helicopters are usually beyond most consumers' affordability, especially if done on a frequent basis. Most of the public now know that drones are used to deliver packages.

It is also obvious that a proliferating LAE translates to congested airspace in our neighborhood. Issues such as noise, invasion of privacy, enforcement of restricted airspace (such as the airport), and midair accidents cannot be ignored and should be addressed by appropriate regulations and strict enforcement. Using package delivery as an example, it is easy to see a drone dropping off packages at the front door or the backyard of a suburban household. But this is not so straightforward in populated cities where high rises are the norm. Innovations by the private sector and government regulations are both needed.

In summary, LAE is here to stay. Its growth is primarily dependent on public acceptance and adaptation together with government regulations.

Reading List

Dauer JC. *Automated Low-Altitude Air Delivery: Towards Autonomous Cargo Transportation with Drones*. Springer; 2021.

Lin YL. *Fujian Province and Low Altitude Economy*. Global Culture Press; 2023.

Torres M. *Low Altitude Economy: Revolutionizing Industries and Shaping the Future with UAV Innovation (Book 1& 2)*. Kindle; 2024.

References

1 What is low altitude economy? Grepow [cited 2025 Feb 21] [online]. Available from: https://www.grepow.com/blog/what-is-low-altitude-economy.html.

2 How much weight can a drone carry? (2024 complete guide). T-Drones [cited 2025 Feb 21] [online]. Available from: https://www.t-drones.com/blog/how-much-weight-can-a-drone-carry.html.

Index

New Space: From Low Earth Orbit to the Moon and Beyond, First Edition.
Edited by Stephen W. Cheung and Chih-Yung Wen.